The Disobedience of Design

Radical Thinkers in Design

Series Editors: Clive Dilnot and Eduardo Staszowski

Not the least of the sins of the obsession with the "research" in the contemporary university is that it unreflectively privileges the appearance of the new. Failing to see the extent to which its own thinking is thereby hobbled and limited, understanding turns in even narrower circles of concern. Design is not immune to this condition. Expansion in practice, and the global increase in numbers of those with design education or who study design, has not necessarily brought with it increased understanding. On the contrary, despite real attempts to counter these impulses, reduced to the crudest understanding of vocation, depth of thought disappears, crises remain untouched; genuinely new practices and conceptions struggle to be comprehended in their full meaning.

Radical Thinkers in Design, a moment of the larger project *Designing in Dark Times*, seeks in a small way to try to address this situation. The project is to bring back into circulation, as provocations and aids to thinking, some key "lost" (out-of-print, untranslated, unjustifiably ignored) texts in contemporary thinking on design(ing). Against the denial of critical thought, these books offer counterviews. In what they open toward, what they explore and present, above all in what they anticipate, they point to the concrete possibilities, as well as to the necessity, of paradigm shifts in design thinking and in our conceptions of what designing today can and should be. They offer approaches, concepts, modes of thinking, and models of practice that can help in thinking not only how design can be rethought and repositioned in its internal momentum, but how it offers an integral mode and capacity of acting in the world. By showing how, at base, designing contains irreplaceable critical and affirmative moments; they point us toward ways of reversing some of the negative and destructive tendencies threatening to engulf the world.

Titles in the series:

John Chris Jones, *designing designing*

Tony Fry, *Defuturing*

Judy Attfield, *Wild Things*

Anthony Dunne and Fiona Raby, *Design Noir*

The Disobedience of Design

Gui Bonsiepe

Edited by Lara Penin

With contributions by
Zoy Anastassakis, Constantin Boym,
Frederico Duarte, Hugh Dubberly,
Ethel Leon, Marcos Martins, Eden Medina

BLOOMSBURY VISUAL ARTS
LONDON • NEW YORK • OXFORD • NEW DELHI • SYDNEY

BLOOMSBURY VISUAL ARTS
Bloomsbury Publishing Plc
50 Bedford Square, London, WC1B 3DP, UK
1385 Broadway, New York, NY 10018, USA
29 Earlsfort Terrace, Dublin 2, Ireland

BLOOMSBURY, BLOOMSBURY VISUAL ARTS and the Diana logo are trademarks of
Bloomsbury Publishing Plc

First published in Great Britain 2022

For legal purposes the Acknowledgments on pp. xviii–xxiv constitute an extension of this copyright page.

Cover design by Andrew LeClair and Chris Wu of Wkshps

A catalogue record for this book is available from the British Library.

A catalog record for this book is available from the Library of Congress.

ISBN: HB: 978-1-350-1-6245-7
PB: 978-1-350-1-6244-0
ePDF: 978-1-350-1-6247-1
eBook: 978-1-350-1-6246-4

Series: Radical Thinkers in Design

Typeset by Deanta Global Publishing Services, Chennai, India
Printed and bound in India

To find out more about our authors and books visit www.bloomsbury.com and
sign up for our newsletters.

Contents

Part One: Thinking Design

Part Two: Design in the "Periphery"

Part Three: Design, Visuality, Cognition

Part Four: Design and Development/Projects

Illustrations

Introduction by Lara Penin

WHY READ GUI BONSIEPE TODAY?

I first encountered Bonsiepe's texts in the late 1990s as a then recent architecture graduate in São Paulo, Brazil, trying to get into industrial design practice. Later, I found my way into a doctorate in design at Politecnico di Milano, in Italy. My trajectory was the reverse of Bonsiepe's, from South America into Europe and from there, to the United States. The Brazilian Portuguese version of Bonsiepe's *Interface* was part of the very small library I took with me to Milan. That book contained a vital key for me as I started my trajectory as a design scholar. It had a language that I could speak—and that spoke back to me. I'm not referring to the language itself, but to the fact that Bonsiepe writes about design through a Latin American lens, recognizing its systems and forces, connecting both theory and practice according to a world-system that I belonged to. Bonsiepe's texts recognized my identity and struggles in ways that other Anglo-centric design theory and literature simply could not. In short, I felt *seen* by Bonsiepe's texts and philosophy.

In my now twenty years as a designer and scholar from the "Global South" living and working in the "Global North," I still hold dear, perhaps more than ever, to Bonsiepe's key as he helps me position myself as someone who needs to bridge worlds by necessity, dialoguing between *periphery* and *center* as a daily negotiation. The process of editing this book involved revisiting Bonsiepe's monumental work not only on a deep intellectual level, but on a personal level; it forced me into revisiting my own trajectory and cultural affiliations.

Gui Bonsiepe is one out of many Europeans designers and architects who migrated to the other side of the Atlantic in the twentieth century. The presence of European designers and architects in the Americas is far from an original route. It's hard not to think about the main Bauhaus superstars, Mies van der Rohe, Walter Gropius, and Marcel Breuer, who established themselves in the United States just before the Second World War. Latin America also received its share of European practitioners and thinkers throughout the twentieth century, some of whom became prominent leaders of local modernisms, such as Gregori Warchavchik, Vilém Flusser, and Lina Bo Bardi, among a myriad of European transplants in Brazil alone.

Across the Americas, from North to South, the language they brought in reflected the dominant Western Eurocentric design epistemologies and cosmologies. The European expats joined local practitioners and thinkers in shaping modern ideals that then manifested in local landscapes. In Brazil, for example, the creative and intellectual avant-garde community tried to define a postcolonial identity, most famously in the Week of Modern Art of São Paulo in 1922, which gathered writers, musicians, painters, and sculptors in a quest for originality. The "Anthropophagic Manifesto" that emerges in the same era proposes cultural hybridization, or the "anthropophagy" as a metaphor to define a cultural practice of eating up or absorbing the ideas and ways of other groups and origins and making them your own, mixing up several aesthetics and practices together combined into one original thing. Remarkable as the anthropophagic approach was, it has been criticized on the one hand for the cultural appropriation of indigenous cultures and on the other for its ties to European cultural movements especially Futurism.

As discussed by Martins and Anastassakis in the postface of this volume, the decolonial discourse in design flourishing today is finally starting to acknowledge and revert the systemic racism, elitism, exclusion, and erasure that have marked the development models and overall cultural identity of Latin America, including its modernity. Bonsiepe's center–periphery model can be understood as an important precedent to the emergent decolonizing design movement. And crucially, Bonsiepe's views and lived experience force us to envision decoloniality in ways that acknowledge the complexities of our lived realities and multiple identities, rather than through a purist aspirational theory.

In Latin America, the more "popular" manifestations of culture, such as music and dance for example, have been, perhaps not surprisingly, areas where its Black and Indigenous roots have been more visible and celebrated. In contrast, the more "technical" professions, such as architecture and industrial design, which are bound to access to higher education, have remained overwhelmingly white and Eurocentric. Only in recent decades have affirmative action policies began to make higher education more accessible to lower-income and more diverse students. Faculty diversity, an equally crucial tool for decoloniality and racial and social justice, remains a change that is still to be enacted in design schools in the whole hemisphere, and worldwide. Bonsiepe's ideas of emancipation have today added meanings and levels of complexity.

Bonsiepe's idea of design education is in itself an idea of emancipation, albeit disciplinary, as he defends that design schools should remain independent institutions, and not be part of universities whose rigid academic systems may become detrimental to design's creative and technical ethos. Based on his experience in ulm, which he presents and discusses in Part I, the design pedagogy he lays out proposes a curriculum based on a unique kind of transdisciplinary on which design is at the center, as a stand-alone discipline independent from the fine arts, and intersects with other disciplines in its own terms, especially science, but also coupling with economics and dialoguing with (political) philosophy, especially the Frankfurt School.

Different from Italian-born architect Lina Bo Bardi, who as an émigré connects deeply with Brazilian culture and whose most famous projects materialize in cultural spheres such as museums and cultural centers, Bonsiepe focuses on the local industrial economy largely bypassing the cultural circuits. He connects straight with South American industrial base, specifically medium-size industry, as Ethel Leon describes, defining thus the industrial world as his medium. In his trajectory from Chile to Argentina and Brazil, Bonsiepe immerses himself in the industrial spheres in these countries and defines his own politics within them. He is convinced that the local industry is the key to self-sufficiency and autonomy. That's where his technical rationality fits perfectly, and becomes the engine of his politics for local development.

Most prominent is his texts is his exasperation with an industrial base that is insufficient in producing goods for local needs and the dependency on imports from central countries. He strongly rejects the logic of economic dependence as a damaging tether binding whole countries to a perpetual state of dependency. He quickly absorbs the Dependency Theory formulated by prominent South American economists in the late 1960s, which suggests that wealth from peripheral countries flows to central countries in an unbalanced and unfair global division of wealth, importing industrialized goods and exporting the much less commercially advantageous commodities, configuring in what Bonsiepe considers economic extractivism tied to old colonial patterns (see, for example, Chapter 17 "Between Favela Chic and Autonomy: Design in Latin America").

One of his most fundamental contributions is defining the dichotomy *center–periphery*, that he uses in lieu of *Third World-First World*, terms largely used in the 1970s marked by the Cold War. The center–periphery terminology is a vastly superior and more sophisticated concept than the geographic determinism of the now-favored *Global South–Global North* as it allows not only its use in geopolitics but also as a concept valid for debates within countries, regions and cities, which have internally their center–periphery relationships dictating priorities, inequality and dependency. (For example: What parts of New York City get their snow plowed first? What neighborhoods have more parks and green areas per resident? What neighborhoods are more polluted?)

Within the center–periphery framework, Bonsiepe's notion of autonomy passes less through cultural identity and more through industrial capacity, from medical equipment to household appliances and, most crucially, machines, as the import of means of production is a main culprit of continual economic dependency. His iconic Cybersyn project in Chile is unequivocally part of his commitment to change the dependency status quo, in Salvador Allende's strive for economic growth and efficiency within a socialist *developmentist* model.

Bonsiepe positions himself thus as a champion for local industrial autonomy as he sees it as a key to emancipation. As noted by Ethel Leon, invited and hired by central and local governments, first the socialist Allende in Chile, later the fading military autocrats in Brazil, hé connects and helps catalyze and activate the nascent local communities of industrial and communication designers as well as engineers. He might have considered himself a *parachutist* in the early years (a term that appears in different texts of this volume, in the James Fathers' interview as well as Ethel Leon's introduction text in Part Two and Marcos Martins and Zoy Anastassakis in their postface); however, as he settles in Argentina and Brazil, his life, both professional and personal, becomes a South American life. He lands and he stays there. It is this crucial fact in his trajectory which allows him to become a "bridge" himself, as suggested by Dubberly. And it is because of his new condition, of one who settled, that he is able to develop his unique "language." This is the key he offers me. With this volume, we hope that a new generation of designers will also be able to access Bonsiepe's key, as this selection of texts and projects were never before presented together in English.

As Duarte writes in his introduction to Part One: "This call for an emancipated, critical consciousness among design professionals and other social agents regarding the social, economic and cultural processes that perpetuate the dependence of peripheral nations is a key aspect of Bonsiepe's original and critical contribution to design discourse."

Bonsiepe's views are indeed original and in contrast to Victor Papanek, the well-known design pioneer who also problematized design in the periphery in the 1970s. Included in this volume is Bonsiepe's critical review of Papanek's bestseller *Design for the Real World* as well as Papanek's heated response. While Bonsiepe agrees with Papanek's premise that design had distanced itself from the "real world," they fundamentally disagree on how to change that. Papanek's appropriated technology model , best illustrated by the tin can radio that is able to pick up only one single station (the project famously commissioned to Papanek by the US Army to be used in developing poor countries) seems misleading to Bonsiepe, who is able to see more nuances and specific possibilities through his connections with local industries and, by the virtue of, unlike Papanek, having settled in the "periphery." The *periphery* for Bonsiepe is not a single panacea. And more importantly, the romantic do-it-yourself designs that emerge from the appropriated technology model seem to Bonsiepe not only insufficiently dignified, but simply not good enough as design responses to people's needs. Furthermore, the one-sided interventionist geopolitics of Papanek is in clear contrast to Bonsiepe's commitment to multilateralism and the idea of state-based design policies.

This is indeed a salient point that appears in several of Bonsiepe's texts and that readers today may connect with current debates about the role of government. For Bonsiepe, design needs to be part of a system on which the state/government has a prominent role in determining economic and social development policies, rather than delegating crucial decisions over society's wellbeing be determined primarily through market-driven or tech-driven forces. Bonsiepe is connected to a world order based on multilateral institutions, as Duarte points out, and some of its most consequential work is precisely related to policy papers and reports. Included in this volume is the historical report "Development Through Design" that Bonsiepe prepared for the United

Nations Industrial Development Organization (UNIDO) in 1973 at the request of International Council of Industrial Design Societies (ICSID), which lays out basic principles of the design profession and how it can be defined as a "tool for development."

Bonsiepe's multilateralism is also very much in line with his refusal of the design individual star system as "*a special field dominated by creatively gifted individuals*" (introduction text to Chapter 7). He strongly rejects design as an elitist practice and despises the idea of design as synonymous of luxury goods. Rather, Bonsiepe firmly believes that design is a force for social emancipation, and he shows that through his own practice. The selection of projects in Part IV showcase his choice of project briefs that are evidence of his views.

We may note that some themes that are crucial in critical contemporary design discourse are largely not discussed in Bonsiepe's texts, for example racial and gender inequalities, prominent issues at the center of any social design agenda today. Our editorial approach was attentive to expand Bonsiepe's own discourse and includes critical voices able to fill some of these gaps. Included here in Part Four is an extract of Eden Medinas' book that dissects the Cybersyn project, which specifically investigates the socialist core of the project while also revealing its highly gendered dynamics.

The book invites new readers to engage critically with Bonsiepe's texts and not to take them as one static body of work, but as a collection of texts that discuss different aspects of design and the world at different places and times. It is worth noting how in several of Bonsiepe's editorial comments he offers for each text included here, he progressively acknowledges some of the failures of the industrial model he once embraced, particularly the environmental crisis. Bonsiepe is in fact an extraordinary learner and an excellent critic of himself. As Dubberly writes "Bonsiepe's career may serve as a signal of where design is heading or even as a model for a new generation of designers." We can sure hope so.

How This Book Is Organized

This volume presents the work of Gui Bonsiepe through a selection of forty of his essays, interviews and projects written between 1965 and 2018. The texts follow Bonsiepe's personal trajectory and provide a unique historical account of transformations of the world, of design and of Bonsiepe himself.

The 1970's and 1980s are Bonsiepe's most prolific period in terms of his practice. The projects presented here in Part Four are all from these two decades. Conversely, the 1990s, 2000s as well as 2010s are the most prolific for Bonsiepe in terms of his writing with the bulk of the texts in Parts One, Two, and Three dating from these three decades, plus a couple of early 1960s texts, especially from the Ulmian period and under the influence of Tomás Maldonado. This timeline reveals how Bonsiepe is both actor and narrator of the transformations happening in and around design in the six decades from the 1960s to present day (2021), traversing key political, technological, and cultural chasms. As Hugh

Dubberly points out, Bonsiepe is himself a bridge between these worlds, but importantly, he is also a critic and proponent of the transformations.

As Bonsiepe notes in his Preface, the texts were written in four different languages (German, Spanish, Portuguese, and English) and for different purposes and audiences, resulting in an editorial project that strives to establish common threads. Rather than presenting the texts following their chronological order, our editorial choice was to group them into main thematic categories that are most salient in Bonsiepe's thought, the common thematic threads indeed of his work. This choice, albeit imperfect as many texts dialogue with more than one theme at the same time, gives the reader the opportunity to engage with Bonsiepe's work in fluid and intentional ways.

The book is therefore organized in four thematic sections, or *Parts*, with ten texts each, clustered in three subsections each. Each *Part* is introduced with a critical essay by a group of outstanding contributors who provide specific and unique insights into the texts of the section: Frederico Duarte, Ethel Leon, Hugh Dubberly, and Constantin Boym.

Part One: Thinking Design opens with Frederico Duarte's essay, *Nostalgia is Futile – Our Future Starts Now*, and gathers ten Bonsiepe essays divided into three sections.

Section *Essays on Ulm* presents four texts focusing on the history, curriculum, and intellectual/disciplinary affiliations of Ulm School of Design, and its overall legacy in relation to a broader project of modernity. Section *Theory and Practice* comprises three texts that discuss design discourse and practice, methods, and how to learn them. Section *Design, Politics, Ethics*, brings three texts that examine the broader role of design in relation to democracy and society, and the virtues of design for the next millennium.

Part Two: Design in the "Periphery" begins with Ethel Leon's essay *A Designer on the Periphery of Capitalism*, and is comprised of a selection of ten essays and interviews covering perhaps the most polemic topic of Bonsiepe's themes, grouped in three sections.

Section *From Europe to South America* offers two important interviews, one with James Fathers on which Bonsiepe shares a deep account of his life and work in Latin America, and the other with Hugo Palmarola that touches upon his brief period in Chile, also noting his passage through Berkeley, California. The interviews are followed by two texts introducing Bonsiepe's model of design and development in the periphery. Section *Design in the "Periphery"* brings three texts on which Bonsiepe problematizes and discusses design practice and role in the periphery, starting with a text born out of a conversation with Argentinian designer and Bonsiepe's wife, Silvia Fernández, and Lujan Cambariere about the history of design in Latin America. Section *The Question of Difference* aggregates three texts that discuss critical aspects of design in Latin America and dependent countries starting with a paper presented at an ICSID Congress in Mexico City in 1979, followed by a text that examines the North–South conflicts especially related to the environmental costs of industrialization, and closes with a text that discuss issues around the ever-thorny issue of Latin America design identity.

Part Three: Design, Visuality, Cognition opens with Hugh Dubberly's essay *Bridge building as Interface Design*, and contains ten texts covering the digital transformations happening on design at the turn of the century offering Bonsiepe's original conceptualization of the interface.

Section *Design and Language* starts with a text that examines how Terry Winograd and Fernando Flores concepts that can inform a design theory, followed by a key text that defines the concept of interface, and then a text whose origin is related to an original course on interface/interaction/information design. Section *Design/Visuality/Theory* texts focus on the larger themes of language and visuality, starting with a text from 1965 that offers a fundamental theory on verbal and visual rhetoric, then jumps into a fundamental text on interface design of computer programs that goes deep into usability issues. The other two texts in this section look into the role of design in relation to language and the role of language especially in the period on digitalization of text. Section *Design and Crisis* brings three texts that zoom out into larger themes and reflections of design starting with a reflection on design in relation to the various crises of the first decades of twenty-first century, followed by a recent (2018) text that ponders on the long arc of design since the Bauhaus and its less well-known director Hannes Meyer. The section closes with the text that gives this book its title, *The Disobedience of Design*, presented in a Symposium in Berlin in 2015, in which Bonsiepe talks about designers as "agents for social change" and how designers may "envisage the possibility of another future."

Part Four: Design and Development/Projects presents a selection of Bonsiepe's practice work. The last part of the book is introduced with an essay by Constantin Boym, who reviews Bonsiepe's design work from his personal lenses rooted in the Soviet Union socialist design but also in Italy and New York from the 1980s onward. As Boym notes, "there is no such thing as Bonsiepe's 'style', each product is different, non-personalist and humble, strictly responding to contextual needs and conditions and yet, dotted with visual quality." His point is that Bonsiepe understands these projects less as "his" unipersonal achievements than, in principle, as the result of a group's collaboration, as result of a teamwork.

Section *Design Policy/Design and Development* includes Bonsiepe's policy contributions for UNIDO, in which he lays a practical vision of what design and development means and a reflective interview pondering on the meaning of design and development forty years after the UNIDO document. As anticipated, included in this section is the famous polemic with Victor Papanek, the two men, despite agreeing on the agenda, disagreed on the means to get there. Section *Gui Bonsiepe: Selected Projects in Latin America* presents projects chronologically reflecting his stay and work on Chile (1972–4), Argentina (1980) and Brazil (1984, 1986). Finally, *Section Case Study of Project Cybersyn, Chile* presents Bonsiepe's most famous and iconic project, the Opsroom for the Cybersyn project, for which Bonsiepe was chief designer. Included in this section are not only Bonsiepe's own voice describing the project in detail but also an excerpt of historian of technology Eden Medina's 2011 book in which she offers a deep historical survey and critical analysis of the Cybersyn project. Both sections are illustrated sufficiently to give readers a sense of each project.

The book concludes in two ways. First, with a postface essay by Marcos Martins and Zoy Anastassakis reflecting on Bonsiepe's work and connecting with their own experience as progressive design educators in a design school (ESDI in Rio de Janeiro) highly influenced by ulm. But we have also attached as an appendix documents from the final issue of the ulm journal (no. 21, April 1968) which announce the resignation of the staff and students of HfG ulm and present Bonsiepe's import of analysis of the context of forces under which the school, and design in general, operated in the late 1960s. The document is drafted just before his move to Latin America.

As suggested by Frederico Duarte, "Nostalgia is futile—our future starts now." In this sense, this book, while bringing in the past, sustain relevance in the present by offering important keys to the future.

A Note on the Making of the Book

When we began work on this book, our intention was in line with how we had begun the series *Radical Thinkers in Design*, that is, by republishing important single texts of design thinking from the last thirty or forty years which had gone out of print or, because of vagaries of fashion and thoughtlessness, had become unjustly neglected. Hence, our first four volumes in the series—John Chris Jones, *designing designing*; Tony Fry, *Defuturing;* Judy Attfield, *Wild Things*; Anthony Dunne and Fiona Raby, *Design Noir.*

We had long considered a volume on Gui Bonsiepe and thought to continue the logic of the series by republishing Bonsiepe's only English-language volume, *Interface – An Approach to Design*, published in a small edition by Jan van Eyck Academie in Maastricht in 1999.

However, as we began to look further into Bonsiepe's work, while the import of this small volume could not be denied—given that, as its title suggests and as Hugh Dubberly makes clear in his introductory essay, this small volume contains in the essential model of design today as interface (a deeper development of Herbert Simon's earlier conclusions in *The Sciences of the Artificial*)—the opening to the full range of Bonsiepe's work both chronologically from the 1960s through to today and in its scope and range made it impossible to confine the project simply to republishing *Interface.*

This was all the more in that much of Bonsiepe's work across his career is untranslated. While Bonsiepe has been extensively published in German, Italian, Spanish, and Portuguese, aside from *Interface*—itself long out of print —there is only a scattering of essays in English (mostly published either through the journal *Design Issues*, which has been a consistent supporter of Bonsiepe's work, or, more recently, in German and Swiss publications on design history and theory).

Accordingly, we were forced to conceive of a different kind of book, one that would make an attempt to present a coherent overview of Bonsiepe's work and writings, both chronologically across his career and in terms of the range of topics and projects considered.

This then explains the logic and character of the book. *Interface* is at its core (ten of the original twenty-one essays are reproduced here) but these constitute only a quarter of the book. The rest of the volume is made up in almost equal parts of a combination of essays published in English, but which are now either out of print (the essays from ulm, for example) or published in mostly German-based texts and newly translated work. Equally, the range of material required some level of systematic organization, hence the organization of the book into four roughly equal parts each of which is designed to give some weight to the themes discussed in each area. It was at the point of organizing this material in this format that we began to work with Gui Bonsiepe, an engagement that took us through the very difficult months of 2020. We remain enormously grateful to him for his patience and extraordinary capacity for detail. Needless to say, it is we as editors who bear responsibility for errors in the work.

Lara Penin, Clive Dilnot, Eduardo Staszowski
Designing in Dark Times/Radical Thinkers in Design

Recognition and Acknowledgments
Gui Bonsiepe

The texts selected for this edition cover a period of over two generations. They have been formulated in geographically, culturally, politically and institutionally different contexts. Original versions of these texts were written in different languages—German, Spanish, Portuguese—which makes their translation difficult because each language has its own flavor and sometimes contains concepts that don't have a precise equivalent in English and thus need to be rendered by an approximation taking into account the different semantic resonance particularly of the terminology in each language. The difficulty increases when a translation has been made from an already-translated version of the original formulation, that is, in case of a second-order translation. I am grateful to the translators for having confronted these difficulties. I did not opt for unifying the terminology because it would have ended in a flattening of conceptual distinctions.

Diverse is the list of friends, colleagues, publishers, participants from industry, and academia (students), whom I would like to thank for their participation in a network that is reflected in this book. Of course, they are not responsible for what I have written. The texts illustrate the contingent and changing character of the issues addressed. In a climate of post-post-isms and the increasing distancing from reason as the supposed cause of all evil of the present technological civilization, they turn against maintaining the status quo and of resigning to accept reality as it is. Here a harsh formulation by the dramatist, poet, writer, essayist, and theatre director Heiner Müller may be quoted: "The world as it is is not possible, one must show it in its impossibility, in the nightmare of its incomprehensible contradictions—in order to be possible it would need to be changed first."[1]

Drawing of the variety and richness of academic sources and attempting to create connections to the issues of design, the established disciplines

---

[1] Helen Müller and Clemens Pornschlegel (eds.), *Heiner Müller: Für alle reicht es nicht – Texte zum Kapitalismus*, SV edition suhrkamp (Berlin: Suhrkamp Verlag, 2017), 92.

can be criticized for their constitutive blindness to the dimension of project ("project" covers a broader range than the increasingly outworn and limited term "design"). From this point of view, the texts can be taken as building blocks of a future project theory to be developed.

Dawn Oxenaar Barrett edited the first English translation (*Interface – An Approach to Design* in 1999, Jan van Eyck Academy, Maastricht), and various chapters were included in this enlarged edition. I owe special thanks to Jan van Toorn for an invitation to give a course on interface design (1997–9) at the Jan van Eyck Academy, where the idea of the first English-language publication gained corpus.

In Argentina, it was Basilio Uribe of the National Institute for Industrial Technology (INTI) who created the conditions for working in the context of industrialization of a country in the 1960s, which explicitly included design. I would also like to thank Carlos and Lala Méndez Mosquera for the platform of publications, as well as Rubén Fontana.

In Spain, I am obliged to Alberto Corazón for the publication of my first book in Spanish, as well as André Ricard and Manuel Lecuona for repeated meetings. I am also grateful to Xavier Llopis, José Luis Martín, and Fèlix Bella of the publishing house Campgràfic.

In Chile, it was Fernando Flores who included a group of designers in the Comité de Investigaciones Technológicas, INTEC, and promoted their participation, among others, in the widely published and emblematic project *CYBERSYN* (1971–3).

In Brazil, the institutional setting for local development activities had been provided by Lynaldo Cavalcanti de Albuquerque, Itiro Iida, and Sergio Gargioni from the National Council of Scientific and Technological Research (CNPq), Brasilia, in the 1980s. I owe special thanks to the editors Edgard and Eduardo Blücher in São Paulo for having published the trilogy of *Design, Cultura e sociedade* (2011), *Design como prática do projeto* (2012), and *Do material ao digital* (2015). I am grateful for their generosity to place this material at the disposal for the editors of the English version. I am thankful to Ethel Leon for the frequent exchange of reflections about design and design education in the Periphery.

In México, the initiative particularly of Fernando Shultz and Simon Sol from the Universidad Autónoma Metropolitana, Azcapotzalco and Xochimilco, Mexico City, permitted me to participate repeatedly in conferences. Also, I want to thank Enrique Cárdenas and María de González Cossío from the Universidad de las Américas Puebla, in the second half of the 1990s, for the opportunity to create a master course in the new area of information design.

Michael Erlhoff as founding Dean of Köln International School of Design (KISD) opened the possibility to teach in Germany again after 25 years of absence. Philipp Oswalt from the University of Kassel who facilitated my participation in various academic symposia in Berlin (2015) and Kassel (2018). I am grateful for his careful editing of various texts, and to David Oswald (Berlin/Schwäbisch Gmünd) for critical reading and exchange of opinions. Dagmar Rinker and Christiane Wachsmann from the archive of the HfG Ulm of the Ulm City offered indispensable help in the search for relevant documentary source material for several articles linked to HfG Ulm issues to which I did not have direct access from Latin America.

In Italy, it was the lifelong friendship with Tomás Maldonado, whom I consider my mentor, always ready to give generous advice. Each meeting with him, each conversation with him, was a source of new stimuli opening new avenues. Having opted for working what is now called the Periphery, he found with a touch of irony an explanation—and as such not to be taken literally—for my decision: anthropological interest. However, that assessment touches one core motif: the possibility to link the notion of emancipation to the notion of design, thus a link between project activity and a sociopolitical context. This concrete possibility I started to perceive in the Periphery. Also I am indebted to Costanza Pratesi and Raimonda Riccini in Milan for the help in the edition and translation of texts.

In Portugal, it was António Gonçalves who as director of the Direcção-Geral da Qualidade (DGQ) invited me to Lisbon in the second half of the 1970s. Later, the contacts with Portugal continued through Raúl Cunca.

My special acknowledgment goes to the members of the different design teams:

— INTEC Design Group in Chile: Guillermo Capdevila, Gustavo Cintolesi, Pedro Domancic, Alfonso Gómez, Fernando Shultz and Rodrigo Walker, Werner Zemp, and Michael Weiss; and to Graphic designers: Eddy Carmona, Jessie Cintolesi, Pepa Foncea, and Lucía Wormald.
— The LBDI team members (Laboratorio Brasileiro de Design Industrial) in Brazil: Maria Regina Álvares Correia Dias, Petra Kellner, Holger Poessnecker, João Rieth, Célio Teodorico dos Santos, and Tamiko Yamada.
— Design studio MM/B Méndez Mosquera/Bonsiepe/Kumcher in Argentina: Hugo Legaria, Miguel Muro, and Héctor Taboada.

Last not least my thanks go to Lara Penin, Clive Dilnot, and Eduardo Staszowski to have included this book in the series of *Designing in Dark Times/Radical Thinkers in Design* and for their patience in the editorial process.

In Argentina, Silvia Fernández provided indispensable support and critical voice as independent researcher. My deep thanks to her.

Gui Bonsiepe
Florianópolis (Brazil), La Plata (Argentina)

Editorial Acknowledgments

All books are complex, even those whose concept naively appeared to its originators as comparatively simple. In the task of realizing this book, our central acknowledgment has to go to Gui Bonsiepe. Conceived from the beginning as one of the key thinkers whose work needed to be included in the series *Radical Thinkers in Design*, the project has been enormously leavened by our sometimes-daily correspondence with Gui. The circumstances of these conversations were not the easiest since they took place largely across several months in 2020 when, because of COVID-19 travel restrictions, the author was in "exile" in Germany, unable to get back to Argentina and thus deprived of his archives and materials. Nonetheless, this editorial interaction has not only strengthened the book beyond what we had first imagined but just as significantly has made it a project of the moment as well as a record of work spanning more than sixty years.

The second acknowledgment concerns permissions. All volumes of collected essays are by definition dependent on permissions to reprint papers and projects. The volume has benefited enormously from the generosity of a few individuals and institutions, most notably Dr. Eduardo Blücher of Editora Blucher in São Paulo, for some of the papers and projects originally published in Portuguese in their three-volume edition of Bonsiepe's works; Philip Ostwald, for assistance with some of key texts originally published in Germany; Herbert Lindinger and Martin Mäntele of the HfG Ulm archive and the editors of *Image and Text* journal in South Africa. More formal permissions for publication have been obtained from MIT Press/*Design Issues* (which over the last decades has been one the few English-language design journals to have published some of Bonsiepe's essays) and from transcript publishers in Germany. However, the main and central thanks for permission to reprint texts go to Dawn Oxenaar Barrett, who was the editor of *Interface*, the first English-language volume of Gui Bonsiepe's writings. It is this text which in many ways forms the initial basis from which this volume was constructed. We should thank here also Jan van Toorn, who, as director of the Jan van Eyk Academie in Maastricht at that time, directly inspired the project (and who passed away, sadly, as this volume was being prepared). Without their work of translation this volume could not have been produced. We also thank Gabriel Patrocínio for sharing the English version of his and José Mauro Nunes' interview with Bonsiepe "Design and Development 40 Years Later." We have reached out to a number of people in different countries searching for specific copies and translations of Bonsiepe's texts, and we thank them all for their help, in particular Tuti Giorgi in Porto Alegre for finding a copy of a magazine with the text "The Relevance of the Ulm School of Design Today" in English.

The third group of people who we need to thank are our translators. Almost every text in this volume has been translated from its original language of publication—beginning with the texts Bonsiepe published in the 1960s in the journal of HfG Ulm and self-translated into English. As it is, around a third of the papers/projects published here are newly translated. For translations from three texts in German, we are indebted to Anke Gruendel, and from Spanish, to Niberca Lluberes Rinicon, Estefania Acosta, and Quizayra Gonzales. From the Portuguese, the editor translated not only a number of papers in the volume but also the critical essays drafted by Ethel Leon and Marcos Martins and Zoy Anastassakis.

Lara Penin

PART ONE
Thinking Design

Nostalgia Is Futile—Our Future Starts Now

Frederico Duarte

The common principles of freedom, peace, democracy, development, and the rule of law guided the creation of the United Nations Organization and a wide range of other multilateral institutions after the Second World War. Their role: to govern an increasingly interconnected planet shared by a growing list of nations. Their unfailing, if perennially unmet, goal: to design a global future free from exploitation, oppression, conflict, and hunger.

The most eloquent expression of this hopeful effort in human history is the Universal Declaration of Human Rights.[1] Proclaimed by the United Nations General Assembly in 1948, this remarkable document provided the blueprint for a global *convivencia*, the historical Castilian word that roughly translates as a living-togetherness of distinct communities. Regardless of how humans had hitherto and would henceforth define, organize, and rule their communities according to biology, ideology, religion, or national identity, the women and men who drafted the declaration's thirty succinct articles set the conditions we all must honor to respect and defend our individual and collective dignity.

This commitment to human dignity is the most distinct hallmark of Gui Bonsiepe's long-life and prolific career. In over half a century of contributions to design practice and discourse, his dedication to building a rules-based, multilateral institution-led world order on the values of this declaration has been expressed in countless of his actions, reflections, and proposals.

In writings such as the nine texts and one interview gathered in this chapter, which date from 1964 to 2005, Bonsiepe expresses his ideas on design with the authority of an experienced practitioner, educator, and policy adviser. His analysis and reflections are informed by a wealth of references and observations on politics, society, and industry, as well as science, art, and technology, which he articulates with the clarity and criticality of a public intellectual.

These essays reveal how Bonsiepe positioned himself within the professional, institutional, and academic establishments of design as a free mover and independent thinker. Six of them were written as lectures and presented in conferences, symposia, and other functions in Italy, Argentina, Switzerland, Chile, and the Netherlands. Each was written not for the eye but for the ear, as a direct appeal to a specific audience. More specifically, to an audience mainly composed of students, to whom the future is dearest.

---

[1] UN General Assembly, *Universal Declaration of Human Rights*, December 10, 1948, 217 A (III), available at: https://www.refworld.org/docid/3ae6b3712c.html (accessed November 1, 2020).

As with many of his writings, most of these essays were delivered in languages other than English. Some were since translated, published, and redacted in other languages such as German, Spanish, Italian, and Portuguese. Regardless of the time, audience, and language they were first delivered in, they remain remarkably relevant.

Bonsiepe's ideas on design have not just been spoken in lecture halls or published in academic journals, trade publications, or books of history and theory. Some of his most relevant analysis and consequential proposals were commissioned by governments, as policy papers or reports associated with design and research institutions. He further contributed to the institutionalization of design, on a local and an international level, by participating in congresses, conferences, and programs promoted by professional organizations such as the International Council of Societies of Industrial Design (ICSID, founded in 1957 and named, since 2015, World Design Organization, or WDO) and international institutions such as the United Nations Industrial Development Organization (UNIDO).

One such congress was the Meeting for the Promotion of Industrial Design in Developing Countries, convened by UNIDO and ICSID in January of 1979. In what was the first design conference sponsored by the United Nations, delegates from all over the world gathered at the Indian National Institute of Design in Ahmedabad, India, to present their positions and discuss the principles and recommendations inscribed in the Ahmedabad Declaration on Industrial Design for Development.

Bonsiepe was actively involved in both the drafting and discussion of one of the twentieth century's most exceptional declarations, which over forty years later has lost none of its relevance and urgency. Especially when taking into account the following three of its seventeen principles: by declaring a "firm belief that designers must have a clear understanding of the values of their own societies and of what constitutes a standard of life for their own people", as well as "that design in the developing world must be committed to a search for local answers to local needs, utilizing indigenous skills, materials and traditions while absorbing the extraordinary power that science and technology can make available to it" and that "designers in every part of the world must work to evolve a new value system which dissolves the disastrous divisions between the worlds of waste and want, preserves the identity of peoples and attends the priority areas of need for the vast majority of mankind,"[2] its signatories proved remarkably conscious of how design, much like development, can and has been employed, often in the name of progress, to provide but also to dispossess humans of their economic, cultural, and even epistemological dignity.

---

[2]   Principles 14, 15, and 16 of the Ahmedabad Declaration on Industrial Design for Development, reproduced in S. Balaram, "Design in India: The Importance of the Ahmedabad Declaration." *Design Issues* 25, no. 4 (2009): 54–79.

In his 1992 lecture *The Cartography of Modernity* (Chapter 1),[3] Bonsiepe claims that in the act of design modernity awakens. "To be radically modern," he pronounced in an event honoring the designer and educator Tomás Maldonado, "is to invent the future, to design and arrange it, and that includes the future of that same modernity" (Chapter 1, p. 13). Such belief in modernity echoed the *ethos* that ought to have driven the multilateral institutions and international networks to which Bonsiepe dedicated much of his life. Yet while claiming to build a more dignified future for all, the world order that these institutions were designed to govern would by the end of the twentieth century reflect the effects of a hegemonic project of Western (or Euro-American, or central) globalization, instead of a more balanced, multipolar world order. Expressed in economic, cultural, and military dominion, this hegemonic project has increasingly been denounced and debated as grounded on the notorious and long-lasting foundations of exploitative capitalism, coloniality, and empire. The Ahmedabad Declaration is, much like Bonsiepe's body of work, a key element in this debate.

Delivered in Milan on April 24, 1992, *The Cartography of Modernity* was heard as the liberal world order was being redesigned. With the dissolution of the Soviet Union, the creation of an enlarged European Union, the application of Washington Consensus measures in Latin America and China's implementation of Socialism with Chinese characteristics, a new world order would henceforth adopt the late-capitalist, neoliberal model, reliant on trade liberalization, the superiority of market-based competition, and individualism. Values such as social cohesion or the defense of the public good were undermined by the financialization of an increasingly connected global economy, corporation-led globalization, and the inexorable quest for profit over purpose. A corrosion of the utopian, and therefore imperfect, effort that drove the creation of the postwar world order and the redaction of the two aforementioned declarations led to the relentless weakening of states and institutions, the progressive commodification of every aspect of human life, and a blatant disregard for social inequality and exclusion in both developed and developing or peripheral nations.

In writings found in and beyond this chapter, Bonsiepe has revealed a recurrent concern with the dismantling or the subversion of this world order. He has been particularly critical of the failure of its institutions to address the economic, social, and cultural divisions between center and periphery. This criticism is also aimed at designers, or rather, of those in charge of design discourse, as they have failed to grasp and manifest how critical this activity is in determining a shared future. In other words, in coming to terms with design's inevitable political dimension, "in that it includes a component of hope—a dream, however vague, with the outlines of the society we want to live in" (Chapter 4, p. 35).

---

[3] As with the other introductory essays in this volume that quote or make references to papers in the book, the pagination will be by the chapter number and page reference.

In his Milan lecture on modernity, Bonsiepe proclaims that "What characterizes the peripheral world is the lack of a design discourse. That is why these countries have not, so far, had a future—for the future is where design unfolds. Only through design is it possible to appropriate the future" (Chapter 1, p. 13). This tension is thus greatly a matter to how a discourse over design is produced, maintained, and eventually challenged.

As Bonsiepe addresses in these and other texts, a dominant discourse on design, buttressed by institutions, organizations, and networks of the aforementioned rules-based international order, aimed toward a professional autonomy and economic, cultural, and political recognition of the discipline. Yet, by the late 1980s this discourse was gradually replaced by another, which reduced design to the symbolic and the ephemeral but also, somewhat paradoxically, to an inevitability of objects and images. Design became less about verbs and more about nouns or, worse still, about adjectives. Considered as a populist, frivolous expression of a consumerist lifestyle or as an expression of artistic or personal style, the practice and consumption of design thus became, in postmodern, late-capitalist, turn-of-the-millennium times, more associated with achieving individual success than with building a shared, global future grounded on human dignity.

The essays in Part I therefore read as appeals to designers to (re)discover the meaning of their practice and activity. This discovery starts with the four texts dedicated to the Hochschule für Gestaltung Ulm (HfG Ulm) gathered in the section *Essays on Ulm*, in which Bonsiepe discusses the aspirations but also the interpretations of the design school where he studied and taught from 1955 to 1968. Built with the support of the Marshall Fund in memory of two members of the resistance to the Nazi regime, HfG Ulm was itself the expression of a desire to build a world *beyond* war.

As Bonsiepe observes, despite having earned a deserved reputation as a citadel of methodolatry (Chapter 2, p. 17), this institution aimed to ground design as an autonomous discipline committed to the enlightenment project. The school's commitment to a critical rationalism (Chapter 3, p. 27) included a utopian component and a certain promotion of disquiet (Chapter 2, p. 17), both of which Bonsiepe defends as essential to design practice and life itself. Refusing to "linger on the side of problems," the Ulm school landed instead squarely on the side of solutions; or, as Bonsiepe rephrases in one of his sharpest observations, "it refused to engage in a purely discursive, theoretical dancing around the problem" (Chapter 3, p. 27).

This is an important point. At Ulm, students and faculty confronted industry and its problems according to a project-oriented approach that unequivocally distinguished design from art. Instead, they focused their attention on the relationship between design and society, which is contextual, contingent, and by no means free of contradiction (Chapter 3, p. 27). This approach considered the teaching and practice of design no less than a critical intervention. Indeed, Bonsiepe sees criticism in design as active criticism, as a practical intervention, which is meant to confront and eventually change the existing system of industrial production. At Ulm, this engagement with and critical attitude toward industry did not however foresee how market problems would replace problems of production (Chapter 4, p.36), especially as the postwar economic miracles of Western Europe gave way to an increasingly complex and interdependent global consumer society.

Nevertheless, Bonsiepe's remarks on the reputation and legacy of the HfG Ulm, as well as on the teaching of design more broadly, are still strikingly critical in 2020. They include considering design an alien body within a design education (Chapter 7, p. 65) that has become both increasingly commodified and consolidated in traditional universities (Chapter 4, p. 35). In the spirit of the 1948 and 1979 declarations, and of Bonsiepe's own cosmopolitan education and intensely peripatetic career, he defends the creation of a new kind of design school, operable on an international level (i.e., beyond the nation-state), where a culturally diverse faculty and student body would offer the necessarily critical curriculum and stimulating learning environment (Chapter 3, p. 27). Pursuing a radical cosmopolitanism is, instead of performing some nostalgic version of modernity, the best way to honor the *ethos* and legacy of the HfG Ulm (and, also, of the Bauhaus). This entails making design political. On this regard, Bonsiepe mentions one of Ulm's unspoken theses: the question of the meaning of design can only be answered in the social context, which also means in the political context (Chapter 4, p. 35).

In the *Theory and Practice* section, Bonsiepe steers the attention of students, educators, practitioners, and those involved or interested in producing and keeping design discourse away from objects, forms, or styles and toward the conditions, intentions, but above all, the methodologies through which design takes place. This particularly rewarding section contains some of his most critical observations regarding the role of theory in design practice and the fetishization of research, innovation, creativity, and methodology itself in what has become as sort of global educational-industrial complex. Furthermore, it claims the roles of visuality and discursivity—how words are brought to images, and images to words (Chapter 5, p. 45)—in design education, inside and outside academia, while proposing to reduce the gap between different societies as "a very relevant subject" for design research: "Assuming, of course, that we consider this at all meaningful and do not dismiss it as the perpetuation of the status quo of a social value system that has no future and needs to be radically renewed and turned upside down" (Chapter 7, p. 65).

This and other of Bonsiepe's calls to action become most acute, and most audible, in the *Design, Politics, Ethics* section. In *Design and Democracy*, a lecture he delivered in Santiago de Chile in 2005 and that has since been published in several languages, Bonsiepe questions how the true meaning and credibility of democracy can be recovered from neoliberal forces. Such forces, he argues, "believe that democracy is synonymous with the predominance of the market as an exclusive and almost sanctified institution for governing all relations within and between societies" (Chapter 9, p. 87). In his interpretation of the term, Bonsiepe emphasizes democracy's sense of participation, which implies that "dominated citizens transform themselves into subjects opening a space for self-determination, and that means a space for a project of one's own accord" (Chapter 9, p. 87), while favoring "a substantial, and thus less formal, concept of democracy as the reduction of heteronomy (i.e., domination by external forces)" (Chapter 9, p. 87).

Bonsiepe returns to the tension between center and periphery, or between autonomy and heteronomy in design, by proposing the term "design humanism," which he describes as "the exercise of design activities in order to interpret the needs of social groups, and to develop viable emancipative proposals in the form of material and semiotic artifacts" (Chapter 9, p. 87). Why emancipative, Bonsiepe asks? Because humanism implies the reduction of domination. In the field of design, he says, this also means "to focus on the excluded, the discriminated, and economically less-favored groups (as they are called in economist jargon), which amounts to the majority of the population of this planet" (Chapter 9, p. 87).

While such a focus echoes the goals and values of the 1948 and 1979 Declarations, Bonsiepe eschews any kind of universalistic attitude, naive idealism, or normative request as of how a designer, but also any other professional, "exposed to the pressure of the market and the antinomies between reality and what could be reality," should act today (Chapter 1.9, p. 87). He more modestly intends to foster a critical consciousness regarding the "enormous imbalance between the centers of power and the people submitted to these powers, because the imbalance is deeply undemocratic insofar as it negates participation" (Chapter 9, p. 87). This call for an emancipated, critical consciousness among design professionals and other social agents regarding the social, economic, and cultural processes that perpetuate the dependence of peripheral nations is a key aspect of Bonsiepe's original and critical contribution to design discourse.

For Bonsiepe, industrialization is the only way for this emancipation process toward autonomy to take place in peripheral nations beyond the whims (and dependencies) of hegemonic powers, but also of national elites. It is the most viable possibility for democratizing consumption and for providing access to the world of products and services in the different areas of everyday life for a broad sector of the population (Chapter 9, p. 87). This entails the investment in industry, technology, and design not just of mass consumer goods, but also of public goods, infrastructure, and services. Such an effort must therefore acknowledge the role of the state in the advancement and institutionalization of design, both as a practice and a discipline, in tandem with investment and intervention by the public sector in industry, innovation, and development. Which may only result in a fairer future for the largest possible number of human beings if driven by long-term, socially minded concerns instead of short-term, market-oriented goals. As Bonsiepe added in his 1992 essay, "The peripheral world, however integrated it may be in the world economy, will only have a future to the extent that it makes design a social practice" (Chapter 1, p. 13).

With the deindustrialization of much of the developed world (and certain developing nations, namely those in Latin America) from the 1980s onward, industrial production ceased to be perceived as an issue of interest for the institutions and individuals in charge of design discourse. More than promoting a postmodern splintering of design practice into a design culture, this approach detached design from its origins as an economic activity. On their approaches to design practice in peripheral nations, agents involved in design discourse such as historians, editors, and curators shifted their attention from the local responses of design and industry, or from the global impact of autonomous thinking and making in the periphery, to the appreciation of the morphological characteristics of a select group of artifacts, its chief practitioners, and the often global repercussion of their lives and ideas.

The postmodern discourse Bonsiepe rightfully criticizes tends to focus on a practice of design that favors representation and style over function, consumption, or use. It claims such a practice is grounded on a kind of individual agency and authorial production that is more akin to artistic practice. Crucially, it stimulates a specific market for design artifacts destined at an elite of consumers or collectors, which are valued for their aesthetic, symbolic, metaphorical, and narrative qualities in detriment of other characteristics, such as commercial application and social impact. In addition, this discourse seeks to find in every design commodity the relevant morphological and semantic elements that express a local/regional adherence to, or divergence from, an international style, trend, idiom, or canon.

A design discourse based on this commodity market leads to two main significant and damaging consequences. The first is how by concentrating interest and seeking value on a limited range of artifacts, disciplines, practices, and agents, this system creates and sustains a skewed, biased, and reductionist discourse on design. This discourse not only misinterprets the potential and social impact of design practice but also subverts the artifact's own use value (as a product) and exchange value (as a commodity), conditioning the understanding and expectation of what design is as activity, profession, and subject matter.

The second consequence of such a discourse is the erasure of any contemporary relevance from the few design artifacts considered. As only a fraction of the results of design practice are appreciated, an overwhelming number of potential instances of design are overlooked. More importantly, the very agency of designers and their contribution to a specific context is ignored. Such a skewed, object-centered, individualistic, apolitical discourse on design has affected the discipline's scope and consequence. No wonder design is not taken as seriously as it should be.

As the twenty-first century enters its third decade, both the twentieth-century liberal international order of multilateral institutions and the postmillennial, neoliberal order of multinational corporations, global finance, and information networks have been shaken by the creative destruction of capitalism, terrorism, and social media. The rise to power of national leaders set to undermine liberal democracy has further endangered the legacy of the women and men committed to designing a better, more dignified future for the planet's inhabitants.

The "war on terror," the 2008 financial crisis, and, to a still unknown effect, artificial intelligence and the COVID-19 pandemic, all urge a collective reflection over the causes and consequences of design. Such a reflection implies inquiring into the social worlds that design both shapes and is shaped by. This also implies acknowledging how designers have been complicit in corroding the foundations of civil society, human dignity, and, in what is a cause dear to Bonsiepe, the dignity of work. Only then we may refocus on what truly matters and take design seriously.

The digital revolution and social media led to an explosion of content available online, but also its progressive devaluation and atomization. This has led to a crisis in the media, which includes publications dedicated to design. More than outlets for uncritical dissemination of works and promotion of their authors and clients, these publications are institutions in their own right. They have over decades offered a kind of barometer of the profession and helped shape a discourse on design as an activity. In fact, the shift in dominant discourse on design identified by Bonsiepe has largely taken place in the physical and digital pages of these publications. As the most impactful of these publications are edited in Western, central nations—and mainly in the English language— they also bear the cultural biases of their editors and contributors toward the periphery—wherever it may be found. Yet like other institutions, design publications are also going through a crisis of authority. This is actually good news, as an increasingly fragmented institutional and media landscape allows for the discovery of new ideas and voices, but also the creation of necessarily collaborative and polyphonic opportunities to enunciate and discuss a design discourse that is fit for our time.

In lieu of conclusion, I would like to advise the reader that many of the ideas, proposals, and tools necessary for the crafting of such a discourse can be found in the pages of this book. They should be read not as nostalgic incursions into a simpler past (for the past has never been simpler) but as clear, coherent, consistent, and consequent calls to action.

Following my own reading of Gui Bonsiepe's texts, I would single out five claims that designers should make if they aim to take their work seriously. The first is that their activity is a vital dimension of how humans dream, build, and keep the ways we want to live together. The second is that their work is a form of political action. The third is that they must claim responsibility toward a defense of the public good as an essential element of individual and collective dignity. The fourth is that they ought to critically revisit the values of a world order Bonsiepe actively contributed to build, expressed in the 1948 and 1979 declarations. Finally, any discourse that contradicts the other four claims should be rigorously challenged by intellectual positions and practical actions.

Essays on *HfG Ulm*[1]

[1]   It was a characteristic of the ulm school of design that the school self-identified itself, from 1953 onward until its closure in 1968, with the use of the lowercase "ulm." This both differentiated the school from the city and became part of the school's identity, used almost exclusively internally and in correspondence and in combinations with uppercase capitals in publications and external presentations. In acknowledging this history, we have stayed with the lowercase for the abbreviation and shorthand ulm, confining capitals to the more formal use of the abbreviation HfG and the full name: Hochschule für Gestaltung Ulm. The original version of this text was published in both German and English in ulm: Zeitschrift der Hochschule für Gestaltung, May 10/11, 1964, pp. 18–29. English translation by Gui Bonsiepe.

"Not only physically speaking"—*friends added this remark when they were talking about the greatness of Tomás Maldonado:* El Grande Tomás. *He was one of the core figures of the ulm school. When I applied in 1955 to study at the HfG Ulm, I had not heard the name of Maldonado who was in charge of the basic course—he was not known at that time in Germany, culturally a flatland recovering from the devastation of the war. Soon he perceived my disconformity with the course. At the end of the first trimester, he invited me for a personal conversation, and after having heard the arguments for my disappointment, he lifted a telephone from the table in the meeting room, put it back with emphasis, and said: When you want to study art, go to Paris; we do THIS here! That was clear. After his move to Italy in 1967, his influence grew with the creation of the first course in industrial design at a public university, in this case at the Politecnico di Milano. His numerous books, first published in Italy, constitute landmarks for the consolidation of the design discourse. The breadth of his cultural interests and his cosmopolitan stance were impressive, as was the insistence on clarity of argument in his texts—endless fine-tuning was the rule until they met his quality standards. Maldonado was decisive for my professional development. His invitation to visit his home country Argentina in 1964 opened a new reality, decisive for the later work in Latin America. We became friends.* (GB, 2021)

1 THE CARTOGRAPHY OF MODERNITY

(1992)

This short laudation puts the focus on the extensive theoretical work of Tomás Maldonado. I should like to raise only one point, but it forms the basic tenor of his writings. In the center of his debate is the design dimension of modernity, the coupling of "design" (in the sense of "project") and "modernity"—terms that are inseparably linked.[1]

The dual concept can be divided into two areas:

— Firstly, modernity as an object of design intelligence;
— Secondly, design as the outstanding feature of modernity, an ontological constant.

The radicalism of that approach can hardly be exaggerated—it is a radicalism expressed both in the boldness of the theses and in the stringency of the conclusions.

The debates on the subject of modernity, from whatever points of view they were conducted, rarely focused on the design dimension. In the chorus of voices, design has been the great absentee, the yawning gap. It was not one of the accredited subjects in the cultural debate, and one is reminded of Braudel's remark about publications on traditional history: in these "man is a being who neither eats nor drinks."[2] Even in writings on the theory of social action, the question has not, so far, been raised of what forces form the arsenal of objective and communicative artifacts. Clearly, these are assumed to be given and never questioned, as if they leaped into the world in a self-explanatory act of magic.

Maldonado has corrected that sin of omission. In his writings he creates, piece by piece, a structure of arguments that permit us to reinterpret the world from the viewpoint of design rationale. Design is not seen simply as an occasional phenomenon, or a subtheme of modernity; it is a driving force of modernity. In the act of design, modernity awakens. To be radically modern is

---

This paper was written for the meeting *Il faut être absolument moderne*, in honour of Tomás Maldonado, Milan April 24, 1992. It was first published in German as "Kartographie des Projekts der Moderne," *form & zweck* 4/5 (1992): 108–9. It first appeared in English in Dawn Oxenaar Barrett (ed.), *Interface – An Approach to Design* (Maastricht: Jan van Eyck Akadamie, 1999), 128–31.

[1]   In the German original of this text, I used the broader and stronger term "Entwurf" (project) in the sense of Project of Modernity as used in social sciences. It is stronger than the increasingly worn-out term "design." This semantic difference and partial overlapping of two terms appear for the first time.

[2]   Fernand Braudel, *La dinámica del capitalismo* (Mexico: Fondo de Cultura Económica, 1986), 18. (First edition, Paris, 1985.)

to invent the future, to design and arrange it, and that includes the future of that same modernity.

So far-reaching a definition goes beyond the borders of the design of material things and tools, their production, sale, use, and obsolescence, although industry is closely connected with the concept of modernity. Maldonado has repeatedly returned to industry as the medium in which modernity is realized, and in doing so he has snatched another theme from the shadows of the past. He looks beyond the traditional design disciplines like architecture, industrial design, graphic design, and fashion design, all of which are only one small sector in the universe of design. But at the same time his discourse can be seen as an undertaking that will liberate these design disciplines from their theoretical morass and raise them to a level where they can be subject to stringent discussion.

Like a surgeon with his scalpel, Maldonado has laid bare the main nerve of modernism. A surgeon needs a sure and sometimes a hard hand. He cannot afford to hesitate or to shake. Work of this nature needs stamina to lift layer after layer of the tissue of traditional discourse and penetrate down to the essence. In my view—and I believe not only in my view—the object of my *laudation* is the inventor of design discourse. No one before him did this in this depth. For that reason, this is not only an achievement with innovatory force, but it is also an achievement of radical innovation; it cuts paths into the future.

As might be expected, the approach is controversial, in that Maldonado clings fiercely to one principle of modernity—the view of the whole. Since the 1960s, a wave of retrogressive rhetoric has spread. The fundamental difference between radical modernism and its critics can be analyzed as follows: although the representatives of radical modernism certainly do not avoid or try to conceal the discontinuities, the contradictions, and paradoxes of modernism, they are never willing to break it down. Radical modernism—and that is the crux of the consequences—postulates the mapping of design rationale on to sociopolitical rationale, so it demands a correspondence between these two basic fields, at least by intent. As we know, the principle of intended concordance is anathema to the supporters of retro-modernism in its many forms.

The debate on modernity has grown in the context of European cultural traditions. But it claimed validity beyond historical contingency. Consequently, it is legitimate, and virtually unavoidable, to ask how relevant the issue of modernity is for all those countries that are on the fringe of the economic, political, scientific, technological, and cultural centers. What characterizes the peripheral world is the lack of a design discourse. That is why these countries have not, so far, had a future—for the future is where design unfolds. Only through design is it possible to appropriate the future. The peripheral world, however integrated it may be in the world economy, will only have a future to the extent that it makes design a societal practice.

All over Latin America one can now hear declarations in favor of modernity. They want to be the First World, or at least catch up with it. Third World is out. They want modern technology, modern industry, modern management, and modern education. However, they appear to be less prepared to accept modern social structures as well, with less of an income differential and less dilacerated social relations. So, there is a danger of an amputated modernity, modernity cut in half in the peripheral world.

If one considers the great open questions of modernity today, one can identify three central themes:

— The crisis of the paradigm of industrialism, embodied in Fordism, and more aptly called fossilism. Today one speaks of sustainable growth. That can be an indication that an ecological rationale is beginning to consolidate, at least in an early form, and the cowboy phase of human history would thus come to an end.
— The conflict-laden North–South relationship, that is, the asymmetrical relation between the small quarter of the world's population who live in the industrial countries and the three-quarters of the rest of mankind.
— The formative contributions from informatics, particularly in networking and digitalization.

Design rationale—modernity—will find, or fail to find, its legitimation to the extent that it can prove how successful it is in finding theoretical and operative answers to these three themes, for they may well dominate the design discourse for the coming decades.

If one looks at the project of modernity in the global view and not only from the standpoint of the small number of industrialized countries, one can say that modernity is actually only beginning now, or it can only begin now, certainly if one sees it as an obligation to make this planet more worth living on.

As is well known, one of the outstanding features of the HfG Ulm was its interest in building bridges between the sciences and design, or rather between the sciences and design education, and in this way to objectify education and transform design into a teachable—and learnable—activity. It was foreseeable that such a project would cause conflicts. Of course, the initiators of the Versachlichung (objectification) of design education—all designers—had not foreseen the severity of the conflicts and occasional excesses of the furor teutonico.

Observing that a one-dimensional interpretation of science and scientific methods could lead to counterproductive results for design teaching—and that means teaching the competence to project—and might even foster reservations against the attempts to integrate scientific disciplines in a design education program, this article offering a panoramic view has been written in 1963. As in any living institution, there existed different approaches and sometimes conflicts. This article permits to see that positions were considerably more differentiated than schematic labels allow. The HfG Ulm was definitely not a monolithic block, although this feature has not always been registered in the external perception of the institution.

2 SCIENCE AND DESIGN

Tomás Maldonado and Gui Bonsiepe

(1964)

Not everything attributed to the Hochschule für Gestaltung Ulm (HfG Ulm) can be legitimately recorded on its credit and debit ledger sheets of achievements and failings. Nevertheless, both fact and fiction surround the HfG Ulm at one point, namely the interest in design methodology, the interest in a relationship between science and design.

The HfG Ulm has in fact deserved the reputation of being the citadel of methodolatry. An important characteristic of its curriculum manifests itself in the emphasis laid on the application of both scientific knowledge and scientific methods in the design process. This rigor is reflected in the various opinions about the HfG Ulm, which has earned for it supporters and opponents. Some, who are inclined from the start to distrust science and scientific thought, consider the ulm approach no more than a new variant of the furor teutonicus—cold, scrupulous, humorless, niggardly, obstinate. Others consider the HfG Ulm a more or less successful model of a synthesis of science and design. Hence, on the one hand, the HfG Ulm methodology—of what is considered to be the ulm methodology—has given rise to a resistance which even reinforces the romantic attitude toward design. On the other hand, it has brought about an altogether indiscriminate, and often unfounded hope in design under the scientific aegis.

The following notes attempt to throw some light on the range of problems indicated, in the title. They are no more than notes of a cursory nature gathered by the authors from their own observations, in discussions, and in a critical assimilation of relevant writings. In this article, sociology and social psychology, two factors which influence design methodology, have intentionally been omitted for the reason that the importance of these disciplines for design is generally accepted and undisputed nowadays. The same applies for product engineering and manufacturing.

*

THE LIMITS OF MATHEMATICAL TECHNIQUES

Methods are determined by goals. Expressed in psychological terms: method is part of a goal-directed behavior; method is part of a behavior directed toward the

The original version of this text was published in both German and English in ulm: Zeitschrift der Hochschule für Gestaltung, May 10/11, 1964, pp. 18–29. English translation by Gui Bonsiepe.

solution of problems. In solving problems, one can employ various methods. If one solves problems methodically, the approach is controlled or planned. This rational factor occasioned Bentham to define it as the implementation of what one could term "tactic faculty"; or according to Buchler's definition: "Method is the strategic dissemination of prudence."[1] Imagination is the dialectic counterpart of method—the rational application of definite techniques within the inventive process. "Invention, free of regulatory restraints on imagination, is flexible. Method . . . introduces fixity."[2] In other words, the function of a method consists in the regulation of unbridled imagination, in its guidance into definite avenues, and in the obtaining of a result in this manner. J. C. Jones argued similarly, saying that method is a means of arbitrating in the conflict between logical analysis and creative thinking.[3] Method operates in the range of possibilities lying between random success and rational determination. According to customary opinion, the more scientific a method, the more that chance can be eliminated—the more that success can be predicted. Methods in science are directed toward two goals: first, the discovering of truth—how are true statements obtained? (speculation, hypothesis, experiment)—and secondly, the checking of their truth value—how does one ensure oneself of the correctness of the statements? (logical compatibility, verification). These methods are characterized by the following features: they are quasi-general (i.e., they refer to more than one case), and they are intersubjective (i.e., they can be repeated by several subjects).

The ensemble of methods employed in designing products, their systematic arrangement, is called the "methodology of product design." This term should not imply the assumption, although the appearance might suggest that there is, or can be, a uniform general design methodology: as though nothing further need be done than to develop a design methodology from scientific methodology—which has still not been defined despite the efforts of the "unity of science" movement. Within this array of methods, some mathematical procedures have attracted particular attention. They stem predominantly from the domain of finite mathematics, that is, the branch of mathematics that does not presuppose the term of continuity nor limit transits or infinite quantities in the treatment of problems.[4] It seems that vector analysis and matrix analysis together with linear programming may be applied to the solution of design problems. But in design the same applies as elsewhere: Which methods are adequate for which aims? There are frequently "good" problems which are approached with "bad" methods, and

[1]   Justus Buchler, *The Concept of Method* (New York: Columbia University Press, 1961).
[2]   Ibid.
[3]   J. C. Jones, "A Method of Systematic Design," in *Conference on Design Methods*, eds. J. C. Jones and D. G. Thornley (Oxford: Pergamon, 1963).
[4]   J. G. Kemeny, J. L. Snell, and G. L. Thompson, *Introduction to Finite Mathematics* (Englewood Cliffs: Prentice-Hall, 1957).

inversely "bad" problems approached with "good" methods.[5] The quality of the questions and the quality of the methods must be correlated.

In the preface to an article by A. Moles dealing with the structural and functional complexity of products[6], a pluralistic methodology for design was envisaged. Insofar as there are various degrees of complexity, appropriate methods must be allotted to them in order to deal with the problems occurring at the various levels. To approach the design of cutlery or cooking utensils or a radio cabinet with an arsenal of techniques derived from operations research is both uneconomical and inappropriate. But there are cases on the boundary between product design and far-reaching product planning. If it confuses the issue to assume that a technique like linear programming may be employed without modification in the design or redesign of products, then the usefulness of this discipline is quite out of the question in the development of new products in the framework of a definite company policy; and, finally, product design is also included in this new development.[7] Moreover, system theory and theory of control could be useful for a designer in combination with mathematical logic as a general background for the understanding of a machine theory. Information theory may furnish the designer with a vocabulary of terms to analyze and to quantify the structural relationships of products. According to experience gathered hitherto, the following mathematical disciplines may be considered useful for the product designer in his practical design activities:

1) Theory of combinations (for modular systems and problems of dimensional coordination);
2) Group theory (in the form of symmetry theory for the construction of grids and three-dimensional lattices);
3) Theory of curves (for the mathematical treatment of transitions and transformations);
4) Polyhedral geometry (for the construction of regular, semi-regular, and irregular bodies);
5) Topology.

DESIGN: A NEGLIGIBLE QUANTITY?

Like all human activities intended to integrate various special disciplines, the specific activity of design must defend itself against each of these disciplines if and for the reason that each wants to dispute the legitimacy of design and expose it as superfluous. In its earliest form, the conflict was contested between the designer and the engineer, whether he were engineering designer or production

[5]   Allen Newell and Herbert A. Simon, "Computer Simulation of Human Thinking," *Science* 134, no. 3495 (December 1961).
[6]   Tomás Maldonado, "Vorbemerkung aus Produkte: ihre funktionelle und strukturelle Komplexität," *ulm* 6 (1962).
[7]   M. K. Starr, *Product Design and Decision Theory* (Englewood Cliffs: Prentice-Hall, 1963).

engineer. This point has already been discussed in detail so often that it is not worthwhile taking up now. On the other hand, it is worthwhile discussing briefly the critical remarks which have recently been brought forward by advocates of mathematical methods against design. According to them, the design of a product scarcely differs from the design of a system, whatever its type: in all cases, the prime task is the solution of a problem or complex of problems. Once all the variables which enter into the solution of a problem have been summarized, a mathematical formulation would be sufficient, whereby all the detailed data or determinative data would be covered. This would then permit an optimal, rational decision divorced from the imponderabilia of the subject. Though the existence of a specific area is admitted, which cannot be so readily reduced to the form of determinative data, this area—the designer's area—is said to be so small that one may consider it a negligible quantity. (That this total technical rationality is only the expression of particularized reason will be dealt with later.)

The thesis of continuous mathematical structuring of a decision space for a design means nothing less than that all design problems can be solved algorithmically, that is, by the employment of a mathematical or logical construction that acts as a program or as an instruction manual.[8] Such algorithm would have to be established with the aid of the following mathematical disciplines that vary from case to case: combinatorial analysis, theory of games, theory of information, mathematical logic, switching algebra, linear programming, system theory, theory of queues, and combinatorial topology. As has already been mentioned, some of these techniques certainly possess an instrumental value for the solution of complex design problems. But this is only if they are not practiced in the sense of a scientific panacea but simply from the aspect of their instrumental value. The mathematical disciplines mentioned should not lead to the erroneous conclusion that creative thought and action in both the fields of science and design can be totally reduced to algorithms. For it is wrong to attempt to simulate the relationship of the designer to the problems facing him with a model in the form of a simply determined system, because, after all, the process here discussed—as in every creative and inventive human behavior—can, if at all, only be simulated with models on the level of complex probabilistic systems.[9] Here the possibility to quantify and formalize design problems in order to optimize or suboptimize the solution of these problems within a definite scheme is not disputed. The relativism implied here of the mathematical instruments of decision theory, in conjunction with design, aims at saving those who behave in a rationalistic manner from a rationalistic belief. However, as Chermayeff said, arty way is definitely not the alternative to reducing everything to mathematics. In the statement "There is no magic in the manipulation of color, anymore than there is in the manipulation of

---

[8]   Georg Klaus, *Kybernetik in philosophischer Sicht* (Berlin: Dietz, 1961).
[9]   Stafford Beer, *Kybernetik und Management* (Frankfurt: Fischer, 1962).

numbers,"[10] he pins down only the plain state of affairs that everything moves in the field of sober feasibility.

SYSTEMATIC METHOD THE EASY WAY

Many advocates of scientific design do actually suggest the application of scientific disciplines in solving design problems. In addition, there are theorists who understand scientific design in a really wholesale sense. They equate science and scientific proceedings with a prescientific rationalism—a kind of observance of Cartesian methodology, of rules for the guidance of the intellect, which has resigned from Cartesian doubt. The advocates of this systematic method rely predominantly on the modern heuristic methods of G. Polya explained in his book *How to solve it* (1945–8). This heuristic method deals with the solution of problems and particularly with the thought processes, which in this operation may be employed in exemplary manner. In a series of papers entitled "Systematic Method for Designers,"[11] B. Archer has given examples of Polya's modernized heuristic method. The results of this systematic method for designers operate as yet within a modest framework.

They remain on the level of reducing the design process to a scheme, that is, a linear sequence of steps, partly in a feedback relation, starting with the gathering of required data and proceeding to the final communication of the design. These schemes indicate what each designer does in any case, and knows already about his activities. Hence, Polya's intentions are led in another direction, since only the sequence of the steps in solving design problems is stated, but not the methods to be employed. Perhaps these schemes have a didactic value; but the systematic method is characterized in that it suffers from an excess of system and a lack of method. The examples cited to illustrate the efficiency of the systematic method remain within the frame of protocol method of stating facts, which are so self-evident that one gathers the impression that the whole matter is being systematically stretched. As long as the argument runs on an abstract level, the comprehensive check lists, seemingly leaving nothing out, behave as paragons of rationalism. The matter becomes self-revealing, however, when concrete problems are tackled, as, for example, design aesthetics. At this moment the entire systematic method becomes remarkably short-winded and lags behind reality. This may be attributed to the fact that no previous efforts were made to define the aesthetic aspects; furthermore, in agreement with idealistic tradition, aesthetics are considered the "theory of the perception of the beautiful." In a concrete situation, questions of design aesthetics are left to intuition. "In the majority of cases it is far, far quicker and cheaper to handle the whole of the aesthetic

[10] Serge Chermayeff, "The Designer's Dilemma," in *Serge Chermayeff, Heinz von Foerster, Ralph Caplan, Sibyl Moholy-Nagy: A Panel Discussion*, ed. Edward J. Zagorski (Chicago: Industrial Design Education Association, 1962).

[11] L. B. Archer, "Systematic Method for Designers," in *Design* no. 172, 174, 176, 179 (London: Council for Industrial Design, 1963).

side of design by intuition, provided that there is an adequate body of prior experience to base it upon." This is truly not a very glamorous end for a rather presumptuous approach to a systematic method for designers. One misses the target if one remains true to rationalism only insofar as he is occupied with entelechies and avoids their profanation. In a precarious manner this position calls to mind the social scientists about whom Mills writes: "They are fully rational, but they refuse to reason."[12]

FECHNER AND THE CONSEQUENCES

It is often maintained that the origin of experimental psychology dates from the morning of October 22, 1850, when G. T. Fechner suddenly realized that "the increase of the mental intensity of a sensation would be proportional to the increase of the living power."[13] In this "sudden enlightenment" lies the thought underlying his "Elemente der Psychophysik" (1860), which, although it was not the first, was the most effective contribution in providing psychology with a modern scientific basis. One must stress, however, that Fechner was an excellent naturalist; but he was not an avowed scientist, at least not according to the customary modern definition. He was occupied with many side-interests—philosophy, mysticism, poetry, and humor—and it is not sure that these side activities were not the main activities in his own eyes. It is very difficult to ascertain to which he attached most value: his experimental work or his philosophic and mystic speculations.[14] It is obvious that experimental psychology—assuming Fechner to be its founder and not J. Müller, E. H. Weber, or W. Wundt—had no satisfactory scientific origin. One can object, of course, that none of the sciences is able to present a satisfactory origin. But in respect to experimental psychology this "original sin" intensified the already-prevailing resistance against this "so-called science which wants to quantify mental processes."

In order to counter the effect of such detrimental conditions, experimental psychology was compelled to pose from the outset as the most scientific of all sciences. For this reason, no nineteenth-century scientific discipline adopted with so few reservations the then scarcely differentiated program of science as did early experimental psychology—a program which suffered from the influence of a doctrinal mechanistic mode of thought. In the interim, the progress of the sciences in our century has contributed to depriving this program of its validity. Nevertheless, experimental psychology still hesitates to separate itself from all of the points of the old program. This is particularly evident in the case of ergonomics, the branch of applied psychology dealing with the study of man-machine-systems. Prejudices that originate from the first program continue to persist in this discipline. F. V. Taylor writes on this subject:

[12] C. W. Mills, *Power, Politics and People* (New York: Oxford University Press, 1962).
[13] Kurd Lasswitz, *Gustav Theodor Fechner* (Stuttgart: F. Frommann E. Hauff, [1896] 1962).
[14] Wilhelm Wundt, *Reden und Aufsätze* (Leipzig: Alfred Kröner Verlag, 1913).

"Although psychologists have become more scientific in their instrumental procedures, using better and better research tools and employing statistics of ever-increasing power, they are still working with pretty much the same old types of syntactically impoverished concepts."[15]

If one points out to the ergonomist that there are factors in man-machine-systems, which, although they can be precisely evaluated, cannot be precisely verified and computed, he takes cover behind the argument that nonverifiable and noncomputable factors cannot be the subject of his investigations. There are many who go so far as to maintain that only one-dimensional problems, or problems that can be solved by linear or scalar means, are scientific problems. Evidently they ignore what has long been practiced in the natural sciences— physics, astronomy, biology—the acceptance of the multidimensional nature of scientific problems and the use of nonscalar mathematics.[16]

It would be a prejudiced simplification to accuse all the opponents of this belated cult, practiced by some ergonomists, of metaphysical, idealistic, mystic, or simply romantic tendencies. To stress the complexity of systems where man constitutes an important component—as in the man-machine-systems, for example—and can no longer constitute the favorite and exclusive subject of the opponents of empirical sciences. Surely there are qualified ergonomists who draw attention to this complexity and to the dangers of underestimating this complexity. A. Chapanis writes:

> A couple of hundred years ago it was the vogue to say that man is nothing more than a system of complicated levers and pneumatic tubes (the nerves and blood vessels) which carry energizing liquids. Fifteen or so years ago, it was popular to say that man is nothing but a servo. Now it is the thing to say that man is nothing but an information-handling channel. Call man a machine if you will, but do not underestimate him when you experiment on him. He is a non-linear machine; a machine that is programmed with a tape you cannot find; a machine that continually changes its program without telling you; a machine that seems to be especially subject to the perturbations of random noise; a machine that thinks, has attitudes, and emotions; a machine that may try to deceive you in your attempts to find out what makes him function, an effort in which, unfortunately, he is sometimes successful.[17]

The trend toward simplification with which Chapanis deals here and against which he warns, may be explained in part by the special nature of investigations to which ergonomists devoted themselves in latter years. They were primarily investigating control systems where man operated under extremely unfavorable circumstances to the limit of his senso-motorical capabilities, that is, in the critical zone "in which human performance deteriorated to an unacceptable

[15] F. V. Taylor, "Psychology and the Design of Machines," *American Psychologist* 12, no. 5 (1957): 249–58.
[16] Anatole Rapaport, *Operational Philosophy* (New York: Harper & Brothers, 1953).
[17] Alphonse Chapanis, *Research Techniques in Human Engineering* (Baltimore: Johns Hopkins Press, 1959).

degree."[18] Investigations of these exceptional machines, offensive weapons or defensive weapons—and of their not less exceptional operators—the soldiers. Because ergonomics has specialized to such a great extent in the solution of critical problems in military equipment, it is sometimes difficult to distinguish between ergonomics and military psychology. "Machines cannot fight alone" was the slogan of the modern ergonomist. The central subject of this discipline was, and still is today: adapt weapons to the soldier, and very often—despite all that ergonomists proclaim—adapt the soldiers to the weapons.[19]

Without doubt the empirical data obtained from investigations of military equipment possess a prototype value for all fields, and even for such fields which are completely removed from military equipment. On the other hand, constant occupation with such issues has no doubt resulted in a certain one-sidedness in the ergonomist—that particular tendency toward a too-abstract version of the human operator. In military equipment, the human component—the component H, as the ergonomist says[20]—had of necessity to be separated from its daily reality, its specific, social, individual, and cultural coordinates. This course is required in the interests of the efficiency of the system, of which H is a part.

THE TWO-SIDED COMMISSION

In our society, neither the world of merchandise can be easily penetrated nor, in many cases, product design which influences this world of merchandise. Despite all these unfavorable circumstances, one thing emerges: in the future the function of the product designer should not consist of designing products according to an outlined demand, as is still the custom in our free economy. Rather more, the product designer should be the one who contributes to the creation of demand; otherwise he will be able to play only a subordinate role and preserve the existing products with only superficial modification. The product designer should not consider his function to keep quiet but to promote disquiet.

---

[18]   J. W. Dunlap, "Bio-Mechanics," in *Handbook of Applied Psychology*, vol. 1, eds. D. H. Fryer and E. R. Henry (New York: Rinehart & Company, 1950).

[19]   W. J. Brogden, P. M. Fitts, H. Imus, and S. S. Stevens, "Human Engineering in the National Defense," in *Notes and Selected Readings on Human Engineering Concepts and Theory* (Ann Arbor: College of Engineering Summer Session, University of Michigan, August 1959).

[20]   F. V. Taylor, "Equalizing the System for Component 'H'," in *Notes and Selected Readings on Human Engineering Concepts and Theory* (Ann Arbor: College of Engineering Summer Session, University of Michigan, August 1959).

It would be presumptuous to assume that an education institution's program could remain valid for decades, especially in an era of intense changes in all domains—technological, environmental (including climate), social, cultural, and economic. In view of a criticism of the project of the Enlightenment which has become known under the term "postmodernism," surges the question arises whether and to what extent the meaning of HfG Ulm and its program have also been affected by this process. The more the antimaterial and spiritualizing tendency of postmodernism and deconstructivism came to the fore, the more the difference and incompatibility with the programmatic position of the HfG Ulm became visible. Recent researches, particularly by French researchers, have revealed the substantially antisocialist drift of postmodernism oriented to reinforce the status quo and its power relations.[1] This does not mean denying merits of having drawn attention to aspects of history that had not been considered before. But in the field of architecture and design, postmodernism and deconstructivism have shown to be more a source of confusion than clarification.

As is well known, one of the cornerstones of the HfG Ulm's program was its focus on industry, without putting into brackets or denying the environmental impact of industrial activities. Also, the creation of energy profiles of industrially manufactured products, and their use and recycling, is relatively new territory for design, but would have fitted seamlessly into the program of the HfG. The attempt to bridge the distance between design (project) and the sciences remains valid to this day. Instead of looking at design in its relation to the sciences, one could today reverse the view and put the sciences in relation to design, that is, assess the sciences from the perspective of project thinking (not to be confused with design thinking).

---

[1] Stéphanie Roza, *La gauche contre les lumieres?* (Paris: Fayard, 2020). The author offers a panorama of the critical and hypercritical theoretical experimentations, known under the term "postmodernism," that constitute the core of the contemporary counter-enlightenment. As far as the domain of design is concerned, they are closed against the possibility of conceiving a socially relevant project because they dismiss the idea of utopia. This explains the distance up to and including the hostility and incompatibility between postmodernism and the program *HfG Ulm*.

3 THE RELEVANCE OF THE ULM SCHOOL OF DESIGN TODAY

(2003)

A look at the still-unwritten history of design education over the past eighty years reveals a number of paradigmatic design schools with international impact, one of which is the ulm school of design. Today, many of the school's innovations in teaching and in methodological and analytical approaches to design are common knowledge and have been absorbed by design education and practice. From this point of view, the ulm school of design was successful.

Differences in the long-term influence of various design schools result from the unique constellations of each and are subject to sociocultural dynamics that unfold beyond the reach of the protagonists. Nowadays, to be sure, the impact and influence of a design school are no longer defined solely by this interplay of contingencies but are purposefully directed to the extent that an institution uses marketing to acquire an influential position for itself—a tendency that will probably increase, above all, in the wake of the privatization of higher education.

The lack of any systematic, comprehensive studies of the ulm school of design, of its approach to design teaching, and of the ways in which its educational concepts developed has contributed to the creation of myths, both positive and negative. Not surprisingly, positions of systematic ambiguity, often cloaked in radical guise, have—and indeed can have—little sympathy for the rational approach of the ulm school. This fact, however, should not blind us to the question that the ulm School once asked of the Bauhaus: Is the ulm school of design still relevant today?

Why was the ulm school able to exert such a powerful influence? Why did it attain the status of a model, even if the normative, Eurocentric, and universalistic connotations of this term make it one to be avoided? In order to answer this question, we must first shed light on the multifaceted context of the 1950s, as characterized in the following observations.

At that time there was no precise sense of the profession that was later to be called "industrial design." In German-speaking countries, the term *Formgeber* ("form-giver") predominated. Nor was there a conception of the profession we now call "information design"; the terms used were "commercial art" and "applied graphics." The concept of "good form," with its socio-pedagogical connotations of aesthetic education, functioned as a point of orientation. Design professionals were educated in crafts schools, with artistic self-images

First published in German and English as "Zur Aktualität der HfG Ulm | The Relevance of the Ulm School of Design Today," in *Ulmer Modelle – Modelle nach Ulm*, ed. Ulmer Museum and HfG Archive, 124–33 (Ostfildern-Ruit: Hatje Cantz, 2003).

that often stemmed from the nineteenth-century concept of "applied" art. Project-oriented instruction, particularly involving work on complex projects, was not given the central role that it deserves. At the ulm school, an unequivocal distinction was drawn between design and art.

In 1950–60, probably no other design school explicitly integrated scientific disciplines in this quantity and selection into the curriculum and assigned them a central place. The concept of design research was nonexistent. The founding of the ulm school of design occurred during a phase of reconstruction in a country whose infrastructure had been destroyed by the Second World War. In addition, its founding must also be understood as a reaction to the trauma of National Socialism. The ulm school concentrated on the materiality of objects, leaving their symbolic-communicative dimension aside, or at any rate not according it the central role it would later acquire. Nowadays, radical technological and industrial innovations in the form of digitalization and the computer industry increasingly affect all areas of life; they influence the content and method of design activity and contribute to the emergence of new professions and fields of activity within the area of design. Competition has been superseded by the struggle for hegemony of the market. Today, designers are expected to produce not solutions, but strategies for solutions. In addition, the spectrum of educational opportunities in the field of design has become strongly differentiated, so that today a broad selection of programmatically diverse courses of study is available. Design education has become consolidated, which is not to say that all of the questions connected with it have been answered. Above all in the areas of design theory and design research, there are still unfortunate deficits, along with considerable differences of opinion regarding the foundations of design, its contents, and its communication.

It is against the background of this profoundly changed context that the question of the relevance of the ulm school of design should be asked and answered—if only in a preliminary way, since a detailed analysis would far exceed the scope of this essay. What made the ulm school of design modern? Positions diverge as to the answer to this question—after all, something more important is at stake than simply the preference for rounded, curving forms as opposed to straight lines and sharp corners. The ulm school accepted industry as a substratum of contemporary society and saw industry and technology as cultural phenomena. The HfG focused its attention on the relationship between design and society, a relationship that is by no means free of contradiction. The HfG accepted the sciences as a central point of reference for design and design education. In addition, the HfG insisted on research in the field of design—with an altogether experimental orientation—in order to create a body of knowledge specific to the field. It insisted on constituting design as an autonomous discipline and rejected attempts on the part of other fields to appropriate design or make it function as their subcategory. The HfG did not focus on individual objects, but preferably on object systems and design programs—that is, it was concerned not with the newest lamp by the star designer, but with the larger complex of which lamps are a part, that is, the question of lighting. The HfG did not linger on the side of problems but landed squarely on the side of solutions—in other words, it refused to engage in a purely discursive, theoretical dancing around the problem.

Although the HfG viewed semiotics as a foundational discipline for design education and even initiated studies in this area, it would have opposed overemphasizing this dimension or even granting it autonomy with regard to design. As is well known, in the 1980s and 1990s the sign character of products gradually attained prima donna status. Moreover, in the course of its commodity-aesthetic popularization, design was reduced to the sign-like and symbolic, to "fun," "experience," and "coolness." This process reached its climax in the "boutiquization" of design. This is also one of the reasons for the tendency to view design as a purely superstructural phenomenon within the framework and art-theoretical categories. The design discourse that arises from this view is hard-pressed to go beyond the mere decoding of signs. This kind of theoretical view fails to see that design involves technology, industry, and economics—in other words, hard materiality. As far as professional practice is concerned, the reduction of design to the dimension of signs and symbols promotes an image of the designer as the creatively inspired outsider of the industry, a beautifier of industrial ugliness, a modern "sign-maker" instead of the erstwhile "form-giver," and, above all, the creator of a new and distinct category of "design objects"— expensive, off-beat, elitist, and, indeed, "designed." This tendency has had its effect even in design education and has given rise to the cliché of design as an easy, playful, fashionable course of study with a lot of "hip" and "hop," sparing the students heavy demands, especially intellectual demands. The process has been reinforced by the collusion of designers themselves, who have delegated the discourse of design and thereby contributed to their own disqualification from it. These critical remarks should not be misinterpreted to mean that design discourse should be reserved for designers alone or that informative and valuable contributions have not and cannot be made by approaches not related to design. Only when they adopt a normative posture should they be examined for the grounds of their legitimacy, especially when the facade of canon-lessness hides a new canon, the canon of the arbitrary, which sometimes goes so far as to accuse the nonarbitrary of authoritarianism.

The aforementioned principles of the HfG—an institution that was considered avant-garde and progressive in the 1950s and 1960s—have today, as was mentioned, been largely subsumed into educational and professional practice so that they no longer seem exceptional. Contours have been blurred, while erstwhile opposites have run together. Yet in view of the establishment of affirmative positions, the fundamental question still arises in all of its variations: What, if anything, continues to make modernity attractive? The answer, as always, is this: it is the promise of self-determination, the lessening of other-directedness, the reduction of power over others in whatever form, imperial or otherwise, for power can be and is exercised through design as well.

In terms of the history of ideas, the HfG followed in the tradition of the Enlightenment, a movement which has lost nothing of its relevance. A look at the current global political situation is enough to persuade one that our present epoch suffers not from too much, but from too little enlightenment. At the social-political level, therefore, the HfG has indeed retained its relevance. With respect to the content of design, on the other hand, a considerably broader panorama exists today, due above all to digitalization. The information and computer industries not only offer new tools for the rapid visualization of design concepts and their speedy transformation into three-dimensional models (rapid prototyping), but also open up new fields of design particularly in the area of the new media, that is,, the internet with networking, e-commerce, e-learning, e-mobilization, and intranets. These areas make it possible, among other things, to use design as a cognitive tool for the presentation and communication of knowledge. In the future, it will perhaps even serve as a tool for knowledge production, for the quality of research depends not only on the answers that are found, but on the perspective from which the questions are asked. Fifty years ago, such things could not have been imagined even in our wildest dreams.

In a future history of design, the effects of the epochal caesura between the predigital and digital eras will probably be the subject of special investigation. The HfG belongs to the predigital phase of design history, although strong affinities *in nuce* to the digital phase can be identified. This is not to suggest in retrospect that digital technologies were anticipated at the HfG; these affinities manifested themselves first and foremost in the basic design exercises, which were aimed, among other things, at the production of continuous figures out of discontinuous elements. The elements used in the grid exercises of that day would nowadays correspond to pixels. Designs which back then were extremely time consuming to produce (in tempera on high-quality cardboard) can now be rendered digitally much more quickly. Digitalization has also opened up fascinating perspectives for design in the time-based digital media, such as interactive animation and the visualization of processes. Thus, today, the exercises in a basic design course—if one wanted to stay with this concept—would concentrate more on cultivating a consciousness of detail and on training competence in generative procedures for the creation of form— that is, design algorithms—than was the case in predigital times.

As is well known, the ulm school experimented with different concepts for its basic design course. The initial one-year Foundation Course, which was initially mandatory for all four departments (Industrial Design, Visual Communication, Industrialized Building, and Information), was later reorganized more in terms of different departments, and it was finally dissolved as a separate teaching unit. Yet the legitimacy of the basic idea was never called into question: the use of specially devised exercises to train formal and aesthetic competency in the area of design. "Invisible design" (as in the slogan: "Design is Invisible," the title of a 1980 exhibition in Linz, Austria) is a dead end; indeed, up to this point no educational concept has been invented that can evade the materiality of the aesthetic, even if the deductive scientific disciplines often encounter this very domain with deep mistrust or naïvely equate it with art and creativity. The exercises in the Foundation Course concentrated exclusively on the visual aspect in two and three dimensions. Today, they would be expanded to include the audio dimension (sound design) as well as the combination of image and sound in the phenomenon of interaction—in other words, the new design parameters and subject matter with which audiovisual studies deal at the analytical level.

Admittedly, the course of study offered at the ulm school as a private institution led only to a diploma and not to a master's degree, though in its final projects consisting of a practical and a theoretical part it could stand comparison to contemporary master courses. Nowadays, the formal approximation of the Anglo-Saxon 3–5–8 model—three years for the bachelor's degree, five years for the master's, eight years for the doctorate—predominates in Europe, including in Germany. From this point of view, the controversial question of the specialization of design education (specialists without "specialism") appears largely irrelevant, for the pursuit of a master's degree is of course intended to deepen a chosen area of study. Fortunately, the once-coddled image of the designer as a coordinator—its contemporary counterpart is the "conceptualizer"—has proven untenable over the years and has yielded to the more modest, but more realistic, conception of design as an integrative activity.

Would it make sense today to recreate a new, updated version of the ulm school of design—naturally without nostalgia—particularly in view of the radical changes in the political and technological-industrial context described earlier? Probably not, for the very reason that the political conditions for it are indeed lacking, with one exception: that a new kind of institution of higher learning was invented in which the category of design was given an appropriate place, one that was organized around problems and not around disciplines. It is certainly no coincidence that two of the most influential educational institutions for design in the twentieth century (the Bauhaus and the ulm school of design) rejected established academic structures. For design is an area of activity and knowledge that does not fit into traditional university structures with their courses of study oriented toward single disciplines. The attempt to integrate design into these traditional structures remains a patchwork at best, producing some notable results, but not permitting the potential of design to fully unfold. What makes design both interesting and difficult as the object of discourse is its seemingly paradoxical and hybrid character: as an activity, it includes essential discursive ingredients and rests on an implicit discursive foundation, yet manifests itself in a nondiscursive result. A design cannot be discursively encapsulated—a stumbling block for a purely theoretical consciousness which, because it clings to the level of reflection, is unable to make the jump into the world of operative consciousness.

A context freed from the ballast of traditional institutions would also make possible the indispensable activity of design research, beginning from the perspective of design rather than from artificial academic criteria and their dehydrated areas of interest. This new kind of design school, moreover, would certainly not be limited to a single national context, but would be operable only on an international level. Only a faculty and student body made up of members of different cultures could offer the diversity that today is more necessary and stimulating than ever in a curriculum and learning environment. It is to Max Bill's credit that he insisted at a very early stage that foreign instructors, the so-called "exotics," be engaged at a German institution of higher learning. This, along with the large proportion of foreign students, lent the ulm school an international character and is probably one of the reasons for its worldwide influence.

Perhaps in the future, the boundaries between science and design will blur; the concern is not, as it was at the time of the ulm school of design, for a scientific approach to the design process (however interpreted) or the integration of scientific knowledge into the design of complex systems. Rather, the concern is for a new kind of science enriched by the category of design. Instead of viewing design from the standpoint of science, one could reverse the perspective and approach the sciences from the standpoint and with the criteria of design. Admittedly, that is a daring speculation for which there is no guarantee, but it is no less plausible for all that. On a new foundation of this type, achieved as it were through an institutional quantum leap, the ambition of the ulm school of design could conceivably be fulfilled: to extend design into the nerve centers of society. For one can scarcely claim that this goal has been attained, which is why the current phase can at best be described as the "prehistory of design."

Drafted more than thirty years after the enforced closure of the ulm school in 1968, this paper reflects on aspects of the pedagogy of ulm and on the politics the school raised in its brief history. The reasons for the closure of the ulm school—and thus for the end of an attempt to renew higher education in postwar Germany in the early 1950s—are complex. Certainly, the experiment had its supporters, in industry and at the international level, but from the very beginning the HfG Ulm was for conservative—and more than conservative—forces a source of permanent annoyance and object of attack. From the beginning of the phase of finding financial support for the project, a countercampaign was launched: "No money for communists." Former Nazis who wanted to dress after the Second World War as defenders of democracy tried to hide their past by a virulent anti-communism.

The reasons for this hostility are manifold. The HfG took up the critical, and not the laudatory, strain of the Bauhaus tradition. From that tradition—as Chapter 3.9 in this volume explores—it focused its attention on a person who was only marginally, if at all, mentioned in official Bauhaus historiography, namely the second director of the Bauhaus, Hannes Meyer, whose concept for design education and the development of a design curriculum for the HfG provided important clues. However, with this orientation the HfG did not make itself friends in politically conservative circles.

Moreover, the notorious internal conflicts, sensationally blown up and spread by influential media in Germany, were in the cultural tradition of German institutions of higher learning an "offence against common decency." University lecturers were generally state employees who were expected to keep calm—the infamous Prussian obedience. Those who did not respect this tradition ran the risk of being disciplined with punitive action. It is not surprising then that attempts were made to blame the reason for the termination of this experiment on the inflexible behavior of the members of the HfG Ulm and their unwillingness to negotiate.

In fact, the "flexibility" or otherwise of the ulm faculty was not the issue. Decades after the termination of the experiment, the purely political motives came to light. One must take into account that the closure of the HfG coincided with the turbulent climate of the 1968 student rebellion. In this atmosphere, it was easy for the politicians who had to decide on the annual approval of the financial subsidies for the private institution to attach unacceptable conditions to this approval—and thus in effect guarantee the end of this promising experiment. Ulm remains an experiment whose potential has not even remotely been exploited and that thus remained a torso. However, for precisely this reason it is important to comprehend more deeply the real complexity and subtlety of the pedagogy that the school explored—particularly so in the light of the necessity to reflect radically on design education today.

The extended statement of 1968 "The Situation of the Hfg Ulm: A Letter of Resignation," published as Appendix 1 of this volume, further develops these arguments.

4 THE INVISIBLE ASPECTS OF THE HFG ULM

(1994)

The debate over the *Hochschule für Gestaltung Ulm* (HfG Ulm) and the role of its protagonists has sometimes been encumbered with arguments directed ad hominem. That has prevented any chance of an objective assessment, free of animosities, and a calm debate.

The HfG Ulm had an impact that was acknowledged and welcomed by some, and by others denied and regretted, if not indeed bedeviled or suppressed. In a well-intended interpretation, the closure of the Institute (probably one of the most stupid cultural policy decisions ever taken in the Federal Republic of Germany) was a fortunate coincidence of internal exhaustion and external political opposition—one that saved the reputation of the HfG Ulm and endowed it with a heroic aura, for by 1968 it was in any case drained dry. Reyner Banham made a similar comment on the Bauhaus, arguing that the productive phase of an avant-garde institution does not last for more than ten years. But the closure of the HfG Ulm was due less to concern for academic quality than to political motives. (As a private institution HfG Ulm nonetheless depended on public financial support. However, in 1968, the government of the state of Baden-Württemberg under the conservative Christian Democratic Union formulated unacceptable conditions for the future support.) It took several decades until the political reasons for the closure were admitted by the later governor Lothar Späth.[1]

Paper given at the conference *Bauhaus, Memphis und die Folgen* organized by the Design Center Langenthal, November 3–5, 1994. It was first published in English (translated by Gui Bonsiepe, revised by *Design Issues*) as "The Invisible Facets of the HfG Ulm," *Design Issues* XI, no. 2 (1995): 11–20. Later published in Dawn Oxenaar Barrett (ed.), *Interface—An Approach to Design* (Maastricht: Jan van Eyck Akademie, 1999), 119–27.

[1] "I believe in the Young Generation—They will do it," in *Jahrbuch 10 – Things Beyond Control*, eds. Nadine Jäger, Jean-Baptiste Joly, and Konstantin Lom (Stuttgart: Akademie Schloss Solitude, 2010). Lothar Späth is a rare and laudable example of a politician who publicly confesses an error. In the interview he told that in 1968 the party leadership of the conservative Christian Democratic Union had decided to cut the financial support for the HfG: The institution seemed to the politicians as "uncanny", and that he took up the task "to tip" the story of the HfG as a left-leaning institution with intransigent faculty and students. The actual political motives for the closure of the HfG had to be erased and a substitute reason of whatever weakness had to be found. That was easy: the story of the supposed unwillingness of the teaching staff to negotiate. But the proposal to negotiate was a farce. At stake was the autonomy of the HfG, thanks to which the HfG could achieve what it had achieved as pioneering institution for cultural innovation. This would have been completely incompatible with one of the offered options to integrate the HfG as subordinate department in an institution for technical professionals (Fachhochschule), whose strength definitely does not lie in standing out for pedagogical-experimental innovation. In 1989, the exhibition about the HfG "The Morality of Objects," after having been shown in Berlin and Paris, was opened in Ulm.(1989) In this context, Otl Aicher asked whether the governor would come to open the exhibition in Ulm, when an exhibition of works by Paul Klee from the Berggrün collection was opening at

The Institute of Design was under permanent innovation and legitimation pressure, and it was not easy to meet those demands, particularly in a climate of growing political polarization. In the time when the waves of political engagement were rising, it was regarded almost as a crime against the spirit of the HfG Ulm—a radicalism that was, if not, incidentally, limited to ulm—to pick up a pencil and start to design. The HfG Ulm also suffered from what could be called media coolness[2] today. The Institute had no knowledge at all of skillful marketing. But it had sufficient reserves to have undertaken the experiment—despite considerable internal conflicts—of changing into a postgraduate institution, with the objective of filling the vacuum of design studies and design research.

That mutation never came about in the end phase of the HfG Ulm, for a number of reasons that cannot be seen as solely due to external factors. One of the reasons why the HfG Ulm remained a torso was that the majority needed for the conversion was not available in the decision-making organs. But as it is idle to speculate on what could have become of the HfG Ulm, I shall turn to my consideration of the Institute and its achievements. My intention is certainly not to wallow in nostalgia; hagiographic intentions are also far from my mind.

Despite the general recognition which it achieved, the HfG Ulm also had a confusing side; this is a difficult subject that has often aroused resentment, in Western Germany in particular, and still does. I should like to examine the reasons for this confusion and throw some light on the paradigmatic weight of the HfG in forming the design discourse.

The title of my paper is a reference to this—to a dimension that cannot be shown through exhibitions of design work, and of which the publications of the HfG Ulm do not give an exact picture, either. At the Institute, not only was design taught, but design was also analyzed and discussed. That aspect has so far been underrepresented in the discussions of the HfG Ulm and its consequences. So, I would like to make a slight correction to the accentuation. As is well known, the use of labels like "ulm," "Bauhaus," "Memphis," "post-modernism," and so on lend these institutions and design directions a monolithic unity that in reality they never had. Many are only abbreviations, and they each need greater differentiation. If, occasionally, I use them nevertheless, it is in the full awareness of sacrificing some nuances. The criticism that is so easily masked in oblique references and the aesthetic dissatisfaction over the physiognomy— be it the unwieldy form of the ulm stool or the lack of psychedelic colors—do little to enrich the design debate, and so I shall leave them aside.

---

the same time in nearby Tübingen. Lothar Späth preferred the art exhibition. Otl Aicher aptly put it "in the question of art or design, he (Lothar Späth GB) decides in favor of art. he senses instinctively, design is possibly an explosive" (Otl Aicher, "Kultur des Staates," in *schreiben und widersprechen*, 202–10 [Berlin: Janus Press, 1988 (1993)], 202–10). A wide swath of the art/ design history discourse seems to have accepted the narrative of HfG's self-inflicted end and has not questioned the officially favoured version of the story of its closure. It is to be hoped that a new, more critical, generation of historians will set out to rewrite the account.

[2] The German word originally used is *Sprödigkeit*, whose meaning in English is unwieldiness. The Ulmians were not very good concerning marketing.

THE PRECARIOUS POSITION OF DESIGN

It is striking that paradigmatic changes in design education in this century have come from institutions that have taken an outsider position and are not institutionally secured. I interpret this fact in the following way: design is an alien body in traditional universities. As an area of human knowledge and activity, it is new, and it bursts the bounds of the tripartite segmentation of the universities into science, technology, and art.

— Design differs from the ideal of cognition[3] and the approach to practice in the universities;
— Design differs from the approach to technology in the technical universities, which is based on the categories of the natural sciences;
— Design differs from the ideal of aesthetic experience in the art colleges.

Moreover, the universities did not have—and still do not have—any scope for integrative studies, as the individual faculties work in isolation from each other. So design had no place in any of these institutions, and if it does find a home, it is at best an alien entity permitted a marginal existence. This may be why both the Bauhaus and the HfG Ulm were able to play such an influential role in training design intelligence. The existing university institutions would scarcely have had the scope for the necessary action and experimentation. But there are signs of a change. In one of the most advanced branches of modern university education, computer science, there are demands for the didactic modality of design-oriented (or project-oriented) study, which is standard practice in architecture and design education, to be introduced.[4] A general design-oriented revision could enrich today's university teaching and liberate design from its fringe position.

INFLUENCES

The HfG Ulm not only influenced the design debate, it also absorbed influences from outside. I see that receptivity as one of the most outstanding attributes of the Institute. The intellectual climate of the HfG Ulm was marked by confrontations with philosophical, scientific, and art-theoretical writings, which included the following:

— Works by the Vienna Circle (R. Carnap, O. Neurath)
— American pragmatism (C.S. Peirce, Ch. Morris, J. Dewey)

---

[3]   This is a translation of the German *Erkenntnisideal*; *Erkenntnisideal* could be translated as ideal of increasing knowledge.
[4]   Terry Winograd, "What We can Teach about Human-Computer Interaction," in *Proceedings of CHI* (Seattle: Washington, April 1–5, 1990; ACM: New York, 1990), 443–9.

— The Frankfurt School (W. Benjamin, Th. W. Adorno, M. Horkheimer, J. Habermas)
— Anglo-Saxon language philosophy (L. Wittgenstein, G. Ryle, I.A. Richards)
— Systems theory (N. Wiener, C.W. Churchman)
— Concrete art and constructivism
— The socio-dynamics of culture and information aesthetics (A. Moles, M. Bense).

A historian who took the trouble to write a standard work on the HfG Ulm would find helpful information in the list of books in its small library; it contained, in the mid-1950s, the first edition of the *Dialektik der Aufklärung* (Dialectic of Enlightenment) published in Amsterdam, the first two-volume edition of Walter Benjamin's writings and the two-volume edition of the fundamental work by d'Arcy Thompson *On Growth and Form.*

If one were to characterize the basic attitude of the HfG Ulm, the term "critical rationalism" would probably be most apt. The HfG Ulm was involved in the enlightenment project, and this contained a utopian component, abandoning which would have ended in cynicism and despair. Admittedly, that utopian attitude is now drawing upon itself the accusation of naïveté. A touch of immunity and indifference to social breaks, coupled with the playful demand that design should be for fun now seems to be regarded as an acceptable attitude. Times have changed. In that regard the HfG Ulm is out of date.

CONTRIBUTIONS

What were the HfG's contributions? In many cases, they have become the common property of professional practice, and they are fully integrated in industry, so their novelty value can now hardly be traced. In keeping with the objective of this address, I shall leave aside the designs for products, objects of visual communication and building systems, and turn to the invisible contributions that are not directly accessible. The sequence of the items in this brief list is not a value judgment. Nor do I claim to do justice to the institution and all its members with such a list. I am simply choosing one segment: the influence the Institute exerted on the design discourse and the effects on education and professional practice.

1) The HfG revindicated and institutionalized design as an autonomous domain that, for precisely that reason, cannot be instrumentalized by other disciplines, so not as a supplement to mechanical engineering, a useful tool for marketing, or as a variant of art in the form of applied art,[5] or as a subcategory of architecture—for none of these areas has the necessary distinctions to cover design.

---

[5]   Adorno called art the abstract negation of the existing. So the domain of design can hardly be seen as an artistic phenomenon. Without wanting to overexaggerate the importance of design, one can say that it is concerned with concrete negation of the existing.

2) The HfG thematized modern industrial civilization as a cultural manifestation. It saw industrial production as an indicator of cultural activity. In doing so, it took up a theme from the Bauhaus and the *Werkbund*, although with a different accentuation. It explicitly relied on industry, so ironic or anti-industrial experiments would have been alien to it.

3) The HfG made training objective with a programmatic approach aiming to build a bridge to the scientific disciplines. It broke with the long tradition of skill-oriented training programs in the *Werkkunst* (Craft) colleges. Design was demystified, and treated as a field that can be learned and taught, using methods that made it unnecessary to fall back on osmotic communication between a master and his pupils.

4) The HfG made a more precise definition possible of the fields of design activity, particularly industrial design, including the typology of investment goods and instruments (e.g., medical equipment). It also extended the field of commercial graphics, which had traditionally been strongly tied to advertising, making it visual communication.

5) The HfG encouraged a pragmatic approach to technology, opposing both a critical attitude to civilization derived from the humanities tradition and too optimistic an attitude to technology, like that held by Buckminster Fuller, for instance. Criticism in design means active criticism; it means intervention. As critical intervention, design differs from discursive criticism. In the HfG's approach to design, art occupies a separate field, partly in order to block the way to applied art and avoid imbuing the design of industrial products with the aura of cultural aristocracy.

THEMES

What themes are predominant in the design debate today, in contrast to the themes discussed in the 1950s and 1960s? A new problem that became increasingly prominent from the end of the 1960s has its origin in the crisis of industrialism, that had been uncontested for 250 years. In teaching and professional practice, this raises the following questions: How can environmentally compatible products be developed? What can the designer contribute to this?

The second new theme derives from technological innovation, and here specifically the spread of informatics, or generally the worldwide process of digitalization. This raises the following question: What new possibilities does informatics offer design?

Thirdly, we have a constellation of themes of a semiotic nature, namely product semantics, the so-called ethnic design and design identity.

Fourthly, after decades of delay, design has finally been incorporated in management debate. This can serve either as an indication of the difficulties of registering new realities in commercial enterprises and bringing the company up to date or as an indication of the inability of designers to enter into the debate between managers and engineers.

Compared with these themes, that are now dominating the design debate today, the following points have become less relevant: design methodology, alternative design, appropriate technology, intermediate technology, and information aesthetics.

RATIONALISM

Can one see the HfG as related to functionalism?

Certainly not, if functionalism is equated with the simplified and untenable thesis that forms can be derived from use, or are determined causally by their purpose, as the shape of a drop of water on a surface, for instance, can be derived from the surface tension. But probably yes, in the thesis that the dimension of use, the use of artifacts for specific purposes, is the main focus of design. No other profession concerns itself with this complex dimension. That is the legitimation of design. Today the word "functionalism" does not generally provoke positive reactions. It does not enjoy a good press. It has been blown up too much, like a horror image to frighten sensitive souls; it has been made the advocate of inhumanity. However, the problem of the origin of the forms of industrial products, which functionalism tackles, has not so far been answered.

The rational approach to design in the HfG implied a tendency to long-life design. Short-lived products and product differentiation, with many formal variants of a product, would have been incompatible with that approach. As far as the historical context was concerned, the HfG, with its fascination with standardization and building block systems, was tied to a phase in which industry found itself confronted primarily with production problems, and not, as later, with market problems.

METHODOLOGY

The HfG is notorious for its interest in design methodology. That interest has, up to now, been misinterpreted as an attempt to try to engage in scientific design, or even change design into a science. That is nonsense, and the HfG was never guilty of it. But its program was intended to open up the rich potential of science for design, in the plausible assumption that scientific knowledge would enrich design and facilitate its analysis.

It should not be denied that interest in what were thought to be rational methods at times took on the nature of caricature, as when the students earnestly set about measuring hundreds of beans in order to experience the statistical distribution of the variants in size in the form of the Gauss bell curve.

POLITICS

Questions regarding the political and social implications of design have now shifted into ecology. Rightly the postmodern realists point out that it is naïve to try to soften social disparities and distortions through the practice of design. But this is to overlook that capping the social ties of design and nestling into one's own special field costs a high price, which probably not every designer is prepared to pay—it is the price of having to suppress the question of the social relevance of design criteria.

It is one of the paradoxes of design that it moves in a social field of tension between conflicting interests, and, despite all the conflicts, has to endeavor to find a balance. Just as the aesthetic dimension is constituent in design, the political component cannot be eliminated from design either, unless the designers are ready to undergo collective lobosectomy. It is not a matter of adding a dash of politics to design. But design is inevitably political, in that it includes a component of hope—a dream, however vague, with the outlines of the society we want to live in. It certainly does not have to be a standardized dream. The question of the meaning of design can only be answered in the social context, and that also means in the political context. That was one of the unspoken theses of the HfG.

A CRITICAL APPROACH

Although the HfG was not prepared to renounce its critical attitude to industrial production, its approach was certainly pragmatic, and in that it differed from other critical attempts, particularly during the period after 1968. These were, in the main, restricted to outlining projects and assigning to them a critical, if not revolutionary, function. In the view of the HfG, these attempts did not progress beyond revolutionary gestures. They remained without influence on industrial production. The system of industrial production cannot be changed by symbolic criticism from outside, only by practical criticism, inherent in industry.

THE INFORMATION DEPARTMENT

Why did the HfG engage in theoretical work as well as practical design work, while the design education centers are generally rather indifferent to theory, if they do not reject it altogether. Presumably because they think it irrelevant for practical purposes or find thinking about design counterproductive?

There are a number of reasons for the HfG's theoretical activities, and I should like to consider only one of them here. The Institute's program included an information department—and it was certainly a highly unusual decision to include language at a design school. The initial aim was to train writers of texts for general use in everyday life and the media, with the competence to reflect on design and write about it from intimate knowledge of the issue. The department never had more than a small number of students. In the course of time, the department developed into a melting pot for theories. Today it would be conceived right from the start as a department for design theory.

As far as I know, that innovation, to see language as a domain of design, has not been taken up and developed by any other design school. That is hardly surprising, for design, which is tied to the visual sphere and located in the retinal space, has an unclear relation to language—and the converse also applies; otherwise the frequency of visual illiteracy would be inexplicable. Possibly its distance from or indifference to language is one of the reasons for the difficulties design experiences in reflecting and articulating in order to take part in the cultural debate. One could accept that. But one should not be content with narrowing design exclusively to practical work. Today, more than ever, it should be clear that professions that do not produce their own specialized knowledge have no future. Design studies, however rudimentary they may be, should come out of the shadows and transform themselves into a constituent part of all design education. The HfG made a fruitful start here. It would be worthwhile taking up the Institute's work.

(b) Theory and Practice

The notion "theory" sounds ambitious and hyped up, particularly when applied to design as a domain distant yet from the strength generally associated with the term "theory." But with the recent expansion and consolidation of master's degree and doctoral degree courses, the term has been used for referring to activities that predominantly don't materialize in a design, but manifest themselves in the production of a text in different literary forms—as essay, research paper, report. Previously, the issue of design theory remained in the background and unnoticed, but today, confronted with the avalanche of papers that make use of the word "design" in their title, one can be tempted to speak of an overproduction of design papers, by far not to the same extent as in the sciences, but remarkable.[1]

To write about design has today become an industry. However, this should not be an excuse to dismiss the design discourse including design theory as superfluous and a waste of time. After all, there are numerous unresolved questions regarding design, especially the indispensable critical discussion about the role of design in society today.[2]

Fundamental contributions to the design discourse have been made by representatives of disciplines that are not considered as design core disciplines. Looking at design from a viewpoint of representatives of scientific disciplines can be very revealing, opening new avenues for understanding the so far underresearched field of project activity for the development of human society. A revealing example for a contribution to the history and theory of artifacts has been provided by Tomás Maldonado.[3]

Written at a point in the middle of the 1990s when the contemporary explosion of research in design was just beginning to consolidate, this paper was an attempt to tease out and reflect on the tensions—creative as well as antinomic—between "theory" and "practice" in designing.

---

[1] Gianfranco Pacchioni, *The Overproduction of Truth – Passion, Competition, and Integrity of Modern Science* (New York: Oxford University Press, 2018).

[2] Sharon Helmer Poggenpohl, *Design Theory to Go, Connecting 24 Brief Theories to Practice* (Estes Park, CO: Ligature Press, 2018).

[3] Tomás Maldonado, "Gli occhiale presi sul serio," *Iride* 2, no. Agosto (2002): 373–86.

5 THE DISCOMFORT OF DESIGN THEORY

(1996)

"Since practice is an irreducible theoretical moment, no practice takes place without presupposing itself as an example of some more or less powerful theory."[1]

In the universal commodity culture, the only thing that counts is that which has a price as a commodity. Anything offered at no price is placed in the drawer reserved for the insignificant. This includes theory. It is produced free of charge, in the groves of the academe, where a surplus of time, however limited, still provides an occasion for such undertakings. Initially, theory is by necessity academic, although this should not be misread to mean divorced from practical life. Practical life is exposed to the pressures of contingency which hardly allow one to nurture theoretical activities. For a lean business, theoretical activities may transpire to be deadweight anyway. Like writing poems, theoretical deliberations do not contribute to growth in the GNP. In line with any strictly economic considerations, theory is pointless; and this is consequently how it is traded and treated. People remember theory at best when it concerns events inflated in the media. Then theory is called for as an entertainment provider or a stopgap. Once it has fulfilled its role as the entertaining after-dinner speaker, it is free to leave again—for the next event. That is one side of theory where no one asks for its use value or exchange value, but only for its show value. Yet there are other sides to theory. And I wish to address them here.

Only at a late date did design become a subject on which philosophy and science reflected. In the 1970s and 1980s, with the wave that popularized design and led to increased efforts to promote it, the discourse on design expanded. At the same time, the predictable danger of a paternalistic relationship emerging between promoter and promoted was increased. Design developed into a theme fit for congresses and became an opportune, easily presented media object. With the dynamics of this process, design became detached from any specific ability to design. In Germany, the concept "design" replaced the older term of Gestaltung (giving shape to) with its antiquated connotations, particularly as it had been impaired under fascism. Yet it is also

Originally published in the journal *formdiskurs*, no. 2 (January 1996): 6–17, as "Über die unerquickliche Beziehung zwischen Theorie und Praxis." It was published in English in Dawn Oxenaar Barrett (ed.), *Interface – An Approach to Design* (Maastricht: Jan van Eyck Akadamie, 1999), 18–25. Translator Dawn Oxenaar Barrett.

[1] Gayatri C. Spivak, *The Post-Colonial Critic: – Interviews, Strategies, Dialogues*, ed. Sarah Harasym (New York; London: Routledge, 1990), 2.

doubtful whether the association of "design" and lifestyle is beneficial. Other competences became relevant, flanked and shored up by traditional values and norms that had nothing to do with the know-how of designing. Design opened up as a field of activity for academically accredited qualifications that were not tied to the design domain—with their own claims to hegemony and an interest in creating a new canon, an interest that increasingly influenced design policy and design discourse by asserting that design was too important to be left to the designers—which is certainly true.

Given this openness and lack of conceptual clarity, the domain of design differs from other disciplines. Discussing theoretical physics requires specific specialist knowledge. This is not the case with the discourse on design. This is an advantage, as it enables unorthodox approaches. However, there is also the danger that the link to the materiality of design gets severed, promoting a bardism of design theory. Theory has not advanced any further, and probably will not get any further until it goes beyond the status of a pastime and is established as a full-fledged field in institutions of education. In order for this step to be taken, new curricula need to be devised that are tailored to the contemporary situation.

As early as the late 1960s, H. A. Simon's fundamental work on design theory positioned within a general theory of artifacts set the standards for deliberations from a precise scientific viewpoint.[2] Approaches from other worlds of discourse have a harder time of it. Reading these you often get the impression that design as an issue is more an annoying irritant than an object of sympathy. In fact, design encounters complacency and arrogance in the wake of a tradition which believes it can get a hold on objects merely by reflecting on them, in other words, purely theoretically in the worst sense. This has to do in part with a lack of familiarity with the object and with deep-seated reservations toward the artifacts (objects, signs) of everyday life and the technical and economic conditions under which they are produced.

Digitalization has brought forth a flood of writings and media-related philosophemes. IT scientists view these with reserve, and they are not alone in doing so. It is as if the distance from concrete experience is directly proportional to the audacity of what are panegyric and, at times, apocalyptic texts. Multimedia and VR, and especially immateriality (not to mention its dialectic counterpart: corporeality/identity), seem at present to exert an irresistible attraction for unfounded speculations. This contrasts sharply with the sober, matter-of-fact overview that is contained in such publications as the collection of articles published by the National Research Council.[3] There, at least, you get a real view of virtual reality.

Theory as contemplative behavior (and there is something of the voyeur about it) turns the object of contemplation into precisely that: an object. It objectifies it and renders it accessible, thus claiming power over it. What Walter

[2]   Herbert A. Simon, *The Sciences of the Artificial* (Cambridge, MA: MIT Press, 1981).
[3]   Nataniel I. Durlach and Anne S. Mavor (eds.), *Virtual Reality—Scientific and Technological Challenges* (Washington: National Academy Press, 1995).

Benjamin said of polemics, namely that they treat an object as lovingly as a cannibal prepares an infant, is also true of objectifying theory. It voraciously consumes actual design. Theoretical discourse is also a discourse of power, a discourse of appropriation. Thus, theory gets caught up in a permanent compulsion to legitimate itself. It emerges in the duality of contemplation and action. Theory presupposes the materiality of what it is theorizing. It consumes its object in order to exist. Initially praxis has priority over theory. In other words, theory leads a parasitic existence and—in a misleading view—always arrives too late. Nevertheless, it affects all design praxis. Conversely, design action all too easily degrades theory to the status of legitimator for a particular form of praxis (in other words, a window-dresser).

John Dewey proposed a way out of this dilemma: renouncing the position of spectatorial vision for knowledge and accepting an open participatory conception of knowledge.[4] This is not a fake reconciliation that simply papers over the cracks and differences. Theory and praxis are different. We would misunderstand both if we were to attempt to render one in a way that functions as an image of the other. In other words, theory needs to avoid the danger of abstractness and head for the so-called lower levels of praxis. And it must do this against the background of the insight that praxis cannot be accessed in a purely discursive manner. Thanks to its pure facticity, a single project—be it a plausibly designed book page, an intelligent metaphor for navigation, or a precisely positioned handle on a medical apparatus—transposed into reality outweighs barrages of verbose speculations.

Praxis, in turn, must not isolate itself in contingency. Precisely action, which wants praxis and only praxis and sets itself as the imperial standard, succumbs to blind opinionating. This is all the more the case when praxis claims not to do this and has a fit when it hears the word theory. Anyone who barks against theory unconsciously falls victim to it. Anyone who thinks that theory is some leisure-time occupation for the intellectual elite, without any relevance for praxis, puts himself on the sidetrack of history called "No Future." Any request that theory should be simple, following the motto "for the rest of us," is likely to take on board a populist prejudice. Theory is as differentiated as the praxis on which it reflects. This is a decidedly complex matter. Were it not to be, then theory would be unnecessary.

Praxis is justified in keeping things at a distance where theory is concerned, if theory raises the suspicion of being directive and denouncing all praxis as narrow-minded. Praxis under the aegis of theory—that would be an off-putting scene, just as theory would be if it were only to follow in the wake of praxis. Theory would be overtaxed if it were expected to provide concrete instructions for action, as if theory could be a toolbox of methodological procedures for design. Conversely, theory would be presumptuous if it were to pose as the regulative agency of praxis and succumb to the temptation of

---

4   David Michael Levin (ed.), *Modernity and the Hegemony of Vision* (Berkeley: University of California Press, 1993), 10.

wishing to influence praxis directly. Such an undertaking would only entangle theory in contradictions between intentionality and operational know-how.

Indifference and aversion are not justified vis-à-vis theory as a domain in which hermeneutic questions are raised. In theory, there is also a tradition of the nondespotic gaze—which perceives opaque areas, discovers complexity, and reflects on contradictions instead of sneaking comfortably offstage.

But then why do we need theory, let alone design theory? Why not protect praxis from all theoretical considerations? From what experiences is theory drawn? Is it somehow a substitute activity for design? Is the prejudice justified that "he conducts theory who cannot himself design"? Must theory be rooted in design practice in order to deserve to be taken seriously? Does design need a theory specific to it? What can one hope to get from it?

We cannot expect there to be clear single answers to these questions. The answers will differ according to interests and career intentions. However much the meaning and purpose of theory may be doubted, there is at least one firm argument in favor of design theory. All practice is embedded in discourse, a domain of linguistic distinctions that form an indispensable part of praxis. Universes of discourse vary in terms of degree of differentiation and stringency. Compared with other realms, design discourse stands out neither through differentiation nor through stringency. One can only speculate as to the causes of this deficient discourse on design. I guess that it stems from the preponderance of the skill-oriented phase in training, as this fosters a basic anti-intellectual stance. Skill-oriented training is gradually being dissolved in institutions of design education. Otherwise, designers will not emancipate themselves, but instead vegetate in the shadows, which sharply contradicts the cultural and economic importance of design as a central domain of Modernity. Let me emphasize that skills are a necessary, but by no means sufficient, condition for design work. Anyone involved in typography cannot survive without being skilled in operating a professional page-layout program. But anyone seeking only to be perfect with the software will be a mere operator or, as the saying goes, a pixel monkey on a rendering ranch.

Theory can be characterized as the domain in which distinctions are made that contribute to praxis having a reflected understanding of itself. Put in a nutshell: theory renders that explicit which is already implicit in praxis as theory. This is why theory is irksome: it casts into question things taken for granted.

In his book *Che cos'è un intellectuale?*[5] Tomás Maldonado made use of the subtle distinction between *pensiero operante* and *pensiero discorrente*. As he admits, it entails all the weaknesses and risks of categorical dualities. Based on this distinction, we can put forward the following interpretation: design praxis as *pensiero operante* acts in the domain of social production and communication. Design theory as *pensiero discorrente (discursive thinking)* acts in the domain of social discourse and, thus, in the final instance, in politics,

---

[5]   Tomás Maldonado, *Che cos'è un intellectuale? Avventure e disavventure di un ruolo* (Milano: Feltrinelli, 1995).

where the central question is: In what sort of a society do its members wish to live? Let me stress that this emphatic concept of politics in design theory has nothing to do with notions of professional politics or party politics, and even less with the simplistic geometry of opposing positions between left and right.

Theory is living in language, and therefore has a contentious relationship to visuality. This is so, in spite of the fact that epistemology has, since the beginning of classical philosophy, always been permeated with visual metaphors—a fact that has been termed the "imperialism of an ocular-centric philosophy." If theory privileges language and declares it the only form of cognition, an anti-visual bias becomes evident. Since the visual turn in the natural sciences, the visual domain has been recognized as a domain that helps constitute cognition. This undermines language's claim to absolute predominance as a primordial basis of knowledge, thus attacking a powerful, institutionally ensconced tradition of discursivity.

Often accused of being inarticulate, designers' statements are assessed in keeping with the standards of discursivity, and rightly so. If you cast a glance at the other side of what is to considerable degree a digital dump, you will find a shameful reversal of this situation: there you have an acute lack of visual articulation. One can only hope that a new university will overcome the division between discursivity and visuality. If Flaubert were to compile a dictionary of commonplaces today, then the following entry would be fitting under the heading of "images": "Images: . . . always preceded by 'beautiful, colorful . . .'. Looks good at the beginning of a lecture, especially if the topic is visual. Serves as an excuse for visual illiteracy and thus aesthetic incompetence."

Design theory could be used for investigating the links between visuality and discursivity. Then words would be brought to images, and images to words.

A new approach to design education would then probably emerge. This would bring the eighty-year-old skill-oriented design training to an end (although I do not intend to belittle those approaches that have created landmarks). To date, all design education has a preliminary character. This will remain the case until, step-by-step, in the unspectacular detailed everyday work of design and design education, the conditions are created from which we can move on from the prehistory of design into the real history of design. Following a retro interlude, design education (and this includes research and theory formation) could begin under the auspices of radical modernity of the twenty-first century.

In the early 1960s, the discussion of design methods took an important place in the design discourse. This overview of the state of the art and a comparison of the different approaches of systematic procedures was an early attempt to test their applicability.

One way they could be interpreted was as an attempt to put design activities on a firm footing instead of hiding behind the black-box term "intuition" or "inspiration," which has only weak interpretative potential if it ever possessed one. However, the limits of rational methods also became apparent when they were only used for ritual purposes. Furthermore, they remained surprisingly silent when the theme of aesthetic quality—a constitutive dimension indispensable to design—emerged: a theme marked by the opposition between esprit de finesse and esprit de géométrie.

6 ARABESQUES OF RATIONALITY

Or the Splendor and Boredom of Design Methodology

(1966)

THE PRESENT STATUS OF DESIGN METHODOLOGY

Methode—Ne sert à rien.
— Gustave Flaubert, *Dictionnaire des idées reçues*[1]

George Nelson had his tongue very much in his cheek when he once described industrial design as a profession which had become a myth before it had achieved maturity. The methodology of design is in very much the same position. The majority of designer's regard disputes about methods of design as frivolous and far removed from actual practice or look upon them as maddening attempts by design methodologists to keep them in leading strings when what they want to do is to get on with the job in hand.

Their job is a tough one, abounding in difficulties. How can they be remedied?

Methodology has no answer to give, and indeed it cannot give an answer. It prefers to speak in terms of design parameters, variables, rational criteria of decision, optimation, systematic procedures for problem-solving, selection of relevant data, and constraints. These are portentous words panoplied in the full armor of scientism.

Anyone who has worked in design for some time without knowing about these things, or without attaching much importance to them, may well regard them with reserve. He may go so far as to suspect they are all bunk. He may find an analogy in language. Just as one can speak a language without explicitly formulating its grammar, he may say, so one can design successfully without necessarily having recourse to an appropriate methodology.

It is no use digging in one's heels and jibbing. A profession oriented on technology and industry is not likely to escape the cold douche of scientism and rationalization. Design methodology can be criticized only from inside. To attack it from outside is like running after a train that left hours ago.

This article is based on a lecture, held at the INTI in Buenos Aires in June 1966. First published in German in the magazine "form" as "Glanz und Langeweile der Designmethode (1)+(2)," *form*, no. 36/37 (1966): 10–14/22–6. Also published as "Arabesken der Rationalität / Arabesques of Rationality," *Ulm: Zeitschrift der Hochschule für Gestaltung* 19/20 (1967): 9–23. Translated by Gui Bonsiepe.

[1] Gustave Flaubert, *Dictionnaire des Idées Reçues* (Paris: Jean Aubier, 1953).

SCIENTIZATION OF DESIGN

Any new idea, if there is anything to it, is easily overvalued and abused and
thus has regrettable as well gratifying consequences.[2]

The idea of incorporating certain scientific disciplines and modes of thought
in design work was put into effect at the Bauhaus by Hannes Meyer at the end
of the 1920s. But even before this, in 1910, the architect Lethaby had argued
for the necessity of such a measure: "We have entered upon a scientific age,
and the old practical arts that work instinctively belong to an entirely different
era. . . . It is high time the scientific side of our studies was given greater
emphasis and the archaeological side less. . . . I would reiterate that the vital
nerve of design lies in scientific method."[3] This statement has lost nothing of
its topicality after more than five decades. Today it still represents a utopian
idea more than a concrete reality. The incipient process of rationalization—
to which all designers are committed unless they are prepared to risk being
left out in the cold—has so far achieved not always encouraging results. More
deformation than formation is often the upshot.

"Those who are no good at architecture go into planning" is the caustic
remark heard in American faculties of architecture. (It may not be many years
before it will have to be turned round the other way.) Certainly, there is more
than just malice in the allegation that design methodology has a special lure for
those who—to use the cant phrase—are lacking in creativity; they use system
in their design work not so much to achieve useful results as to dissemble their
paucity of design ideas.

The motives for adopting rational methods and incorporating scientific
methods and knowledge in the design process are many, various, and
contradictory. First, there was and is the desire to use scientific results to
humanize the environment—a task which has so far been criminally neglected.
And then this inclination towards the "scientific" fulfilled, and fulfills, a placatory
function in the process of integrating the designer with society. Adaptation to
prevailing conditions is of dubious merit, even when the conditions are those
of "science," whose conservative functions all too easily stifle its formerly
critical impulse.

And so it comes about that those who make a great fuss about rational
criteria of decisions and dazzle with optimum design solutions also make out
a case for themselves by demonstrating the sort of solid utility an industrial
system requires. Design on scientific principles carries two implications: first,
an instrumental interest, and, second, a quietistic deference to science—or
what designers take to be science.

Anyone who dutifully gives anxious thought to the rationality of design
methods inadvertently runs a risk of stultifying his awareness of the rationality
of design purposes.

---

[2]   F. J. Anscombe, "Some Remarks on Bayesian Statistics," in *Human Judgments and Optimality*,
      eds. M. W. Shelly II and G. L. Bryan (New York: John Wiley & Sons, 1964).
[3]   W. R. Lethaby, "Architektur als Wagnis," in *Anfänge des Funktionalismus*, ed. Julius Posener
      (Berlin; Frankfurt; Vienna: Birkhäuser, 1964).

Rationality can generate liberating forces but it can also encourage repressive tendencies.

Rationalization can obscure as well as illuminate. It is no mere coincidence that in psychoanalysis rationalization means the adduction of evidence for a specious purpose under conditions of stress. "One must be ready to call sour grapes sweet."[4]

The discomfort engendered by design methodology is due primarily to the fact that—although the necessity for design methodology is not disputed—this radical weeding out of the design process may blur our view of the aim of design or obscure it altogether unless measures are taken to correct the calm divagations of method.

DESIGN AND FORM OF A PRODUCT

Form is the ultimate object of design.[5]

It is about time to recall the almost discredited idea of "form"—an idea which is about as easy to disjoin from design as the idea of healing from medicine.

Orthodox design methodology sometimes gives the impression that the form and shaping of a design is a necessary evil, a distasteful incubus on the designer from which it is best to hold aloof since form cannot, as it were, fail to emerge from the coordination of the design parameters. It would, of course, be a splendid thing if coordination were to yield a final form with such promptitude. How very convenient if designing a product—and that is creating a form—were like twisting various yarns into a thread! It would be like working to a pattern. Now there is something rather odd about the final forms which arise from the coordination of design parameters: the often astonishingly unsophisticated quality of the results is something that cannot be accounted for in terms of the various factors systematically integrated. That is a pity and must sadden every design methodologist. One cannot master form by abjuring it.

STYLING, ANTI-STYLING, PRESTIGE DESIGN

Of all the criticism of our artefacts, which is the most important? I believe it is their lack of coherent form.[6]

Design . . . is an attempt to make a contribution through change. When no contribution is made or can be made, the only process available for giving the illusion of change is "styling."[7]

4   A. Mitscherlich, *Die Unwirtlichkeit unserer Städte* (Frankfurt/M: Suhrkamp Verlag, 1965).
5   Christopher Alexander, *Notes on the Synthesis of Form* (Cambridge, MA: Harvard University Press, 1964).
6   R. Latham, "The Artifact as a Cultural Cipher," in *Who Designs America?* ed. Laurence B. Holland (New York: Anchor Books, 1966).
7   George Nelson, *Problems of Design* (New York: Whitney Library of Design, 1962). Originally published in 1957.

The deep-rooted aversion of many European designers from styling has had some curious consequences. The stylists' preoccupation with appearance or form alone has made the concept "form" almost suspect in Europe. The stylist who beautifies the surface to titillate the consumers' prestige appetite is disparagingly dismissed as a "product cosmetician." The serious designer, it is said, is concerned with more significant matters; he is concerned with the concept of the product, with improvement of its characteristics of use (in German *Gebrauchsweisen*), with ease of assembly, with low production costs—in brief with meeting genuine needs. He produces a proper design and not a "prestige design" like the stylist. The concept "prestige design" is so vague and elastic that the only definite thing one can say about it is that a designer uses the term to indicate that a colleague's work is not to his liking— even though this work may have formal qualities which are not to be sneezed at. "Prestige design" is a dirty word, "styling" is a dirty word, and "form" is on the way to becoming one. Now it would be a serious mistake to imagine that discarding the design philosophy of styling is an open sesame to form and formal quality in design. A stylist who draws gingerbread scrollwork by the yard is no doubt more deserving of sympathy than an equally well-meaning anti-stylist who, through sheer narrow rectitude, never gets a detail of the stern on paper. Doubts about the stylists' design procedure and his philosophy of design are entirely warranted. On the other hand, a strategy which virtually eliminates responsibility for the form of a product or whittles it down to the coordination of design factors cannot be entertained either. The stylist's aesthetic imagination tends to hypertrophy and the anti-stylist's to atrophy.

Much the same thing can be observed in kindred design professions. Charles Colbert, formerly Dean of the Faculty of Architecture at Columbia University, wrote: "It is very possible that the creative architects of our time, either those infatuated with external shape, whom I shall call stylists, or those obsessed with everyday convenience, whom I shall call anti-stylists, have been so debilitated by a society fraught with seemingly insoluble problems that they have retreated into a nihilistic aesthetic."[8] As far as anti-stylists in industrial design are concerned, this shot is wide of the mark. They have no aesthetic.

METHOD, PLAN, PROGRAM, SYSTEM

We shall never revolutionize society through architecture, but we can revolutionize architecture—and that's exactly what we are called upon to do.[9]

Before comparing several design methods, it is advisable first of all to revert to a general theory of method in order to obtain precise definitions of concepts such as "method," "plan," and "systematic behavior." Such a general theory

---

8   Charles Colbert, "Naked Utility and Visual Chorea," in *Who Designs America?* ed. Laurence B. Holland (New York: Anchor Books, 1966).
9   Vittorio Gregotti, *Il Territorio dell' Architettura* (Milan: Feltrinelli, 1966).

is to be found in praxiology, that is, the science of efficient action, which purposes to organize techniques of sound, effective work aimed at maximum efficiency.[10]

It is the business of this science to elaborate a grammar of actions which is based on the analysis of planned, purposive behavior. Design methodologists focus their attention primarily on the way in which an action—designing— leads to a result—the product. The question "how is something done?" can be rephrased as the question: "what method, what process is used?" This question in turn is best explained by reference to the concept of composite action (action bundle). This forms either an action chord (concurrent actions) or an action sequence (consecutive actions). Within action sequences, special attention should be paid to the preparatory acts preceding a main action and simultaneously causing or facilitating it. Tests (here meaning practice, experiment) are a subclass of preparatory acts. Common to all tests is the intention of doing something. This, whatever it is, can be done systematically or unsystematically. Systematic procedure serves to eliminate all arbitrary actions, whereas unsystematic procedure runs haphazardly through the whole gamut of possibilities. Systematic behavior—and hence systematic designing— thus means planned or controlled behavior. Planning is itself a preparatory act. Instead of a plan, we can speak of a project or a program. All three are alike in that they refer to the possible selection and composition of actions oriented to a common objective. If the plan provides a description of a specific selection of actions, the method is neither more nor less than this planned selection. Method is accordingly a special characteristic of an action bundle. The difference between action bundles which are methodically constructed and those which are not can be expressed thus: the agent—the designer— knows that he should proceed precisely in this manner in his action. Awareness that specific procedures have to be followed is linked up with systematic behavior. Methodical behavior and systematic behavior are synonymous. Method—systematic procedure—is expressed in the deliberate selection and arrangement of subactions; it must also possess the characteristics of a plan and be applicable more than once. This objectified concept of method is too rigorous for both design and other applied disciplines. Granted, this recourse to reflective thought is fruitful, but only if this reflective thought is assigned a regulative function and not a totally determinative one.

The man who knows that he must act thus and not otherwise—who is aware, that is, of the inherent necessity of the sequence of actions—is amenable to argument. He recognizes the rules of the rationality game. He does not leave the course of action to his idiosyncrasies. Yet here again there is a latent factor which might develop into a restrictive component. If methodical (systematic) behavior is yoked so firmly to conscious thought, one might be inclined to conclude that each step of an action must be determined and that the methods must be—to use the jargon—highly structured. But this is in flat contradiction

[10]   T. Kotarbinski, *Praxiology* (Oxford: Pergamon Press, 1965) and K. Alsleben and W. Wehrstedt (eds.), *Praxeologie* (Quickborn: Schnelle, 1966).

to the peculiarities of designing which have so far proved verifiable. The concept of method described here should therefore be more freely formulated, and Abraham Moles has done precisely this.

"All these methods are aleatory: their success is never guaranteed. Methods are not recipes which lead one infallibly to a result; there is no such thing as an inventing machine. . . . By and large these methods are not highly structured and that is how they must remain. If they were too highly structured, they would turn into recipes and would lose applicability in proportion as they gained precision."[11]

It is advisable to maintain this critical attitude toward methods in general and design methods in particular. The rigor and perfection of the method spell its own end. A strict design method, however, has value in one place: a museum. Only old men are perfect.

EMPIRICAL FACTS AND NORMS

Technique of design, however, cannot serve in lieu of a committed point of view, or faith. This is a point that needs to be stressed in days when it is all too easy to succumb to oversimple scientism.[12]

Anyone wishing to evolve a methodology might try polling practicing designers to find out the peculiarities of the design process and then distil from the replies an extract of method. She/he would sort out the essential from the incidental, arrange and organize it in the light of his own understanding, and thus obtain an idealized picture of existing practice which—short-circuited with herself/himself—would become a measure of herself/himself. Setting aside the practical difficulties of such an enterprise—for how ready would designers be to watch over their own shoulders and set down in words what they perceived—the empirical-statistical approach is vitiated by the fact that it postulates the existence of what it is seeking. But that is precisely the point at issue. Design methodologists have pondered less on the existent than on the nonexistent which should serve as an exemplar to guide designers. A design methodology has not so much a descriptive as a normative content. It provides a framework within which design must accommodate itself if it is to take account of changed conditions in the technical and industrial sphere and thus be something more than a secondary factor in shaping the environment.

Arguments about the design process and the changed conditions under which it operates have prompted a number of publications which, although widely different in pragmatic content and theoretical basis, nevertheless have starting points that are very close. Both Bruce Archer in his series of articles "Systematic

---

[11]   A. A. Moles, "Le contenue d'une methodologie appliquée," in *Methodologie—vers une science de l'action* (Paris: Gauthier-Villars, 1964).

[12]   Serge Chermayeff and Christopher Alexander, *Community and Privacy: Toward a New Architecture of Humanism* (New York: Doubleday, [1963] 1965).

Methods for Designers" (London 1963/64) and Christopher Alexander in his book *Notes on the Synthesis of Form* (Cambridge 1964) start with much the same subject matter for their meditations on the methodology of design.

Alexander puts forward four arguments for fortifying the design process with an admixture of methodology:

1) Design problems have become too complex to be handled by intuition alone.
2) The amount of information required for the solution of design problems has increased by such leaps and bounds that no designer alone can by his own efforts collect it, let alone evaluate it.
3) The number of design problems is growing rapidly.
4) The type of design problem has changed at a brisker tempo than in former times, so that there is less and less possibility of having recourse to empirically established practice, that is, tradition.

These four arguments are rooted basically in the idea of complexity. If a design problem comprises a series of variables, its complexity increases with the number of variables; the design of an aircraft seat presents the designer with a larger group of variables than the design of a stool.

It would all be plain sailing for the designer if each variable could be handled in isolation from the others. But this is impossible by reason of the fact that the variables are more or less closely interlocked. The solution of one variable influences the solution of another variable, favorably and unfavorably. An optimal design—for all the pious optimism inherent in the expression—does not represent the sum total of separate optima but a complex of subsolutions bundled together or—in other words—forced into compromises. The variable "economic production" is not entirely compatible with the variable "practical quality" or "aesthetics"; the variable "use of semi-finished products" may clash with the variable "small number of parts." To reconcile these incompatibilities is the hard and intractable nub of the designer's task.

COMPLEXITY AND VARIABLES

The first thing to do when confronted with the complexity of design problems is to seek a method of partitioning complex entities into simple ones. Splitting up a complex problem means hierarchizing it; the various groups of variables are thus weighted according to their relative importance. It will be apparent at once that personal judgments and prejudices inevitably creep into the design process at this point. The process of dividing up a problem can be visually represented in the form of a graph, or, more specifically, a "tree" consisting of elements (= variables) and connecting lines (= reciprocal relations between the variables). At the top of such a tree stands the problem in an undifferentiated and thus insoluble form. As the ramifications increase in a downward direction, the subproblems are arranged at various levels. Analyzing a problem in this way does represent an important step forward but it stops short of the product form, which is to say that the product has not yet been designed. In essence the form is contained in the "tree"; it must therefore be decoded from the diagram and converted into an object. This process of conversion—the actual design work—has hitherto formed the arcanum of all design methodologies. Suffice it to say—without attempting any premature explanation of the fact— that so far no design methodology, not even in its most sophisticated form, for example that of Ch. Alexander, has proposed techniques for successfully accomplishing this process of conversion from an analytical diagram to a form. It is at this point therefore, that future efforts to inject methodology into the design process must begin.

SPECIFICATIONS, DESIGN CRITERIA, FORM, AND CONTEXT

Design begins with the recognition of need. One of the most important steps of all in the design process is the determination of the job—the variables and constraints—sometimes called "defining the problem."[13]

Alexander suggests that a design problem should be regarded as consisting of two elements, namely a form and a context belonging to it. The context— largely identifiable with the sum of requirements and constraints—receives its rational complement in a form adequate to it, whereas the form embodies the sum of the characteristics which satisfy the context. A form is appropriate to a context if it is conducive to the smooth coexistence of both. As it is exceedingly difficult, if not impossible, to describe a form which conforms to its context (e.g., a comfortable chair), it is preferable to enumerate the possible types of nonfit between form and context. The design process, then, can be interpreted as a scheme designed to neutralize, eliminate, or eradicate those factors which cause the undesired nonfit between a form and its context. It is indeed only when a nonfit—a deficiency state—is recognized that a design problem crops up in the mind at all. Hence, in the initial phase of his work the designer should concentrate on those factors which might upset the desired equilibrium between the product and its context. Design originates in a clash with the negative.

If a list of design criteria is set up in terms of a number of potential interference factors, the designed form will have to be assessed in the light of its harmonization or otherwise with a particular set of requirements, or the context. If there are standards with quantified comparative values (e.g., prescribed tolerances), it should not be difficult to ascertain to what extent a design satisfies requirements. But there is every indication that, as a problem-solving activity, designing involves a preponderance of variables for which there is no comparative scale. On this point Alexander writes: "The importance of these nonquantifiable variables is somewhat lost in the effort to be scientific.

---

[13]   J. N. Sidall, "A Survey of a modern Theory of Engineering Design," *Product Design and Value Engineering* 11, no. 9 (September 1966).

A variable which exhibits continuous variations is easier to manipulate mathematically and therefore seems more suitable for a scientific treatment. But although it is certainly true that the use of performance standards makes it less necessary for a designer to rely on personal experience, it also happens that the kind of mathematical optimization which quantifiable variables make possible is largely irrelevant to the design problem. A design problem is not an optimization problem."[14]

PAY-OFF OF RATIONALITY

Where there are preservers at work, a corpse is to be expected.[15]

Whereas Ch. Alexander bases his design methodology primarily on mathematics and specifically on set theory, B. Archer borrows his systematic methods of problem-solving from organization and planning techniques and from computer programming. The incorporation of these organizational methods undoubtedly perfects design methodology in the sense that it makes it rational throughout; but doubts were soon voiced whether these loans from working methods of the kind generally used today in implementing technical and scientific programs of development constituted anything more than an approach to the problem of design or did more than merely trim it out with a methodology (if the word applies here at all) that leaves the core of design itself untouched. Undoubtedly, the design process can be suitably organized with the aid of network and arrow diagrams, provided that the scope of the work makes the use of these techniques seem reasonable. Finally, a distinction must be made between a material necessity to use modern planning techniques and the mere wish to "apply" them. Organizational precisianism in the field of design satisfies ritual needs and as such is harmless enough, provided it does not get in the way of the design process. But, apart from the fact that organizing the design process can make it easier, all the minutiae of organization are very useful ploys in an argument. Anyone presenting his client with a design elaborated according to sophisticated checklists, and put, as it were, on a sound empirical basis, is more likely to bring him down to the ground of rational argument—or at least to his knees—than a designer who produces a model pure and simple and leaves it at that. He who argues that hundreds of thousands of documents have been sifted in the course of work on a design and millions of data on the functional efficiency of a product processed in a computer might create an atmosphere that can spellbind even reluctant negotiators and conditions them in favor of a design, provided they are the type that will yield to the brute force of facts.

As a tactical means of enhancing the credibility of a designer, network techniques can be used in design, more especially while they still have a

[14] Alexander, *Notes on the Synthesis of Form.*
[15] H. Heissenbüttel, *Über Literatur* (Olten: Walter, 1966).

certain novelty value. Being quasi-scientific additives, they help to banish the—wrongly or rightly—tabooed artistic element in industrial design. As a technical aid, they objectify design and bureaucratize it. And as psychological stabilizers, they function rather like a superego which the ego, lacking the courage of its convictions, can obey.

PHASES OF THE DESIGN PROCESS

Probably the most startling feature of twentieth-century culture is the fact that we have developed such elaborate ways of doing things and at the same time developed no way of justifying any of the things we do.[16]

In spite of the gross simplification to which (in this instance) complicated processes must be submitted when they are converted into block diagrams, the phases of some design processes or design-related processes are shown here.
    Bruce Archer divides the design process into six stages:

1 Programming
2 Data collection
3 Analysis
4 Synthesis
5 Development
6 Communication

Kjetil Fallan also divides the value analysis process into six stages:

1 Preparation phase (stating the problem)
2 Information phase (getting the relevant facts)
3 Evaluation phase (definition of the function of the product)
4 Creative phase (find less costly ways of performing the same function)
5 Selection phase (scanning a set of alternatives)
6 Implementation phase

Sidall—himself a designer—distinguishes between thirteen stages in the design process:

1 Definition of problem
2 Scanning of all possible design and environmental variables acting on a machine
3 Definition of constraints (legal requirements, standards) writing of design specifications
4 Creating the basic concept

---

[16]   C. W. Churchman, *Prediction and Optimal Decision* (Englewood Cliffs: Prentice-Hall, 1961).

5   Analysis of the evolutionary basis of the machine
6   Optimization
7   Detailing
8   Calculation
9   Procurement
10 Prototypes
11 Testing
12 Final changes for production

Staged plans of this kind, which are of limited value as statements both of the relevant facts and the extent to which such facts should determine our future course of action (they represent the beginning rather than the end of methodological endeavors), still have their place in the standard theory of problem-solving behavior, according to which problem-solving is the specifically human activity. Yet here are signs that emphasis is shifting from this aspect to the recognition (creation) of problems. Certainly, these divisions of the design process are a gesture in the direction of objectivity. Moreover, they create an impression of serious habits of thought. Even if schemata should be disowned by practice, to which they should nonetheless be adjuvant, and even if they should contain a compulsive element, it is a compulsion which is intent on transcending the merely regulative. Methodologies—or at least the best of them—point to a multiplicity of courses rather than insist on one particular track. Whether they are conceived in a critical or a credulous spirit, the difference between such methodologies is bridged over by one factor: they mold the designing process in advance and thus obviate reckless and unconsidered designing which proceeds as if there were a spontaneity that was not covertly introduced. Methods cut back the impulsive and unpremeditated activity that is obedient to the truism, false like all truisms, that designing begins in a welter of busyness. The rationality of method is needed even by him who thinks he can dispense with it. This rationality comes through primarily in the analysis, in the elucidation of the structure of the problem, in the identification of those desiderata that constitute a problem, and, finally, in their systematic satisfaction.

Today, design methodology is in the same position as psychology in the nineteenth century when it hankered after the status of a "true" science. The method of science continues to be the idol of scientism. Care must be taken that design is not subjected to a heteronomous methodological ideal in which it will no doubt receive the label of approved scientism but will virtually nullify itself. Only when design methodology liberates itself from its often-parasitic relationship with other disciplines can it move a stage higher. It would gain in independence and rigor which are not to be acquired from any other source. Whereas those sciences which prepare "hard data" have long been mandatory for design methodology, it will have to expand in future to embrace precisely those branches of knowledge concerned with the more diversified "soft data." It will be no more able to close itself to a broader technological scope than to go chasing after it out of sheer blind enthusiasm; for as said in another context—it is quite conceivable for technical modernity to be grafted onto a provincial habit of mind.[17]

Design methodology can be the target for a great deal of criticism, ranging from allergic reaction to anything smacking of the rational to the charge that it represents die-hard pedantry and gratuitous bogus rationality. Methodology would be supererogatory if each design process would emerge spontaneously out of its own method of design; for methodology stands and falls by the hypothesis that there are invariables in design and that out of these can be constructed a framework for designing. This undialectical separation of a general action scheme from a particular action content testifies against any methodology in its previous form. This is a contradiction which will no doubt have to be worked out to a satisfactory conclusion.

[17]   Marianne Kesting, *Vermessung des Labyrinths* (Frankfurt: s. Fischer Verlag, 1965).

In view of the inflationary use of the term "design," it may be appropriate to point out a difference between design and "projecting," especially in those languages which have the equivalent of the term "project," but which have imported the term "design" and thus refer to a certain class of products which have "something special" about them.

All too often, this "something special" refers to a phenomenon dominated by marketing interests and faithfully reflects a system-compliant interpretation of design, that is, a special field dominated by creatively gifted individuals.

Liberating design from this association can be a goal of design research.

Of course, at the moment when design researchers allow design to be decided upon in a grandly inquisitive gesture—especially in design education—the question arises as to the legitimacy of this claim. For this reason, there is a latent possibility of tension between design research and design practice, since they correspond to opposing interests between the interest in knowledge and the interest in intervening in the everyday living environment. How to think this tension creatively rather than reductively was one of the aims of this paper.

7 THE UNEASY RELATIONSHIP OF DESIGN AND DESIGN RESEARCH (2004)

ON THE LIMITS OF DESIGN SCIENCE

In 1848, a thin book appeared with the provocative-sounding title: *Die Wertlosigkeit der Jurisprudenz als Wissenschaft* [The Worthlessness of Jurisprudence as a Science]. The author was the well-known lawyer Julius Hermann von Kirchmann.[1] In this work, he analyzed the part played by jurisprudence in improving the practice of law and arrived at a conclusion that did not go down very well with lawyers. In order to allay any suspicions that he was trying to start a futile dispute, he began his exposition with the following sentence: "The subject of my paper today might easily lead some to suspect that I am only interested in a piquant sentence, with no concern for the deeper truth of the matter."[2]

He explains the ambiguity of the title, which may mean that jurisprudence is indeed a science, albeit one without any influence on everyday activity. Or, conversely, it may mean that jurisprudence is not a science, since—as he writes—it does not "fulfill the requirements of a true concept of a science."

Why this reference to jurisprudence and legal practice? What do these concepts have to do with the dialectics of design, and the related question of design research, which are at issue here? For all their difference in content, parallels can be drawn. The conceptual model in Kirchmann's comment may, when transferred to design, mean that although design science is a genuine science, it has no influence on design practice; or that design science is not a science because (as the philosophers put it) it does not fulfill the requirements of a true concept of the latter. It is the task of science to "understand its subject, discover its laws, with the aim of creating concepts, of identifying the relationship and connections between the various phenomena and, finally, of assembling its knowledge in a simple system."[3]

We shall leave aside the question as to whether today's scientists (and this includes design researchers too) would accept so unreservedly the goal of

---

This text is based on a lecture presented at the first research symposium of the Swiss Design Network, May 13–14, 2004 at the HGK Basel. Published under the title *Die Dialektik von Entwerfen und Entwurfsforschung* in Erstes Design Forschungssymposium published by the SwissDesignNetwork, 2004. First published in English in Ralf Michel (eds.), *Design Research Now*. Board of International Research in Design (Basel: Birkhäuser), 25–40.

[1] Julius Hermann von Kirchmann, *Die Wertlosigkeit der Jurisprudenz als Wissenschaft* (Heidelberg: Manutius Verlag, 2000). Originally published in 1848.

[2] Ibid.

[3] Ibid., 12.

amassing their discoveries in a simple system. Here, the first goal is to create free space for reflection and thus avoid making premature characterizations of what design research and design science are and what they ought to be doing. In this situation, a fluid physical state is preferable to a solid one.

"DESIGNING" AND PROJECTING

The English term "design" does not allow for the differentiation, made in German, between design and *Entwurf* (project). Consequently, it is difficult to grasp this distinction in the English language. It may even seem incomprehensible that the term "design" (in contrast to "project") is used in this context with a certain degree of detachment. However, there are reasons why the term "design" should be used carefully in both languages. The popularization of the term "design" during the past decade—not only in English-speaking regions— and its more or less inflationary usage have turned the word design into a commonplace term that has freed itself from the category of projecting and has now attained a sort of autonomous existence. Everyone is entitled to call himself or herself a designer, especially as people generally equate design with the things they see in lifestyle magazines. Not everyone would suddenly call herself/himself a project-maker (in Daniel Defoe's sense of the term[4]) because this carries an overtone of professionalism that the word design has lost. As an alternative, we could use the German expression *Gestaltung*. The only problem is, of course, that it has no equivalent in other languages. For although it refers to design, it does so primarily from the perspective of perception (Gestalt psychology) and aesthetics.[5] However, the German term *Gestaltung* has taken a few blows, from which it has not yet recovered.

If we examine the relationship between design education and design science, it becomes apparent that they appeared almost simultaneously in the 1920s: in the Dutch De Stijl movement and at the Bauhaus. After the Second World War, design research gradually began to establish itself. There are various reasons why this happened; these will be considered later.

In 1981, Bruce Archer, who became well-known through the publication of his Systematic Methods for Designers, characterized design research as a form of systematic inquiry performed with the goal of generating knowledge of the form/embodiment of—or in—design, composition, structure, purpose, value, and meaning of human-made things and systems.

This definition of design research is clearly tailored to industrial design and does not therefore touch on communication design. Archer then goes on to explain his definition, ending with the plausible conclusion that: "Design research is a systematic search for and acquisition of knowledge related to

---

[4]   T. Maldonado, "Das Zeitalter des Entwurfs und Daniel Defoe," in *Digitale Welt und Gestaltung* (Basel: Birkhäuser Verlag, 2007), 257–68.

[5]   It has strong cultural connotations as opposed to civilization with technical connotations.

design and design activity." There ought to be no dispute about this, especially as the statement comes close to being tautological.

In English-speaking regions in particular, the main representatives of design research formerly came from the fields of engineering science and architecture. Interest correspondingly focused on developing rational design methods and on procedures for evaluating buildings and products. Graphic design was barely mentioned. It is hardly surprising then that practicing industrial and graphic designers viewed events dealing with design methods and design science as well as any papers (if they registered the latter at all) published in this context as esoteric glass-bead games[6], played—with no noticeable impact on practice— in academic "reservations" shielded from the constraints and exigencies of professional practice. As a result, discourse on design science found itself cast— rightly or wrongly—in a bad light, since it appeared to have been usurped by a network of concepts irrelevant to design practice. This may be due, in part, to the fact that design research was carried out under the aegis of systems theorists, computer scientists, operations research specialists, and mechanical engineers, whose categorical conceptual systems bypassed industrial and graphic design. Furthermore, they often had no experience—or, at best, very little—in product design or visual communication. The autonomization of method research thus also motivated Christopher Alexander at quite an early stage to distance himself from such research projects because he felt that they had either forgotten or lost sight of the goal of producing better designs.

How has the theme of design research/design science come to assume greater significance? There are two possible reasons for this:

— First, complex design problems can no longer be solved without prior or parallel research. It should be noted that design research cannot be equated with consumer research or variations of it that take the form of ethno-methodology, that is, an empirical science that examines the behavior of consumers in their everyday environments and thus refrains from carrying out laboratory research. Whether or not we are prepared to designate such activity (which accompanies design) as "research" is a question of judgment and depends on which criteria are applied to research. We cannot rule out in advance the possibility that design activity will raise questions that will, in turn, yield new knowledge as a result of the research involved in answering them.

— Second, the consolidation of design education at universities and colleges creates pressure to adapt to academic structures and traditions. Anyone who seeks to pursue an academic career is expected to acquire the appropriate qualifications in the form of a master's degree or doctorate. Anyone who does not possess this "symbolic capital" (Bourdieu) may find himself/herself unable to fill certain key positions in hierarchically structured institutions. In Turkey, for example, it is impossible to gain a

---

6   The use of the term "glass-bead games" refers to the Hermann Hesse's novel with the same title, published in 1943.

professional qualification in architecture or industrial design without a doctorate.

We can therefore identify two reasons for the emergence of design research: one linked to professional practice and the other to academic activity. The tension between the two can and does lead to controversies and divergences.

Designing is initially a free and independent form of activity unconcerned with the existence of design science. However, this form of design has a provisional character. After all, it is quite possible that this activity will increasingly come to depend on the existence of design science: in other words, that design science will become the precondition for practicing design. This trend will obviously have dramatic consequences for design courses, especially in the case of industrial and communication design, as well as all the new fields of study, such as interaction design and information design, which have arisen and are arising in the wake of digitization.

SCIENCE AND DESIGN

In general, scientific activity and design activity are—rightly—distinguished from one another, for each pursues its own mundane interests. The designer observes the world with an eye to its designability, unlike the scientist who regards it from the perspective of cognition. It is thus a question of divergent points of view with different contents in terms of innovation. The scientist and the researcher generate new knowledge. The designer gives people an opportunity to have new experiences in their everyday lives in society, as well as with products, symbols, and services; experiences of an aesthetic character, which, in turn, are subject to sociocultural dynamics.

The tension between cognitively related activity (research) and noncognitively related activity (designing) becomes apparent here. To avoid any misunderstandings, however, it should be pointed out here that design activity is increasingly permeated by cognitive processes. This also raises the issue of mediating between these two areas, something that has been done with varying degrees of success since the 1920s. The unavoidable revision and updating, which are now on the agenda, of traditional study courses in the field of design and planning, inevitably raises the question of how students' cognitive competence can be improved. And this also touches on the intimately linked part played by language in the teaching of design, among other things.

Despite the difference between design and science, there is also a hidden affinity and structural similarity between the approaches of the innovative scientist and the innovative designer: both are engaged in "tinkering," as the American philosopher Kantorovich puts it.[7] They both try things out in

---

[7]   A. Kantorovich, *Scientific Discovery—Logic and Tinkering* (New York: State University of New York Press, 1993).

accordance with the motto: let's see what happens when we do this or that. Both proceed experimentally.

A glance at contemporary design problems clearly shows that the cognitive demands on design have grown. For this reason, neither design studies nor design practice can ignore the sciences and research. One example should make this clear: nowadays, when an industrial designer is commissioned to design sustainable packaging for a carton of milk for a client, she or he will need access to scientific information about energy profiles and ecological footprints and, if necessary, to systematic experiments on material combinations to place design activities on a scientific footing. It is no longer possible to tackle a task of this nature intuitively. As an example, from the field of communication design shows, it is impossible to develop an interface for courseware without engaging in subject-related research. Anyone who relies on their inner voice and supposed creativity will go to the dogs.

REFLECTION/THEORY AND DESIGN

With the introduction of design courses at universities of applied science[8], education programs are now expected to stimulate students' capacity for reflection. In other words, design students must learn to think—a demand that may sound totally normal, but which has by no means been fulfilled. As an American graphic designer wrote: "Design has no heritage of or belief in criticism. Design education programs continue to emphasise visual articulation, not verbal or written. The goal is to sell your idea to a client and/or a hypothetical audience. Design in relation to culture and society is rarely confronted."[9]

Reflective behavior is discursive thinking: thinking that manifests itself in language. Although the idea of including language in design courses goes back to the 1950s, teaching programs generally have a lot to catch up on when it comes to language and texts, especially in the field of visual communication. The anti-discursive tradition and predisposition of design education remains powerful. We need to admit and recognize that design's image frequently attracts the wrong students. Hip-hop and cool are qualities to which design— fortunately—cannot be reduced.

What is reflection? Reflection means establishing distance to our own activities and thematizing our interdependences and contradictions, especially those of a social nature. Theory points beyond what exists. With regard to the emphasis on design research, it should be noted that free space must be set aside for theoretical activity in the future as well: anyone who only considers the direct application of an idea will suffer from a narrowing of their horizon and the degeneration of their speculative consciousness. In *Lob der Theorie* Gadamer mentions "the closeness of theory to the realm of pure play, to

---

[8]   This remark refers to the situation in Germany, where design education formerly linked to Applied Art Schools changed in the 1990s by moving design courses to universities of applied sciences.

[9]   K. Fitzgerald, "Quietude," *Emigre* 64 (2003): 15–32.

purely contemplating and marvelling, far removed from all customs and uses and serious business." Furthermore, he establishes a relationship between theory and those "things . . . that are 'free' from all the calculating attitudes associated with need and use."[10] When we speak of theoretical activities in the field of design, we are certainly not issuing a kind of *carte blanche* for people to start speculating about designing and design in a manner that is totally alien to design, in which speculation occasionally serves scientists as a welcome vehicle for distinguishing themselves academically when they treat design as an object of research. Such strategies are quite tempting, because the subject of design, with its complex ramifications and interconnections, is virgin territory for scientific activity. It seems that people easily forget that talking about a subject demands a minimum degree of knowledge about it, and that, no matter how good people's intentions, speculative theoretical studies are no substitute for specialized knowledge. Hence, when such discussion contributions on design—which are frequently full of preconceived ideas and interpretation models—serve as norms aimed at standardizing practice in the guise of scientific dignity, then it is time to cut those displaying such presumptuousness down to size.

DESIGN AS AN OBJECT OF CRITICISM

For many decades, design was not thematized in scientific discourse. It was a nonissue. Despite its presence in everyday life, it hardly awoke the interest of scientific disciplines. Now that design has become a media topic, however, the situation has changed so much that there is no lack of critical discussion on the subject. One pertinent publication worth mentioning here is the latest work by the art theoretician Hal Foster, entitled *Design and Crime*.[11] Foster argues from an anti-conservative, culturally critical perspective. The title of Foster's book alone, an allusion to Adolf Loos' *Ornament und Verbrechen*,[12] which attained fame because of its polemical tone, speaks volumes. Foster writes:

The old debate (infuse art into the utilitarian object) takes on a new resonance today, when the aesthetic and the utilitarian are not only conflated but all but subsumed in the commercial, and everything—not only architectural projects and art exhibitions but everything from jeans to genes—seems to be regarded as so much *design*.[13] (Emphasis in original)

To which we can only reply: all of this is design. Foster continues:

the old project to reconnect Art and Life, endorsed in different ways by Art Nouveau, the Bauhaus and many other movements, was eventually accomplished, but according to the spectacular dictates of the culture

---

[10]   Hans-Georg Gadamer, *Lob der Theorie* (Frankfurt: Suhrkamp Verlag, 1991).
[11]   Hal Foster, *Design and Crime (and Other Diatribes)* (London: Verso, 2003).
[12]   Adolf Loos, *Ornament und Verbrechen*, in *Trotzdem. Gesammelte Schriften 1900–1930* (Vienna: Prachner Verlag, new edition 1997), 78–88.
[13]   Foster, *Design and Crime (and Other Diatribes)*, 12.

industry, not the liberatory ambitions of the avant-garde. And a primary form of this perverse reconciliation on our time is design.[14]

Here we find a distinct example of a mistaken understanding of design (applied art). Design has long since ceased to involve the aestheticization of everyday life, if it ever did in the first place. We can hardly get to the roots of design using art-theoretical concepts. Design is an independent category. Located at the interface of industry, the market, technology, and culture (living practice), design is eminently suited for engaging in culturally critical exercises that focus on the symbolic function of products. Foster uncritically adopts a theorem formulated by Jean Baudrillard that asserts that design is basically limited to the symbolic dimension of objects, to the "political economy of symbols." Design is thus dematerialized and degraded to a sign exchange value. Those very positions that view themselves as anti-conformist show a remarkable tendency to pour blanket criticism on modern design for being pure ideology. There were times in which avant-garde positions in philosophy (the Vienna circle, for example) and modern design regarded one another with mutual respect. Nowadays, such an attitude is rare indeed. Today, design serves as the compliant stooge for critics of the commodity society: for critics of pan-capitalism.

RESEARCH IN CLASS

When and how should students be taught how to reflect and do research? So far, universities and colleges have failed to provide a unanimous answer to this question. Reflection and research should not be reserved for students in the more advanced classes but should be taught and practiced from the beginning of the first year. Design studies would then no longer be limited to master's degree courses but encouraged and required in bachelor's courses too. This does, of course, also entail risks. Every design instructor has experienced students who try to steal their way out of doing designs by performing rhetorical acrobatics and concealing weaknesses in the field of design with the aid of verbal gymnastics. Such discourse, which is a strategy for avoiding design, must be prevented. It has nothing to do with the kind of cognitive competence envisaged here, which is anchored in design. Teaching programs must take into account cognitive competence, especially if students display theoretical interests that have hitherto been tolerated, at best, but not explicitly encouraged; for this defect is one of the reasons for the oft-criticized speechlessness of designers.

In the field of design, it is possible to distinguish between two different approaches to research:

1. Endogenous design research, that is, research initiated spontaneously from within the field of design. This primarily proceeds from concrete

---

[14] Ibid., 19.

experiences in designing and is frequently integrated into the design process, thus signifying a primarily instrumental interest. It may be hoped, however, that in the future a form of endogenous design research will be pursued that goes beyond its immediate application in the design process. This would create a pool of knowledge that the field of design still lacks. (The complaint about the lack of a pool of knowledge specifically related to design is well known.) Designers should definitely be involved in this kind of research in order to counteract the danger of other-directedness in design discourse. Should the profession fail to address this need, it would put the future of industrial designers and graphic designers in doubt. These two professions might then find themselves members of a dying species.

2. Exogenous design research, which views design as an object of research and other disciplines as meta-discourses, so to speak. We should proceed with caution here, however, for the further removed texts and research exogenous to design are from concrete experience with the contradictions, paradoxes, and the aporia of design, the greater the danger that they will be at the mercy of sweeping judgments. The last thing that designers need are scientific high inquisitors who, with one finger raised, try to drum norms into their students' heads, telling them what they should and should not do.

As far as the content of research work is concerned, a rhizome table can be drawn to illustrate the broad range of themes and arrange them in a distinct order. It goes without saying that this classification, like every other, contains subjective moments and is subject, above all, to certain plausibility criteria. The map outlined here is subdivided into six thematic groups:

— History
— Technology
— Form/structure
— Media
— Design/daily practice
— Globalization/the market

Each of these themes is, in turn, subdivided into a series of subthemes.

Within the framework of a historical research project, it would be possible to draw a timeline of the subjects of the discourse on design that shows the emergence and duration of certain themes that appear in design discourse. The timeline would show the ups and downs of the discourse too. Certain themes vanish and new ones—whether under familiar or novel names—appear, while old ones may experience a revival. (This would make a fruitful field for research on design history.) As far as design education is concerned, it would be very interesting to examine how the dominance of discourse subjects has left its mark on diverse curricula.

INNOVATION AND PROJECTING

The various branches of economics distinguish different strategies that a company can pursue to assert itself in competitive markets are:

1. Technical innovation (e.g. a new chip)
2. Quality (reliability, durability, finish)
3. Rapid delivery
4. Design

There is, of course, also the strategy of competing via lower prices. However, this is likely to play an ever-smaller role. Markets demand quality products, technically advanced products, and products with high-quality designs.

How is innovation manifested in industrial design and communications design? How does innovation differ in the fields of engineering science, management, and the applied sciences? In other words, what is design innovation?

How do the results of design activity look in terms of product design and visual communication, inasmuch as the latter are intended to be innovative? Before this question can be answered, the main characteristics of design must be outlined, concretized—from an integrational perspective—with the aid of nonpropositional knowledge. The following list shows areas in which innovative design leaves its mark:

— Innovation in the form of improving the usability of a product or information
— Innovation in the form of improving the production process for manufacturing a product
— Innovation in the form of inventing new affordances
— Innovation in the form of sustainability
— Innovation in the form of the accessibility of information or a product (social and nonexclusive design)
— Innovation in the form of finding possible applications for new materials and technologies (solutions looking for a problem)
— Innovation in the form of greater comprehensibility in dealing with information or a product
— Innovation in the area of formal aesthetic quality (sociocultural dynamics)
— Innovation in the sense of strategically extending a company's product range (e.g., citing a manufacturer of agricultural machinery who has broadened his range—as part of strategic branding—to include services in the form of optimal fodder compositions)

From an economic point of view, there is evidently a relationship between an economy's competitiveness and its design ranking at an international level. A survey by the New Zealand Institute of Economic Research reveals that the twenty-five countries with the world's most competitive economies are also world leaders in design.[15] The survey is quite instructive, despite reservations about the marketing criteria underlying the ranking system. The survey could be broadened to find out which countries are leaders in the field of design research and whether there is a correlation between this and general economic competitiveness.

THE FOUNDATIONS OF DESIGN

Another unanswered question in design education concerns the foundations of design and related research into the foundations of design. Very divergent views prevail on this matter among design instructors. There is, for example, the question as to what actually constitutes the foundations of design, and, taking this a step further, whether design can be said to have foundations in the first place, that is,. whether design isn't an activity without foundations. To anyone who subscribes to this position, any insistence on foundations would indicate nothing other than a pious, outmoded, and unfounded wish. In this context, the sciences are often cited for the purpose of comparison, as they are generally considered to rest on foundations, and are therefore upheld to serve as a model and benchmark for design.

If we consult the sciences, however, we learn otherwise: that the sciences do not rest on foundations either. In a lecture in 1941, Max Planck said:

> if we . . . subject the structure of the exact sciences to precise analysis, we
> very soon become aware that the edifice has a dangerously weak spot, and this
> spot is the foundation. . . . for the exact sciences there is no principle of such
> general validity and, at the same time, of such great importance with respect
> to content that it can serve science as an adequate basis. . . . We can, therefore,
> reasonably draw only one conclusion from this, namely, that it is absolutely
> impossible to place the exact sciences on a general foundation composed of
> definitively conclusive content.[16]

No matter what position we adopt regarding whether design has foundations or not, it must be noted that, in the field of design education, the teaching of foundations has always aimed to solve an undeniable problem: providing the students with a formal-aesthetic education that seeks to cultivate not only their receptive differentiation skills but also, and above all, their generative differentiation skills. A glance at the history of design education reveals that heated controversies raged over the Bauhaus basic course, on which design courses across the globe would subsequently model their identity (distinguishing them from courses in other fields of study). When the basic

---

15 Building a Case for Added Value through Design. Report to Industry New Zealand, February 2003. New Zealand Institute of Economic Research (NZIER). This report is based on indicators from the World Economic Forum's Global Competitiveness Report.
16 M. Planck, *Sinn und Grenzen der exakten Wissenschaft* (Leipzig: Johann Ambrosius Barth, 1942).

course was being developed, there was a debate on whether formal-aesthetic generative competence should be allowed to develop independently and as an organizational unit within the curriculum, or whether the basic course should be simply abolished as a relic from a hazy, romantic period of design education. Terms such as "basic course" and "basic design course" sometimes irritate people and cause them to adopt rigid positions that block all discussion from the start. Hence, it might be advisable to refrain from using these terms. Such a move would not do away with the problem of educating students to develop formal-aesthetic competence, but it would at least diffuse the situation. Instead of talking of basic courses and foundations of design, we could use Christopher Alexander's term "patterns," which refers to recurring phenomena that exist in a context relatively free of the influence of economic factors, production technology issues, etc. This would make it possible to avoid the immanent danger of academicizing basic courses and thereby transforming design exercises into formal recipes that assume the form of canons or "style-bibles."

FROM DISCOURSES TO VISCOURSES

For some years now, there has been talk within the social sciences of an iconic turn. This notion denotes a new epistemological constellation that ends the primacy of discoursivity as the privileged domain of cognition. The term "iconic turn" signifies the recognition of the visual plane as a cognitive domain in contrast to the centuries-old tradition of verbo-centrism. This turn is determined by technological innovations, especially in digital technology, which have made possible the new processes of image generation and image production. In the words of Günter Abel: "Here (in the basic processing of the images) it is not a question of merely passively, illustratively or graphically reproducing something that already exists in finished form, but of an original, active process of revealing or showing something visually [Ins-Bild-Setzen]."[17]

It is well known that training the ability to reveal or present an object or an idea visually is central to graphic design and visual communication courses. Thanks to the iconic turn in the sciences and to digital technology, it is now possible to explore the cognitive potential of visual design and adequately to characterize the indispensable role that visual design plays in the cognitive process. This opens up a fascinating new field of activity and research for traditional graphic design. Even so, it is difficult—in the beginning at least— for a mode of thinking in which the discursive tradition has been dominant to acknowledge the cognitive status of images and, above all, of visuality. The deeply rooted prejudice against images is evident in the fact that they are so often downgraded with the adjective "beautiful," revealing a visceral distrust of anything that betrays even a trace of aesthetic sensitivity. The anti-aesthetic

---

17 Günter Abel, "Zeichen- und Interpretationsphilosophie der Bilder," in *Bildwelten des Wissens. Kunsthistorisches Jahrbuch für Bildkritik*, eds. Horst Bredekamp and Gabriele Werner, 1, no. 1 (Berlin: Akademie Verlag, 2003), 89–102.

attitude, or at least the indifference of a scientific tradition that is fixated on language, is sufficiently well known. For centuries, Plato's allegory of the cave contributed to the contempt for visuality and its being situated outside the mainstream. The epistemological constellation based on enmity toward pictures is the counterpart to a design tradition that adheres exclusively to images and disdains language. Günter Abel characterizes visual knowledge thus:

> In contrast, non-linguistic and non-propositional knowledge refers to a form of knowledge that one can possess without having the corresponding linguistic predicates and concepts and without having learned these.[18]

Digital technology will bring about far-reaching changes in epistemological traditions and indicate a new role for visual design. One media analyst wrote in this context: "Writing and reading will certainly not lose their meaning immediately; however, they will be come to occupy a less-central position among the broad range of cultural performances." He went on to say that the claim "that only a printed monograph can represent, for instance, the standard of knowledge achieved by a scientific discipline is generally viewed as one of the 'myths of book culture' these days."[19]

If it is true that designers can no longer design the way they did one or two generations ago, then it must also be acknowledged that researchers can no longer do research as they did one or two generations ago—that is, orienting themselves primarily or exclusively by texts. This newly emerging trend can be summarized in four words: from discourses to viscourses. Under these circumstances, an iconic turn in the sciences might correspond to a cognitive turn in the design disciplines. It is still the early stages.

POSTSCRIPT

Since the 1980s, when the term "globalization" found its way into the social sciences, design, too, has been called upon to reconsider its role vis-à-vis this process. However, the term itself must be treated with a pinch of salt, especially in view of the fact that the alternative is not between globalization fundamentalists and globalization phobics (the expression unjustly used by critics in the conservative media to stigmatize them). In a recent interview, Kenneth Galbraith criticized the naïve use of this term, which he exposed as a means of camouflaging the process in which world economic policy was subjected to the hegemonic economic interests of the United States. That said, there are purely practical reasons why it will not be so easy to exclude the term from discourse. If we consider the adverse impacts of this development,

---

[18] Ibid.
[19] Frank Hartmann, *Mediologie—Ansätze einer Medientheorie der Kulturwissenschaften* (Vienna: Facultas Verlag, 2003).

we cannot avoid seeing a tendency toward social exclusion and the ruthless plundering of our planet's resources. Considering design in this context, we are entitled to ask after design practices that oppose this trend and refuse to unthinkingly fall in line with or subordinate themselves to this process. Of course, not everyone wants to occupy themselves with these questions. How would a design practice look that presented an alternative to a form of design that excludes people; if it no longer restricted itself to addressing a mere 10 or 20 percent of the world's population in the highly industrialized countries, or in enclaves within zones formerly known as the Third World? It seems that there has been no unanimous answer to this question so far, especially as it is a highly explosive issue with an inescapably political character. Design research could certainly find a very relevant subject here if it aimed to reduce the gap between the different societies, assuming, of course, that we consider this at all meaningful and do not dismiss it as the perpetuation of the status quo of a social value system that has no future and needs to be radically renewed and turned upside down.

(c) Design, Politics, Ethics

As becomes evident in many of the papers in Part II of this collection, design practice in peripheral countries, especially in the public sector, is characterized by a high degree of instability and unpredictability. From one day to the next, the political starting conditions and goals of a project can change, which provides reason to question the coherence of design and gives ideological patrols a chance to demonstrate their vigilance. The weak structure of political institutions is susceptible to all kinds of influence and instrumentalization of particular interests. This different institutional context of design activity in peripheral countries with insufficiently consolidated political institutions is reflected in design practice, so that it seems questionable whether it makes sense to establish generally or globally valid criteria for assessing design. Design practice in peripheral countries can be characterized as continuity of discontinuity—as the continuity of instable framework conditions.

As a brief reflection on two decades of practice in Latin American countries, this interview was an attempt to explore the deeper—and often unstated—political and economic circumstances to which practice, because it itself is always economic and often political, is inevitably subject.

8 DESIGN, NOMADISM, AND POLITICS

Interview with Alejandro Lazo Margain

(1992)

Alejandro Lazo Margain (LM): Why did you decide to work in Latin America?

Gui Bonsiepe (GB): Latin American societies are still in the process of formation—they are a mixture of immobility and chaos, with a correspondingly high degree of unpredictability. There is scope for new possibilities, despite the notorious restrictions in the form of endemic political, institutional, and economic instability.

LM: What aroused your interest in these societies in the process of industrialization?

GB: I am interested in how they tackle modernization—it is a process in which design plays—or could play—a crucial part.

LM: Why did you leave the Federal Republic of Germany in 1968?

GB: I wanted to learn something outside the European context, although I was not clear what it was. My move to Latin America coincided with the politically motived closure of the HfG Ulm but it was not the direct result of this politically conditioned fact. In general, there is a tendency to migrate from the periphery to the center; in my case the direction was the reverse.

LM: What induced you to decide to study industrial design?

GB: I came to design from an artistic background. In ulm I studied what one can call the theoretical aspects of design, in a department that bore the name "information." That department was intended to train specialists who could reflect on the broad and little-understood theme of design and write about it.

LM: What do you understand by industrial design?

GB: For me, design means taking on the concerns of social groups of users, sometimes mediated through a commission by a client.

Interview with Alejandro Lazo Margain, published in the newspaper *Excelsior*, Mexico, March 17, 1992. The German version of this text was translated by Anke Grundel.

LM: What does ulm mean to you?

GB: An uncompleted project of radical modernism. Ulm faced up to the challenge—and the tensions and conflicts it brought—of the relation between the world of sciences and design and thematizing them. As we know, this aroused opposition. The critical rationalism of the ulm variety was a constant source of annoyance and polemical conflict, and even resentment in the FRG. But in my view the opinion generally held of the HfG is far too monolithic, and this prevents the more differentiated reality from being perceived.

LM: Who were your tutors in ulm?

GB: I would like to name Tomás Maldonado, for I regard myself as one of his pupils; then the philosopher Max Bense, the sociologist Hanno Kesting, the critic Erich Franzen, the literary theorist Käthe Hamburger, to name only a few. I was strongly influenced by the Frankfurt school, especially the writings of Walter Benjamin, Theodor W. Adorno, Max Horkheimer, and Jürgen Habermas. On the other side, I had the opportunity to concern myself with the neopositivism of the Vienna Circle, with the analytical philosophy of language and with pragmatism, especially C. S. Peirce. That stimulating intellectual climate is one of the most remarkable qualities of the HfG Ulm in my eyes, although that side of the design institution is not publicly known.

LM: What do you see as your professional contributions?

GB: I regard it as an achievement to have introduced industrial design as a new field to the technology and industrial development policy of a number government institutions of Latin American countries, and to have helped to form the professional profile of design in peripheral countries.

LM: What are the most important objectives in order to orient design to the international level?

GB: I assume that no one will presume to try and formulate directives for design on international level. Having said that, I regard it as a task that cannot be postponed to create a design discourse, to institutionalize design research and promote it intensively.

LM: Do you agree that designers' work should have a national identity to reveal its country of origin?

GB: I have repeatedly been asked about national identity, since I came to Latin America.[1] It is almost an obsession here. But I ask: Is that really where the problem

---

[1]   In other words: "The problem of a Latin American identity is inseparable from the destruction of its historical culture and its historical awareness." Eduardo Subirats, *El continente vacío* (Madrid: Anaya & Mario Mushnick, 1994), 27.

lies? Do you believe that Japanese designers care about "Japanese" design, as long as they can sell their cars, video cameras, and television sets to the rest of the world? Designing means listening to others, and not staring at one's own navel. Furthermore, I confront with distrust manifestations of essentialism and of attempts to look for roots of Latin American design in the past, mainly in the preindustrial period. Essentialism does not lead into the future.

LM: What do you say to the political events in Russia?

GB: "Real existing socialism"—to use a term that is unwieldy and aims to discredit a socialist utopia generally, beyond historical contingency—proved unviable, both as a production system and as a distribution system. As far as the ideological consequences of this implosion are concerned, I doubt whether the economic doctrine of real capitalism being practiced today in Latin America with draconian severity can provide answers to the global problems, especially the bipolarization between and within societies, and the North–South conflict. History is certainly not at an end.

LM: How do you see the future of Cuba?

GB: According to official pronouncements, the Cold War is over. Consequently, it seems contradictory to me to try to pursue a Cold War policy. And one may ask on what international legal norms the economic blockade of a country, that has lasted for decades now, is based. The subordination of law to contingency interferes with the law and annuls the right to speak in the name of that law. What idea of "right" is it that interprets what is right according to criteria of opportunism?

LM: What project in industrial design has given you the greatest personal satisfaction?

GB: The agricultural machines that were designed by a team in Chile in 1971–3.

LM: And in graphic design?

GB: The interface of a computer program which I helped to design in the United States in 1988–9. I should like to add that it blurred the traditional borders between industrial design and graphic design.

LM: What interests you in computer graphics?

GB: Firstly, the interplay of graphics and text. Secondly, the development of new tools to learning in the form of interactive hypermedia.

LM: How do you judge the position of industrial design in Latin America?

GB: The existence of design depends less on the designers than on the managers and companies, who will decide whether to invest in design or not. It would certainly be meaningful to move away from a self-referential approach and star reputations; for they make it more difficult to integrate design into industry.

LM: What should our industry in Mexico concentrate on—assembly (maquiladoras), redesign, or the development of new designs?

GB: In the assembly factories along the border to the United States, with their low wages, the share of design, as the name *maquiladoras* shows, is virtually zero. Real industrial dynamic now feeds on innovation. In most professional work that is redesign, that is, the improvement of an existing product. If firms want to be active on the international market in a significant way, they have no choice but to make innovation a permanent activity. Otherwise, they will be left on one side and wither away.

LM: On many occasions, it has been said that the Latin American countries are developing design for the broad mass of the people and so have to meet mass requirements. Do you agree?

GB: It is not a matter of agreeing with that or not. It is easy to agree with it, but difficult to move from the level of well-meaning intentions and high-sounding declarations to the level of effective action in design.

For decades, the history of design in Chile during the politically and socially dynamic period from 1970 to 1973 remained hidden, as if it had never existed. However, in the long run this process could not be silenced, especially since one of the conceptually most daring projects of this period known by the name Cybersyn—began to provoke international interest, if less on the part of design history, but on the part of the history of computer science and cybernetics.

As the papers at end of this volume by myself and Eden Medina make clear (see Chapters 4.10 (i) and (ii)), Cybersyn was a revolutionary project. The Opsroom, and its design, was a component of this complex system. The photos of the program that have gone into the media served then and now to make the rather abstract economic project of Cybersyn experienceable and that was instrumental for the resonance of the project and publication.

An invitation to Chile in 2005 offered the opportunity to present broader considerations of the role of design in relationship toward democracy.

9   DESIGN AND DEMOCRACY
(2005)

I shall present a few thoughts about the relation between democracy and design, about the relation between critical humanism and operational humanism. This issue leads to the question of the role of technology and industrialization as a procedure for democratizing the consumption of goods and services, and finally to the ambivalent role of esthetics as the domain of freedom and manipulation.

The main theme of my lecture is thus the relation between design—in the sense of projecting—and autonomy. My reflections are open-ended and do not pretend to give quick and immediate answers. The university—still—offers a space to pursue these questions that will not generally be addressed in professional practice with its pressures and contingencies.

Taking a look at the present design discourse, one notes a surprising—and I would say alarming—absence of questioning design activities. Concepts like branding, competitiveness, globalization, comparative advantages, lifestyle design, differentiation, strategic design, fun design, emotion design, experience design, and smart design prevail in design magazines and the—all too few—books about design. Sometimes one gets the impression that a designer aspiring to two minutes of fame feels obliged to invent a new label for setting herself or himself apart from the rest of what professional service offers. I leave aside coffee-table books on design that abound in pictures and exempt the reader from intellectual efforts. The issue of design and democracy doesn't enjoy popularity—apart from a few laudable exceptions.

If we look at the social history of the meaning of the term "design," we note on the one side a popularization, that is, a horizontal extension, and on the other side a contraction, that is, a vertical reduction. The architectural critic Witold Rybczynski recently commented on this phenomenon:

> Not so long ago, the term "designer" described someone like Eliot Noyes, who was responsible for the IBM Selectric typewriter in the 1960s, or Henry Dreyfuss, whose clients included Lockheed Aircraft and Bell Telephone Company . . . or Dieter Rams, who created a range of austere-looking, but very practical products for the German company Braun. Today, "designer" is more likely to bring to mind Ralph Lauren or Giorgio Armani, that is, a fashion

This is a slightly abbreviated translation of a speech given in Spanish at the palace of the former national congress, Santiago de Chile, on occasion of an academical distinction by the Metropolitan University of Technology in Santiago, Chile, June 2005. It originally appeared in English in *Design Issues* 22, no. 2 (spring 2006): 27–34. It has been published also in *Civic City Cahiers 2: Design and Democracy* (London: Bedford Press, 2011. Kindle Edition).

designer. While fashion designers usually start as couturiers, they—or at least their names—are often associated with a wide variety of consumer products, including cosmetics, perfume, luggage, home furnishings, even house paint. As a result, "design" is popularly identified with packaging: the housing of a computer monitor, the barrel of a pen, a frame for eyeglasses.[1]

More and more design moved away from the idea of "intelligent problem solving" (James Dyson) and drew nearer to the ephemeral, fashionable, and quickly obsolete, to formal-aesthetic play, to the boutiquization of the universe of products of everyday life. For this reason, design today is often identified with expensive, exquisite, not particularly practical, funny and formally pushed, colorful objects. The hypertrophy of fashion aspects is accompanied and increased by the media with their voracious appetite for novelties. Design has become thus a media event—and we have a considerable number of publications that serve as resonance boxes for this process. Even design centers are exposed to the complicity of the media running the risk of failing to reach their original objective: to make a difference between design as intelligent problem-solving and styling. After all it is a question of a renaissance of the tradition of the Good Design Movement, but with different foci and interests. The advocates of Good Design pursued socio-pedagogical objectives, and the Life-Style Centers of today pursue mainly commercial and marketing aims to provide orientation for consumption patterns of a new—or not that new—social segment of global character, which can be labeled with the phrase: "We made it."

The world of everyday products and messages, of material and semiotic artifacts has met—with rare exceptions—in cultural discourse (and this includes the academic discourse) a climate of benign indifference that has its roots in classical culture in the medieval age when the first universities in the Occident were founded. This academic tradition did not take note of the domain of design (in the sense of project). However, in the process of industrialization one could no longer close one's eyes to technology and technical artifacts that more and more made their presence felt in everyday life. But the leading ideal continued to be cognitive character in the form of the creation of new knowledge. Never design achieved to establish itself as parallel leading ideal. This fact explains the difficulties of integrating design education in the institutions of higher learning with their own traditions and criteria of excellence. This becomes evident in doctoral programs in design that favor the production of discursive results and don't concede projects the same value or recognition as the production of texts. The sciences approach reality from the perspective of cognition, of what can be known, whereas the design disciplines approach reality from the perspective of projectability, of what can be designed. These are different perspectives, and it may be hoped that in the future they will transmute into complementary perspectives. So far

---

[1]   Witold Rybczynski, "How Things Work," *New York Review of Books* 52, no. 10 (June 9, 2005): 49–51.

design has tried to build bridges to the domain of the sciences, but not vice versa. We can speculate that in the future design may become a basic discipline for all scientific areas. But this Copernican turn in the university system might take generations, if not centuries. Only the creation of radically new universities can shorten this process. But the decision space of government institutions is limited due to the weight of academic traditions and due to the bureaucratization with emphasis on formal procedures of approbation (title fetishism). Therefore, the new university will probably be created outside of established structures.

Relating design activities to the sciences should not be misinterpreted as a claim of a scientific design or as an attempt to transform design into a science. It would be foolish to design an ashtray with scientific knowledge. But it would not be foolish—and even mandatory—to tap scientific knowledge when designing a milk package with a minimal ecological footprint. It is no longer feasible to limit the notion of design-to-design disciplines such as architecture, industrial design, and communication design because scientists are also designing/projecting. When a group of agricultural scientists develops a new sweet from the carobbean that contains important vitamins for school children, we have a clear example of a design activity.[2]

Now I want to focus on the central issue of my lecture: the relation between democracy and design. Indeed, during the last years, the notion of democracy has been exposed to a process of wear and tear so that it is advisable to use it with care. When looking at the international scene, we cannot avoid stating that in the name of democracy colonialist invasions, bombardments, genocides, ethnical cleaning operations, torture, and breaking international laws have been—and are—committed, almost with impunity, at least for the moment. The invoice for this lack of humanity is not known. Future generations will probably have to carry the burden. With democracy these operations have nothing in common.

According to the neoliberal understanding, democracy is synonymous with the predominance of the market as an exclusive and almost sanctified institution for governing all relations within and between societies. So, we face questions: How can the notion of democracy be recovered? How can the notion of democracy gain credibility again? How one can avoid the risk of being exposed to the arrogant and condescending attitude of the centers of power that consider democracy as nothing more than a tranquilizer for public opinion in order to continue undisturbed with business as usual?

I am using a simple interpretation of the term "democracy" in the sense of participation so that dominated citizens transform themselves into subjects opening a space for self-determination, and that means a space for a project of one's own accord. Formulated differently, democracy reaches farther than the formal right to vote; similarly, the notion of freedom reaches farther than

---

[2]   "Crean un nuevo alimento para escolares en base a algarroba," Monday, May 9, 2005, http://www.clarin.com/diario/2005/05/09/sociedad/s-03101.htm.

the possibility to choose between a hundred varieties of cellular telephones or a flight to Orlando to visit the Epcot Center or to Paris to look at paintings in the Louvre.

I favor a substantial, and thus less formal, concept of democracy as the reduction of heteronomy, that is, domination by external forces. It is no secret that this interpretation fits into the tradition of the enlightenment that has been criticized so intensively by, among others, Jean-François Lyotard, when he announced the end of the grand narratives. I do not agree with this approach or other postmodern variants. Without a utopian element, another world is not possible and would remain the expression of a pious ethereal wish without concrete consequences. Without a utopian ingredient, residual though it may be, heteronomy cannot be reduced. For this reason, the renunciation of the project of enlightenment seems to me the expression of a quietist, if not conservative, attitude—an attitude of surrender that no designer should be tempted to cherish.

In order to illustrate the necessity to reduce heteronomy, I am using a contribution from a linguist, a specialist in comparative literature—Edward Said. He characterizes in an exemplary manner the essence of humanism, of a humanist attitude. As a philologist he limits the humanist attitude to the domain of language and history: "Humanism is the exertion of one's faculties in language in order to understand, reinterpret, and grapple with the products of language in history, other languages and other histories."[3] But we can extend this interpretation to other areas too. Certainly, the intentions of the author will not be bent when transferring his characterization of humanism—with corresponding adjustments—to design. Design humanism would be the exercise of design activities in order to interpret the needs of social groups and to develop viable emancipatory proposals in the form of material and semiotic artifacts. Why emancipatory? Because humanism implies the reduction of domination. In the field of design, it means to focus also on the excluded, the discriminated, and economically less-favored groups, as they are called in economist jargon, which amounts to the majority of the population of this planet. I want to make it clear that I don't propagate a universalistic attitude according to the pattern of design for the world. Also, I want to make it clear that this claim should not be interpreted as the expression of a naïve idealism, supposedly out of touch with reality. On the contrary, each profession should face this uncomfortable question, not only the profession of designers. It would be an error to take this claim as the expression of a normative request of how a designer—exposed to the pressure of the market and the antinomies between reality and what could be reality—should act today. The intention is more modest, that is, to foster a critical consciousness when facing the enormous imbalance between the centers of power and the people submitted to these powers. Because the imbalance is deeply undemocratic insofar as it negates participation,

---

[3]   Edward W. Said, *Humanism and Democratic Criticism* (New York: Columbia University Press, 2003), 28.

it treats human beings as mere instances in the process of objectivization (*Verdinglichung*) and commodification.

Here we come to the role of the market and the role of design in the market. In a recently published book, the economist Kenneth Galbraith analyzes the function of the concept of the market that according to him is nothing more than a smokescreen for not talking openly about "capitalism"—a term that, not in all social classes and in all countries, enjoys a high rating on the popularity scale. Galbraith inserts design in the context of techniques of corporations for gaining and consolidating power:

> Product innovation and modification is a major economic function, and no significant manufacturer introduces a new product without cultivating the consumer demand for it. Or forgoes efforts to influence and sustain the demand for an existing product. Here enters the world of advertising and salesmanship, of television, of consumer manipulation. Thus, an impairment of consumer and market sovereignty. In the real world, the producing firm and the industry go far to set the prices and establish the demand, employing to this end monopoly, oligopoly, product design and differentiation, advertising, other sales and trade promotion.[4]

Galbraith criticizes the use of the term "market" as an anonymous and impersonal institution and insists instead on talking about corporate power. Against this use of design —after all a tool for domination—stands the intent not to remain fixed exclusively on the aspects of power and of the anonymous market. In this contradiction, design practice is unfolding and resisting a harmonizing discourse that is camouflaging the contradictions. One can deny the contradictions, but one cannot bypass them.

The issue of manipulation has a long tradition in design discourse, especially in advertising. I remember a popular book that at its time provoked a wide resonance, *The Hidden Persuaders* by Vance Packard (1957). But one should be on one's guard against a critique with declamatory character that merely denounces. More differentiation is required. Manipulation and design share one point of contact: appearance. We design, among others and certainly not only, appearances. For this reason, I once characterized the designer as a strategist of appearances, that is, phenomena that we perceive through our senses, above all visual senses, but also tactile and auditory senses. Appearances lead us to the issue of aesthetics—an ambivalent concept. On the one side, aesthetics represents the domain of freedom, of play—and some authors claim that we are only free when we play; on the other side, aesthetics opens the access to manipulation, that is, the increase of outer-directed behavior. When designing products and semiotic artifacts, we want to seduce, that is, foster a positive— or according to context, negative—predisposition toward a product and sign

---

[4]   John Kenneth Galbraith, *The Economics of Innocent Fraud* (Boston: Houghton Mifflin Company, 2004), 7.

combination. Depending on intentions, design leans more to one pole or the other, more to autonomy, or more to heteronomy.

At this point I want to insert a few reflections on technology. The term "technology" in general is understood as the universe of artifacts and procedures for producing merchandises with which companies fill the stage of everyday practice. Technology implies hardware and software—and software implies the notion of design as a facet of technology that cannot be dispensed with. Here in Latin America, we face the problem of technology policy and industrialization policy. Research on these issues reveals interesting details about progress and setbacks. But these seem to me to favor a reductionist interpretation of technology. Only in exceptional cases do texts mention the question of what is done with technology. The question for the design of products remains unanswered. This presents a weak point without wanting to underestimate the efforts by historians. But one cannot defend them against the reproach of being blind to the dimension of design, the dimension of projects, or at least of facing this dimension with indifference. The motives for industrialization include the wish to diversify exports and not to remain an exporting economy of commodities without added value. But behind this plausible argument is hidden another generally not explicitly formulated motif. I am referring to the idea that apart from the growth of the GNP, industrialization is the only possibility for democratizing consumption to provide for a broad sector of the population access to the world of products and services in the different areas of everyday life: health, housing, education, sports, transport, and work, to mention only a few.

However, to mention today the role of government in promoting industrialization can appear almost as an offense of good manners. The role of public intervention has been demonized with one exception, paying the debt of a bankrupt privatized service. In that case public resources are welcome, thus reinforcing the idea that politics is the appropriation of public goods for private purposes. But when the history of industrialization and technology of this subcontinent will once be written, one shall see with clarity that the role of government has been decisive, though the detractors of the public sector with their bellicose voices have belittled its function and contributions. If we look at the recent history of Argentina—a country that until a few years ago followed in subservient manner the impositions of the International Monetary Fund and that in a moment of delirium enthusiastically praised its "carnal relationships" with the leading military and economic power—then we see that this country didn't fare very well with this policy of relentless privatization and reduction of government presence. This process plunged a great part of the population into a situation of poverty unknown until then. It led to an income concentration with the corresponding bipolarization of the society divided in two groups: the excluded and the included. Privatization in this context is synonymous with de-democratization because the victims of this process have never been asked whether they approved the credits and sales of public property that led the country into bankruptcy. Relentless privatization and reduction of the role of government, the unconditional opening of the economy for imports initiated a process of deindustrialization of Argentina, thus destroying the foundations for productive work, including work for industrial designers.

The industrialization policies in various countries in which I have participated, above all Chile, Argentina, and Brazil, concentrated exclusively on hardware, leaving the communication and information industries untouched. Today the constellation has changed radically. An updated industrialization policy would need to include the information sector of the economy, for which graphic design and information design can provide essential contributions. Here new problems show up that confront designers with cognitive demands that in design education programs generally are not taken into consideration. The expanding process of digitalization fostered a design current which claims that today the important design questions are essentially of symbolic character. As second argument for the semantization of products—and thus for semantization of the designer's work—miniaturization is mentioned, made possible by printed circuits and cheap chips. These do not allow us to see how the products are working—functions become invisible. Therefore, the designer's task would consist in rendering these invisible functions visible. Though it would be blind to deny the communication and symbolic aspects of products, their role should not be overvalued as some authors do. Between the alternative to put a nail into a wall with a hammer or the symbolic value of a hammer, the choice is clear. The material base of products with their visual, tactile, and auditive conformation provides the firm base for the designer's work.

With concern, one can observe the growth of a generation of designers that obsessively focuses on symbolic aspects of products and their equivalents in the market—branding and self-branding—and that doesn't know anymore how to classify joints. The search for a balance between the instrumental/operational aspects of technical objects and their semantic aspects constitutes the core of the designer's work, without privileging one or the other domain. As the historian Raimonda Riccini writes: "The polarity between the instrumental and symbolic dimension, between internal structure and external structure is a typical property of artifacts, insofar as they are tools and simultaneously carriers of values and meanings. Designers face the task to mediate between these two polarities, by designing the form of products as result of an interaction with the sociotechnical process."[5]

It is revealing that Riccini does not speak of the form of products and their interaction with functions, that is, the affordances, but that she alludes to sociotechnical development. In this way she avoids the outdated debate about form and function. The once-secure foundations for arriving at the configuration of products have been dissolved today—if ever they existed. It would be naïve to presuppose the existence of a canon of deterministic rules. He who defends such a canon commits the error of essentializing Platonic

[5] Raimonda Riccini, "Design e teorie degli oggetti," *il verri*, no. 27 (febbraio 2005): 48–57.

forms. At the same time, it would be equally naïve to claim a limitless fickleness of forms that would arise from demiurgic actions of a handful of creatively inspired designers. We face here a paradox. To design means to deal with paradoxes and contradictions. In a society plagued by contradictions, design too is affected them. It might be convenient to remember the dictum of Walter Benjamin that there is no document of civilization that is not also at the same time a document of barbarism.[6]

---

[6]   Walter Benjamin, "Über den Begriff der Geschichte," in *Walter Benjamin - Gesammelte Schriften*, eds. Rolf Tiedemann und Hermann Schweppenhäuser, volume I.2 (Frankfurt: Suhrkamp, 1991), 691–704.

Teaching at the Jan van Eyck Academie in Maastricht in the second half of 1990s opened an opportunity for reflecting on the design and on desirable properties of this activity in the future. The intellectually stimulating climate at this institution fostered the formulation of ideas about the future of design—necessarily very personal reflections. Looking for examples as starting point or points of reference, I found the incomplete Harvard lectures of Italo Calvino. They provided a brilliant example for reflecting about the future of writing—avoiding the high-sounding tone of a manifesto that runs the danger of ending in a confrontationist militant pamphlet: noisy, but not more than that. Furthermore, the quality of public design in the Netherlands provided a concrete example for taking this area seriously: a counterfoil to the otherwise domination of consumerism and private interests.

10 SOME VIRTUES OF DESIGN

(2007)

AN UNFASHIONABLE TERM

I chose to focus on the issue of virtues of design when I was reading—once again—the *Six Memos for the Next Millennium* by Italo Calvino. In this remarkable small volume, he speaks about the values he would like to see maintained and brought into the next millennium as far as literature is concerned. These shared values he calls virtues. Taking his approach as starting point, I want to talk about the shared values of design for the next millennium.

VIRTUE 1: LIGHTNESS

The *Six Memos for the Next Millennium* include the following: Lightness, Quickness, Exactitude, Visibility, Multiplicity and Consistency. Without wanting to push the issue, several of these values for literature can be—with due corrections—transferred to the domain of design. A literal transfer certainly would be naive and inappropriate. But parallels and affinities seem to exist. For instance, when Calvino defines *Lightness* as the attempt to remove weight from the structure of stories and from language, are there not analogies in the field of design? Lightness in design might be a virtue to be maintained, especially when we reflect on material and energy flows and their impact on the environment and when we confront the mundane issue of congested lines cloaked with digital trash in the Net. When later on he refers to the "sudden agile leap of the poet-philosopher who raises himself above the weight of the world, showing . . . that what many consider to be the vitality of the times—noisy, aggressive, revving and roaring—belongs to the realm of death, like a cemetery for rusty old cars"[1] lightness acquires a critical dimension and dissipates wrong associations of easygoing aloofness and superficiality.

Definitely I would include under the term "Lightness" the notions of humor, wit, and elegance for which we have, particularly in Italian design, such well-known examples (e.g., Castiglioni's tractor seat mounted on a flat elastic steel profile).

Originally given as a contribution to the symposium *"Design beyond Design . . ."* in honor of Jan van Toorn, held at the Jan van Eyck Akademie, Maastricht, November 1997.

Published in English in Dawn Oxenaar Barrett (ed.), *Interface—An Approach to Design* (Maastricht: Jan van Eyck Akademie, 1998), 152–9. Translated by Dawn Oxenaar Barrett.

[1] Italo Calvino, *Six Memos for the Next Millennium* (Cambridge, MA: Harvard University Press, 1988), 12.

VIRTUE 2: INTELLECTUALITY

On the occasion of the Aspen Congress 1989, dedicated to Italian Design, Ettore Sottsass surprised the audience by presenting himself—quite naturally I would say—as an intellectual and cultural operator. Only an Italian or a Frenchman can say that. Italy and France are two countries in which the notion of intellectual does not produce a lifting of the eyebrows and a climate of suspicion. In Germany, in the United States, and I assume also in the Netherlands the world "intellectual" carries negative overtones, and certainly many of the practicing design professionals would accept but with reluctance the self-interpretation as intellectuals. Rather they would say that they are practitioners, and they want to distance themselves from the neighborhood of the intellectual; they do not share Gramsci's notion of the organic intellectual, who uses his technical competence within social institutions like private companies or public administration.

Intellectuals are—rightly or wrongly—characterized as wordsmiths because they play a decisive role in shaping the discourse of domains—political, cultural, scientific, and technological. In the field of design, intellectual formation does not have a strong history, because design education grew out of craft training with a deeply engrained mistrust of anything "theoretical." Recently, however, we can observe some promising signs of a shift away from an indifferent, if not openly hostile, attitude toward an interest in articulation and theoretical issues. Designers start to write, particularly graphic designers—for me a promising symptom to overcome a period of collective muteness of the profession.[2] Design and writing about design are no longer seen as sterile and mutually exclusive opposites. On the contrary, a design historian in the year 2050 who looks back at the design scenery at the end of the twentieth century might be surprised about the binarism between action and contemplation. In two generations this opposition might appear as out-of-date as for us the debate about types between Muthesius and van der Velde nine decades ago.

Intellectuals have repeatedly reflected on their role in society.[3] The most salient characteristic seems to me the stamina to reveal contradictions, to rock the boat of self-complacency, to compare what is to that what could be, and in particular to ask for the legitimation of power. This is a business that is not wholeheartedly welcome to the powers that be, whatever they are and wherever they are.

I do not want to heroize the role of the intellectual, and even less to overestimate his possibilities of influence, above all in the field of design. Neither I do want to stylize her or him as a permanently resentful protester driven by the drive of "being against". But I would not like to see this ingredient of a critical stance in the design culture missing or abolished. An antidote to

[2]   Michael Bierut, William Drenttel, Steven Heller and D. K. Holland (eds.), *Looking Closer* (New York: Allworth Press, 1996); Ellen Lupton and J. Abbott Miller, *Design Writing Research* (New York: Princeton Architectural Press, 1996); A1G, *Essays on Design 1* (London: Booth Clibborn Editions, 1997).
[3]   Edward W. Said, *Representations of the Intellectual* (London: Vintage, 1994).

intellectual acquiescence does not only seem to me desirable, but indispensable, if one wants to avoid the danger of falling into the trap of indifference and accommodation.

As a second conclusion, I would like to see *Intellectuality* maintained as a virtue of design in the next century: readiness and courage to put into question the orthodoxies, conventions, traditions, agreed-upon canons of design—and not only of design.

That is not only a verbal enterprise, an enterprise that works through the formulation of texts, an enterprise of linguistic competence of a critical mind. The designer acting as designer, that is, with the tools of his or her profession, faces the particular challenge of an operational critique. In other words, she or he faces the challenge not to remain in critical distance from and above reality, but to get involved and intervene in reality through design actions that open new or different opportunities for action.

VIRTUE 3: PUBLIC DOMAIN

The Netherlands possesses a great tradition in civic virtues that manifests itself in the care for the public domain.[4] A foreigner visiting the Netherlands is struck by the attention given to detail in such simple everyday objects as an address label for post parcels or a timetable for trains. Moreover, he is struck by the apparent naturalness with which caring for the public domain is taken for granted and considered one of the noble tasks and outright obligations of public administration. This care for details and quality of public service is a result of a political commitment that might be traced back to the civic history of the Netherlands. Certainly, it is not the result of a single short-term action, but rather the outcome of a steady practice rooted in the political body of Dutch society. Politics is the domain in which the members of a society decide in what kind of society they want to live. Politics therefore goes far beyond political parties. Care for the public domain, though a profoundly political commitment, is at the same time transpolitical insofar as it exceeds—or better should exceed—the interests of the government in turn.

As the third design virtue in the future, I would like to see maintained the Concern for the Public Domain, and this all the more so when registering the almost delirious onslaught on everything public that seems to be a generalized credo of the dominant economic paradigm. One does well to recall that the socially devastating effects of unrestricted private interests have to be counterbalanced by public interests in any society that claims to be called democratic and that deserves that label. The tendency toward Third-Worldization even of richer economies, with a programmatic binary system of a small group of haves and a majority of excluded have-nots, is a phenomenon that casts shadows on the future and raises some doubts about the reason

---

[4]   Paul Hefting, *Royal PTT Nederland NV* (Royal PTT Nederland NV, 1990).

in the brains of the people that find utter wisdom and desirability in such a dilacerating scheme of social organization.

VIRTUE 4: OTHERNESS

As a fourth virtue I mention *Otherness*, or better Concern for Otherness. This issue is linked to the discussion about Self and Identity, about Presentation and Representation. It plays a strong role in discussions about feminism, gender roles, and race and ethnic diversity. It has virulent political implications because it is rooted in the question of autonomy, that is, the power to participate in the determination of one's own future. This leads us to put into focus the— as Edward Said formulated it—blithe indifference to a good three-quarters of reality.

Today design and design discourse reflect the interests of the dominant economies that, under the banner of globalization, are engaged in the process of modeling the world according to their hegemonic interests and imagery. Globalization as a new economic fundamentalism is the name for the actual planetary project or drift, a process that seems to advance with inexorable ruthlessness, like an objective force passing over the heads of individuals, governments, and societies. Tapping the conceptual repertoire of anthropological discourse, globalization can be interpreted as an attempt to incorporate Otherness and to subject Otherness. That might not be to everybody's taste. It should not come as a surprise that the victims of this process that euphemistically and cynically are labeled with the term "social costs" resist the attempt of incorporation and prefer to enter the arena with better preparation. When fight and competition are the order of the day or the supposed inexorable divine imperative that not to accept would be Quixotesque romanticism, one might agree; but the entrance conditions into the arena should be less distorted. So my fourth virtue of design is Respect for Otherness, leaving behind the racist distinction between developed and underdeveloped countries. This virtue implies the acceptance of other design cultures and their inherent values. It definitely requires a critical stance against ethnocentric, messianic visions of whatever type, European, North American, or Asian. This virtue can counteract the propensity to focus exclusively on the one-quarter of humanity that, according to international statistics, forms part of the industrialized rich economies.

VIRTUE 5: VISUALITY

As an equivalent to Italo Calvino's virtue of Visibility, I take *Visuality* in the field of design. He characterizes visibility as "thinking in terms of images." That is an assessment with radical implications, because in our culture thinking is associated with linguistic competence, with dealing with texts, whereas the visual domain is put into the subaltern role of quacks, trickery, treachery, superficiality, shallowness, appearance, *Schein, bloßer Schein,* something not to be trusted, that is, the opposite of macho-style thinking, at best a second-rate kind of thinking, but definitely an intellectual nullity.

The denigration of vision and visuality has its philosophical origins in Plato's well-known cave simile. We can call this deep linguistic bias against visuality and its cognitive potential the "imperialism of the word." The possibility that the visual domain has cognitive power and is not a simple subordinate or corollary to text has been perceived sometimes, but it never got a strong foothold in our educational system and has been filtered out in academe where mastery of texts is institutionally consolidated. Nobody would doubt that literacy is a prerequisite for higher learning, but graphicacy, as it has been called—the competence in dealing with images—is far from being recognized as a competence of equal importance. That might change in the future, putting an end to visual illiteracy that is disfiguring and disbalancing university education, producing masses of visually and esthetically atrophied graduates.

There are symptoms of change provoked by technological innovations. I refer to the process of digitalization. To an increasing degree, sciences and cognition depend on the power of the visual domain, of images and visualizations, not in the traditional ancillary role of providing illustrations for the higher glory of texts, but in its own right. The still-fledgling imaging science is a new branch that deals with the multifaceted phenomena where images are not taken as examples of mimesis, but in which images reveal realities that are not accessible through words and texts.

The theory of post-structuralists based on the assumption that reality is a "text" that has to be "read," that architecture is a "text," that cities are "texts," that our designed environment is a "text" to be deciphered by the master decoders, will have to be revised. This text fundamentalism has to be relativized by showing that the deeply engrained predominance of the word in the Judeo-Christian tradition (In the Beginning was the Word) is now starting to be technologically undermined and that its claim of the word as the exclusive and predominant domain of cognition is simply that: a claim that today shows signs of corrosion.

The antivisualism, the logocentrism counts on a long and strong tradition that—save a few exceptions—has passed with Olympic indifference over the visual domain. Therefore, a change will not occur from one year to the next; the shift might stretch over a period of generations.

For design, undreamed and radically new possibilities open up. But so far, apart from dispersed initiatives to tap the potential of design for visual cognition, the profession of graphic designers pursues well-trodden tracks. Here then is the challenge for design education to explore this new domain and to loosen the strong association between graphic design and sales promotion—from detergents to political candidates.

We do not yet have a name for this new domain that would correspond to imaging science. Perhaps in the future the notion of "image design" or "visualization design" will become popular, though I would prefer the term "information design," because the binarism between word and picture should be avoided. The emerging field of information design would not only require a considerable collective effort to get outlined and established as a promising field of expertise, it would furthermore contribute to a problem-oriented approach to design issues that differs from the self-centered design approach that gained attractiveness in the eighties.

The fifth virtue then I would like to see maintained and increased in the next millennium I call Visuality. Let me quote a scholar of visuality to reinforce my argument: "The history of the general move toward visualization thus has broad intellectual and practical implications for the conduct and the theory of the humanities, the physical and biological sciences, and the social sciences— indeed, for all forms of education, from top to bottom."[5]

VIRTUE 6" INTEREST IN THEORY

Coming to an end of this panoramic tour of the domain of virtues let me now look at the question of design theory—a question that is related to the general issue of design discourse and design research. As I have argued elsewhere, I do not see any future for the design profession if within the next years we don't overhaul all our design education programs and open an institutional place for design theory. There are two reasons for this declaration: first, every professional practice takes place in front of a theoretical background; this holds even for practice styles that vehemently deny any theoretical involvement. Second, professions that do not produce new knowledge do not have a future in technologically dynamic societies. Therefore, design theory should and—according to my assessment of the future—*must* become part of our educational programs. Design theory still leads a marginal existence. It is considered the pastime of some eccentrics in academic settings protected from the harsh realities of professional practice in the labor market. That is a somewhat biased view that does not reveal a particularly perspicuous vision.

Theory is not a virtue. But concern and cultivation of *Theoretical Interests* is a virtue that I would not only like to see continued into the next millennium, but brought to full blossoming.[6]

---

[5]   Barbara Maria Stafford, *Good Looking – Essays on the Virtues of Images* (Cambridge, MA; London: MIT Press, 1996), 23.
[6]   After one generation since the original version of this text and in front of the addictive expansion of social media I would today be less optimistic about the use of digital technologies, at least as far as social media are concerned.

PART TWO
Design in the "Periphery"

A Designer on the Periphery of Capitalism

Ethel Leon

True tradition is not a testimony to the past, but a living force
that animates and informs the present.

—*Igor Stravinsky*

In the first half of 2020, we witnessed with astonishment the international piracy skirmishes to obtain lung respirators made in China. In this context, Bonsiepe's writings stand out as almost prophetic, when he insistently denounced the deindustrialization of Latin America countries and defended an autonomous form of industrialization as the only possible alternative to reduce the heteronomy of peripheral countries.

As deindustrialization has reached many regions of the world, the debate on reindustrialization makes the texts of this collection remarkable, which is not restricted to the countries of the south. This observation leads me to see two critical issues of "design in peripheral countries":

1) Much of Gui Bonsiepe's practice and theory took place in peripheral countries: Chile, Argentina, Brazil, apart from consultancy work in other countries in the region as well as in India. In the long years he worked in these countries, Bonsiepe thought about issues such as visual rhetoric, the interface, teaching, and, more broadly, design theory—themes that are present in the texts of the different sections of this publication—while he also designed consumer items and capital goods. Some of his reflections were perhaps informed by his South American experience, as he thought that Latin America in the 1970s was at the forefront of design worldwide. A project such as Cybersyn, extensively presented in this volume, was carried out in Chile and not in Germany or the United States.

2) The question of periphery itself becomes global, insofar as there is a periphery in the center, emerging from the negative effects of globalization, and its crises and policies, especially related to that of migratory movements in 2008 toward central countries.

Thanks to Luiz Antonio de Carvalho, for reading the text and offering suggestions.

I am paraphrasing the title of the book *A Master on the Periphery of Capitalism* by Brazilian theorist Roberto Schwarz.

THE MOVE TO LATIN AMERICA

The publication of this volume in English resonates with a growing interest in anti-hegemonic thinking built from the South and the decolonial discussion happening in the social sciences. This interest might explain the recent attention given to the strategic projects of the Chilean Unidad Popular/Popular Unity (UP) government from 1970 to 1973 in which Bonsiepe participated.

In his usual understated manner Gui Bonsiepe responds that his work in Salvador Allende's Chile was a lucky spell, being in the right place at the right time. Discounting the role of chance in his trajectory (chance is always an antidote to a false sense of control over one's life), it is possible to trace some of the "reasons" for this happy synchronicity.

What would be the perspective of Bonsiepe, a former student and professor at Hochschule für Gestaltung Ulm in his country of origin? When the turmoil of the 1968 barricades arrived at the school, the then conservative state government used it as an excuse to end its activities. This was at the peak of postwar recovery, with West Germany becoming the ultimate example of capitalist regeneration in the world. What would be Bonsiepe's perspective if he had remained in Germany[1]—or even Italy[2]?

Bonsiepe saw two possibilities for the practice of design in his country of origin. The first would be totally adapting to the demands of local industries in their advance toward internationalization.[3] The second would be renouncing the design activity itself, a kind of immobility that our author connects to the (seeming) radical criticism that associated design with absolute submission to capital, resulting however in a kind of paralysis among young designers. Many of them opted for nondesign and, eventually got closer to the artistic sphere, as if the latter was purer, and equally not part of a marketplace.[4]

The arrival of Gui Bonsiepe in South America,[5] more specifically in Chile, where he worked for five years (1968–73), proved to be strategic and was carried out at institutions such as the United Nations Industrial Development Organization (UNIDO), which had then established various technological cooperation programs. The Alliance for Progress, a program launched by John F. Kennedy in 1961, as a kind of Marshall Plan for the poor in the Americas, presented $20 billion aid package to Latin America. After the Cuban Revolution

---

[1]   The conservative climate of postwar Germany is well outlined by Paul Betts in *The Authority of Everyday Objects: A Cultural History* (Berkeley; Los Angeles: University of California Press, 2004).

[2]   At that time, Tomás Maldonado invited Bonsiepe to live in Milan.

[3]   It is worth noting that the paradigmatic Braun, often connected to the Ulm school, was partially acquired by Gillette in 1967.

[4]   A key text on the apocalyptic and reductive dimension of design as a manifestation of the capital is Wolfgang Fritz Haug, *Crítica da estética da mercadoria* (São Paulo: Unesp, 1997). Bonsiepe criticizes it in his book *A Tecnologia da Tecnologia*.

[5]   Bonsiepe first visit to Brazil and Argentina was in 1964 as a vacation trip, stimulated by Tomás Maldonado. He went to South America again in 1966 when he taught a course on packaging in Buenos Aires. He moved to Chile in 1968.

and, crucially, after the 1968 anti-American and anti-imperialist demonstrations in South America, the American tactic was to assist the industrialization processes by governments such as Argentina, Mexico, and Brazil.

Bonsiepe arrived in Latin America coming from the international atmosphere of the ulm school, in which the Argentine Tomás Maldonado had become director, and which had an international student body.[6] Anti-parochialism and anti-nationalism had been programmatic foundations of the Hochschule für Gestaltung. In addition, Bonsiepe went on to learn Spanish (and later Brazilian Portuguese), but also quickly got acquainted with the main theories and political movements emerging on the continent at the time. We can also see in his texts an appreciation for South American literature. His ulmian training helped him to join multidisciplinary and interdisciplinary teams in collective projects that required more teamwork capacity than individual creativity.

One of his activities in the Chilean capital was to teach engineers at the Catholic University of Santiago. Designers are often considered by engineers as somehow superfluous, decorative, arriving at projects after the main design decision have been made.

Breaking with this prejudice would mark Bonsiepe's practice. Engineers played an important role in leading teams in several countries in Latin America, with a strong presence in government policies linked to the so-called public works, following the French tradition of the École Politechnique's public works. In South America, engineers had a prominent role in design culture.[7]

In the three South American countries where he lived and worked as an industrial designer and consultant to government agencies, engineers were the ones who had the power to implement industrial design policies: Fernando Flores in Chile, Basilio Uribe in Argentina, and Lynaldo Cavalcanti de Albuquerque in Brazil. It was through them that Bonsiepe landed on cutting-edge institutional projects in these three countries.

WORKING IN CHILE 1970-3

Bonsiepe's ability to interact with engineers proved to be essential for him to join the Cybersyn (or Synco) Project.[8] In the Cybersyn Project, design took on the task of defining the interface/s—which would become a central element of his practice and theoretical work. The development of the interface meant not only designing the Opsroom, and the control room and its chairs with buttons (and not keyboards, as shown by Eden Medina), but also, and foremost,

---

[6] See Silvia Fernández's "The hfg ulm: On the Origins of Design Education in Latin America: From the hfg in Ulm to Globalization," *Design Issues* 22, no. 1 (Winter 2006): 3–19.

[7] The rapprochement between designers and engineers was nurtured in Ulm.

[8] For more on Cybersyn, see Eden Medina's 2011 *Cybernetic Revolutionaries* (MIT Press), and Chapter 4.10 "Socialism by Design" by Eden Medina in this volume.

it involved the gigantic task of processing overnight a massive amount of data coming via telex from the Social Property Areas (APS) comprised of a hundred companies that were nationalized by the Allende government). This processed data would be the basis from which the management team could make their decisions. Here, Bonsiepe has the experience of transforming raw data into processed information that would feed decision-making. This will be a fundamental theme in his thinking: design as a cognitive tool.[9]

Based on his Chilean experience, Gui Bonsiepe made a life commitment to Latin America, working later in Argentina and Brazil, with incursions in Mexico and Cuba. In addition, he played a fundamental role in proposing in 1980 the formation of the Latin American Industrial Design Association (ALADI), which did not submit to the directives of the International Council of Societies of Industrial Design (ICSID), an international but Eurocentric society of industrial designers.

The ten texts chosen for this section of the book reflect Gui Bonsiepe's permanent commitment to this part of the world. The first text is from 1979 and the last from 2018. The texts refer, in one way or another, to some of the issues that are central to his thinking. One of them is the concept that the world consists of *the center and periphery*, and the technological divide between the two into the so-called world-system.[10] Another postulate that runs through his work is *design as interface*, which makes principles such as "form follows function" sound anachronistic. Another frequent point of Bonsiepe's thought is related to teaching, a fundamental issue for the entire world of design, even more critical in the world's peripheries.

Latin America was unquestionably central in Bonsiepe's professional growth and defined his way of being in the world. His Chilean years were marked by disciplinary convergence bringing together cybernetics and emancipatory social transformation, industrial design practice, and the struggle to break from dependency relations between countries.

DEVELOPMENT AND DEPENDENCY: WORKING WITH THE STATE

Although a reader of Adorno, Horkheimer, and Benjamin—texts from all of which were present in the library of the Hochschule Ulm (and absent from other German educational institutions of the period as per Bonsiepe's account)— he identifies himself with Jürgen Habermas and the second generation of the Frankfurt School. Just like Habermas, Bonsiepe is a defender of modernity and his unfinished emancipatory project. A self-proclaimed constructive pessimist, he moves into the field of critical rationality in the sense that he recognizes the

---

[9]   See Gui Bonsiepe, "Design as Cognitive Tool: The Role of Design in the Socialisation of Knowledge," in Silvia Pizzocaro, Amilton Arruda, and Dijon de Moraes (ed.), *Design Plus Research Proceedings of the Politecnico di Milano conference*, May 18–20, 2000, p. 23. See also in this volume Chapter 3.3 "Design as Tool for Cognitive Metabolism: From Knowledge Production to Knowledge Presentation."

[10] I am here referring to Immanuel Wallerstein's concept, adopted by Bonsiepe.

need to impose limits on instrumental rationality, without denying rationality per se,[11] as well as the power of the law and the government working toward the common good.

A second matrix of Bonsiepe's thinking comes from the proponents of the Dependency Theory. These are Latin American social scientists who in the 1960s[12] studied the situation of the continent based on the notion of dependency. Bonsiepe embraced the negative foundation of the Dependency Theory: the refusal that the misery and inequality of the so-called underdeveloped countries are because they haven't yet taken the right steps toward their full development. This idea of economic/technological process as progressive, mimetic, and stage oriented was denied by Bonsiepe, who advocated for a break with patterns of economic and technological dependence. He understood that that the peripheral (rather than underdeveloped) countries were in fact the product of an international division of labor. His alignment with this school of thought is clear in much of his work. In the text "Identity and counter-identity of design," he writes, "The central thesis said: Latin America is not underdeveloped by the lack of social capitalist structures, but, on the contrary, by the predominance of these structures" (Chapter 20, p. 195).[13]

One of the consequences of the application of the so-called Dependency Theory in Chile was the strengthening of the role of the state. In order to break with the political, economic, and cultural links of dependence, it was crucial for the state to become itself a direct driver of change. In the absence of companies producing artifacts for the population (tableware, toys, furniture, stereo systems, among others) as well as adequate machinery for local agricultural produce that are different from the ones that grow in the cold countries in the North, it would be up to the state to provide for the design of these artifacts. An industrial design program promoted by the government would also promote the local industry and become a major task force for the strategic management of the economy. In the text "Between Marasmus and Hope," Bonsiepe notes: "Given the weak local industry, the support provided by public and semipublic institutions is of crucial importance for the implementation of a design policy" (Chapter 18, p. 181).

It is possible to bring Bonsiepe's view of the state closer to what the economist Mariana Mazzucato advocates, suggesting that the state should be involved in the creation of value, in mission-driven innovation, leading the economic scene and creating wealth.[14] In the Chilean socialist government,

---

[11] But for Bonsiepe, the completion of the modernity process is tied to the reduction of heteronomy or the search for autonomy by peripheral countries.

[12] Many of whom were involved in the Chilean context through the CEPAL, the United Nations Economic Commission for Latin America and the Caribbean.

[13] As with the other introductory chapters in this volume that quote or make references to chapters in the book, the pagination will be by the chapter number and page reference.

[14] Economist Mariana Mazzucato comments on the great innovations shaping the world today: "The state may act as a principal investor and catalyzer, triggering a network of knowledge creation and dissemination. The state can be a creator rather than a mere facilitator of the knowledge economy. Mariana Mazzucato, *O estado empreendedor. Desmascarando o mito do setor público vs setor privado* (São Paulo: Penguin, 2014).

this direction was clear: the control of about 100 strategic companies was part of the emancipatory project.

STRATEGIES

Although he worked for state agencies in Argentina and Brazil, the deepest involvement Bonsiepe had with governments was in Chile, whose political and economic planning was far from the strongly criticized Soviet model. The Chilean path to socialism, sought by the democratically elected government of Salvador Allende, not only kept the Constitution intact, but also pledged for freedom of the press and freedom in terms of labor organizing. Their version of democratic socialism foresaw a multiparty model in political-institutional life and control over the economy that still granted autonomy to factories and state properties managed by their workers and monitored by the central government.

The notion of autonomy, so dear to Bonsiepe, was in this model a structural part of the ongoing political project, whose thinking was influenced by the Marxist Rosa Luxemburg, a defender of democratic freedoms, essential for the makeup of an egalitarian society. There is a clear parallel between Rosa Luxemburg's statement, "The emancipation of the working class must be the work of the working class itself,"[15] and Bonsiepe's maxim, "The problems of the periphery must be solved by the periphery itself," present in so many of his texts.

In defending the spontaneity of social movements that emerged in revolutionary processes, Rosa Luxemburg understood that several organizational forms were liable to coexist and that the revolutionary process could contain them.[16] This flexibility and even spontaneity in action is opposed to the Leninist model of party and government and resonates with the cyber management approach that Stafford Beer helped to implement.

Finally, Bonsiepe's training at Ulm proved to be adequate for this open, dynamic, and flexible process. It is possible to acknowledge today the adaptability of the ulmian teaching model that was brought into so many parts of the world, especially in its peripheral corners, not as replicas, but as adaptations to local contexts.[17] Bonsiepe himself understood that the designer's tasks in the periphery were not the same as those of their peers in the central countries where the industrial infrastructure had been in place for a long time.

---

[15]   Rosa Luxemburg, "O que quer a Liga Espartacus?" Available at: https://www.marxists.org/portu gues/luxemburgo/1918/12/14.html, accessed on August 20, 2020.
[16]   A text by Hernán Ouviña about the Italian theorist Lelio Basso, a guest of Chilean UP socialist government, covers the peaceful path toward socialism, a sign that Rosa Luxemburg's thoughts were present within the intellectuals and activists at the base of the socialist government of the UP. Hernán Ouviña, "Lelio Basso y el debate sobre el Estado y la transición al socialismo en Chile. Apuntes en torno a su pensamiento político en el contexto de la Unidad Popular," *Revista Ciclos Número* 55, Buenos Aires, 2020.
[17]   See Fernández, "The hfg ulm: On the Origins of Design Education in Latin America."

On the periphery, he says, the designer must make production feasible, acting as a link between different structures.[18]

In the text "The Ulm model in the Periphery," Bonsiepe writes that the Ulm is born international, aimed at industrialized countries but "That does not mean that the HfG claimed universal significance," but "also reached those countries that saw industrialization as a tool to reduce their technological dependence" (Chapter 13, p. 141).

SCHOOLS

In several texts, Bonsiepe considers that one of the key structural characteristics that allowed the pedagogical innovation and the solid and multidirectional education of the Hochschule für Gestaltung was the fact that the school was not under existing university structures. For him, any innovative pedagogical project tends to be short-lived; Bauhaus and Ulm are clear examples of that. Both schools have reinvented themselves a few times, although each only lasted for about fourteen years. At Hochschule für Gestaltung, the project was the axis around which foundational and ancillary disciplines were organized, without any clear hierarchy between them. The study of philosophy and social sciences was made *pari passu* with the study of disciplines such as cybernetics and physics, in an effort to give its students the skills to deal with the complexity of the multiple military to civilian conversions happening in the industrial and technological worlds in the 1950s.

His view of the study program was not that of the sum of disciplines. In a recent response to a student, he observed: "It has long been known that the criterion adopted to organize higher education institutions in disciplines is questionable and that it would be more convenient to organize them according to problem fields, that is, to overcome the traditional criterion existing since the creation of universities in the Middle Ages."

His criticism of education encompasses also engineering courses in peripheral countries, which formed and trained more managers of imported technology rather than innovators. If we consider, for example, the importance that engineers had in Brazil in the implementation of railroads in the nineteenth century with English technology, aimed at enabling the transportation of coffee beans to the exporting ports, we understand the engineers' role in maintaining the technological hegemony of central countries.

This kind of industrialization would be, for Bonsiepe, just a simulation of industrialization, a reflexive and not autonomous process, whose protagonists

---

18   Gui Bonsiepe, "Technological Asymmetry: A Dilemma of the Periphery," *Revista Brasileira de Tecnologia*, Brasília 12, 92 (April / June 1981): 54. The Brazilian design firm Questto Nó developed a lung respirator, articulating several companies and using the technology of credit card machines to replace electronic components produced by few countries. This is a concrete example, in the current pandemic situation, in which designers make production feasible, by articulating several companies and acting as managers, not relying on the latest technology, established as the dominant matrix.

understand technology as "things" and not as social processes. Breaking with these patterns of dependence requires political positioning and excludes any form of paternalism from the well-intentioned philanthropic attitudes of the central countries.

Bonsiepe has been defending the study of the practice of design from the anthropology of everyday life, but his approach, unlike that of anthropologists who examine the life of objects from their realization, includes design decisions.

The education model he advocates for is project based. It has some parallels with what has developed in medical education as "problem-based learning." The parallel between design and medicine is interesting: both are activities based on scientific knowledge, but are not science; both are based of clinical practices—of diagnosis and application of knowledge—and both are activities in which theory and practice form a feedback loop, the basis for critical operationality.

FROM CHILE TO BRAZIL

With the military coup in Chile on September 11, 1973, Bonsiepe headed for Argentina, where he was invited to transform the Centro de Investigaciónes de Diseño Industrial (CIDI), in which he remained for less than two years, until the Argentinian military coup in 1976. He stayed in Buenos Aires however, working as a designer at the MM/B studio, with architect Carlos Méndez-Mosquera and engineer Felipe Kumcher. This studio was in charge of designing the equipment and signage for the stadiums for the 1978 Soccer World Cup, held in Argentina.

In 1981, the engineer Lynaldo Cavalcanti de Albuquerque, president of the National Council for Research and Technological Development (CNPq), invited Bonsiepe to set up the Laboratory for Product Development in the city of Florianópolis, Santa Catarina state, in Southern Brazil.

At that time, the Brazilian military regime, which had ruled the country since 1964, was in a deep crisis, and would end in 1984. Bonsiepe settles in Florianópolis in January 1983, and at the beginning of the year he would coordinate a first course taught at the School of Mechanical Engineering at the Federal University of that state, where most students had an engineering degree. Three other courses were taught by teams formed by the Laboratory, only two of them aimed specifically at designers. The fourth course was aimed at industrial technicians. Bonsiepe's anti-corporate thinking can be seen in these initiatives: he did not choose to build a disciplinary/academic field based on the Laboratory, and yet had a profound repercussion in Brazilian university circles.[19]

[19]   After Bonsiepe ends his directorship in 1986, the Laboratory focused on the task of training teachers and strengthening professional activity. See Ethel Leon, *Canasvieiras, um laboratório para o design brasileiro* (Florianópolis: UDESC / FAPESP, 2014).

HISTORY AS RECOGNITION OF INNOVATION

An aspect to be emphasized in Bonsiepe's thinking, especially in recent years, concerns how he understood research and teaching in the history of design. Echoing the break with the past as a fundamental element of its constitution, the German paradigmatic schools, Bauhaus and ulm did not have history courses in their programs. In ulm, history was integrated into the humanities-based disciplines such as philosophy, psychology, and political science.

Bonsiepe started to look at the history of design as a critical exercise. The biographical and hagiographic narratives that dominated, for a long time, the research and teaching of design history, came into his sights. They reinforced the mythology surrounding the designer as a creator, a notion opposed by Bonsiepe, who began to advocate for the object of the history of design as a research of innovation, as a way of building a heritage capable of pointing to the future. Bonsiepe defends the history of design as totally connected to social-political history, industrial technology, and everyday life.

At the same time, he realized that the history of design in Latin America suffered from a discourse deficit, which made him take the decision to embrace Silvia Fernández's ambitious project to organize a history of design on the continent, calling on researchers from different countries to undertake empirical research. The book was published in Spanish in 2008, and is still the only design history manual on the continent, covering Argentina, Brazil, Colombia, Cuba, Chile, Ecuador, Mexico, Uruguay, and Venezuela.

This work is described in the text "History of Design in Latin America," of this collection, coauthored with the designer and researcher Silvia Fernández, and has its outline in a previous text, "Industrialization without Design," also in this volume, in which Bonsiepe tried to tackle the evolution and problems of design in the continent based on his experience as a consultant in several countries (see Chapters 14 and 15). In it, Bonsiepe also mentions his work in India. His attention to history forced him to understand the correlations between countries at different times, as a reflective act of recognition.[20]

CRAFTS, ANCESTRAL CULTURES, AND DESIGN

The characterization of artisanal production in Latin American countries is a delicate point. Publications on the theme often equate precraft work, rudimentary practices recently acquired by diverse social groups, artisans/artists with signature work, commercialization to luxury markets with the handicraft production of the pre-Columbian peoples.

Gui Bonsiepe is a staunch critic of many of the programs that promote connections between designers and communities of artisans. Many such official programs and philanthropic initiatives rarely result in elevating the

---

[20] 'for the architect—and for the designer—to know is to always to recognize," writes Julio Katinsky. In: Ethel Leon, *Design brasileiro, quem fez quem faz*, ed. Mosly (SENAC, 2005), 12.

economic and political-cultural conditions of the populations involved into autonomy. Many programs are focused on the manufacture of artifacts that maintain existing characteristics—labor intensive, materials, and techniques. Changes are generally introduced either in the sphere of taste—to cater to the tourist and decoration markets for the elite and middle classes—or to increase the productivity of artisans, through procedures aimed at simplifying/speeding up the production process.

Recognizing the richness and depth of artisanal practices such as those practiced in some Latin American countries, such as Mexico, Peru, Ecuador, or Colombia, Bonsiepe refuses their appropriation as meaningless style, as a marketing tool, facilitated by designers to help sell products with a local flavor in the world market.

In this sense he strongly criticizes those who apply indigenous graphics into industrial products, as if the ornaments of pre-Columbian popular cultures had the power to overcome technological dependence. He asks: "How could a pocket electronic calculator or a water tap be designed with formal elements of the Maya culture? Nostalgia is not an effective way to prepare for the future" (Chapter 16, p. 163). This problematic attitude of designers and industries is connected, for example, with the misappropriation of indigenous decorative patterns, without any form of remuneration to their creators, which is the case for several fashion companies.[21]

Bonsiepe's criticism of the misappropriation of elements of the visual culture of pre-Columbian peoples does not mean contempt for their history and for the concrete contributions that these traditions offer. Bonsiepe points out the importance of knowing the history of Latin America and its ancestral peoples: "greater familiarity with Latin American and pre-Hispanic civilization will enable us to understand the present better and from there look into the future" (Chapter 16, p. 163).

The issue of (cultural) identity is obsessively discussed in several spheres and also among designers in different Latin American countries, especially in the early years of this century. For Bonsiepe, identities are not essential, ontological, but always relational, historical. Bonsiepe rejects the fundamentalism of a certain mythical ancestry and bets on universality (but not on Eurocentric universalism). To refuse inequality, and not accept it as "difference" or weak multiculturalism, often made into a marketing strategy, is to refuse domination.

---

[21] In June 2020, the Carolina Herrera brand was accused by the Mexican government of cultural appropriation of visual elements of peoples in the region. Other fashion brands have received similar accusations. Many traditional knowledges of pre-Columbian peoples were expropriated in actions of biopiracy, one of the first perhaps being the cochineal domesticated by the Aztecs and producer of top-quality dyeing material, stolen by Spanish colonizers. See, in this regard, Amy Butler Greenfield, *A Perfect Red: Empire, Espionage, and the Quest for the Color of Desire* (New York: Harper Collins, 2005).

THE ENVIRONMENTAL ISSUE AND UNIVERSALITY

The text "The environment in the North-South conflict" written shortly before the United Nations Conference on the Environment and Development held in 1992 in Rio de Janeiro, demonstrates the scope of his vision that aims at changing the system of economic relations starting in the periphery, but including the industry of the central countries.

Bonsiepe reports the distrust of sectors in the peripheral countries when they were charged with environmental actions, as the highest rates of pollution and environmental problems come from the countries of the North, but also because without radically changing the economic geopolitics of the world, it seems impossible to change the unsustainable living standards with the necessary radicalism.

Bonsiepe takes a stand against the patterns of development and industrialization reached by the North, that is, by the central countries. And while he argues that design should take its share of responsibility for changes, he does not advocate for peripheral countries to aim at a similar pattern of development of the central countries; rather, he pushes for changes on a global scale. He writes: "The irrefutable evidence of environmental destruction calls into question the paradigm of industrialization as an instrument to create worldwide prosperity and happiness" (Chapter 19, p. 187).

Bonsiepe pushes here for a universal voice, criticizing the contradictions of a supposed planetary environmental agenda, which however imposes restrictions on peripheral countries, while not fully committing with solutions that might have real impact for the future.

Notwithstanding, he is also critical toward those who, on behalf of North–South inequalities, pitch the fight against poverty against environmental issues, a common narrative of conservative sectors in peripheral countries, falsely pushing for "progress" as an excuse to destroy water resources, forest, and minerals.

The parochial view and some ad hominem criticisms resort to a chauvinistic argument which "accuses" Bonsiepe of being a "Foreigner" in Latin American territories. This claim is unsustainable and quite curious especially when Tomás Maldonado, once he left Ulm, settled in Italy, where he lived until the end of his days. While his German disciple chose to remain in the southern part of the world.

As it happens, Bonsiepe—teased by Maldonado, who said his option for the South was driven by anthropological curiosity—had embodied the premise of his mentor found in a text from 1966: "At first sight, the new world which is taking shape is associated with what was conventionally called the third world; but as a matter of fact it is not exclusively with the third world but with all the worlds where food, housing, and cultural misery exist and must be overcome."[22]

---

[22] Tomás Maldonado, "La formación del diseñador en un mundo en cambio,. *Summa*, nos. 6/7 (December 1966).

(a) From Europe to South America

In the course of working in different cultural and political contexts, especially in Latin America, a corpus of experiences was formed that made it possible to characterize the particularities of design activity in the periphery. This interview, made in 2000, explores these experiences, both biographical and professional. It touches on a number of themes discussed in more detail in the papers below and therefore provides a useful introduction both to this part of the volume and to Bonsiepe's career and trajectory as a whole.

11 PERIPHERAL VISION AND DESIGN EMPOWERMENT

Interview with James Fathers

(2002)

This article documents an interview with Gui Bonsiepe conducted by James Fathers and given as part of the program of the *Mind the Map* conference hosted by Istanbul Technical University and Kent Institute of Art and Design in Istanbul, Turkey, in July 2002. The interview was an attempt to shed some light on Bonsiepe's career as a figure who has been at the heart of the discourse on design in a developmental context, but who, twenty years ago, while extensively published in Germany, Italy, and Spain and across Latin America remained little known in English-speaking countries. Trained at the Hochschule für Gestaltung in Ulm (HfG Ulm) in the second half of the 1950s, Bonsiepe went on to teach and design there, from 1960 to 1968, alongside his friend and mentor, Tomás Maldonado. When the institution closed in 1968, he decided to move to Chile. Latin America has remained Bonsiepe's base ever since. The interview explores some of the thoughts, methods, and motives behind a career spanning the last forty years devoted to addressing the challenges of design in the "periphery," the term that Bonsiepe developed in the early 1970s as a then more appropriate way of describing those geographic areas and peoples that were usually referred to at that time as "developing countries."[1]

THE INTERVIEW

Q1: You are well known for your writings and experiences designing in developing countries, especially in the 1970s and 1980s. Can you describe why you first became interested in the role of design in development?

I studied at the HfG Ulm in the 1950s, when we had a considerable number of foreign students, particularly from Latin America. So this was my first contact because, similar to other Europeans, at least at that time, I didn't know anything about Latin American history or culture. Then, in 1964, I was invited

Originally published in English in *Design Issues* 19, no. 4 (Autumn 2003): 44–56.

[1]   See Gui Bonsiepe, "Precariousness & Ambiguity: Industrial Design in Dependent Countries 2," in *Design for Need*, eds. J Bicknell and L. McOuiston (London: Pergamon Press, 1976).

to Argentina by my teacher, friend, and intellectual mentor, Tomás Maldonado, whom I considered one of the most important design theoreticians of the twentieth century, a real giant, though his works weren't widely known outside the Spanish and Italian language context.

I arrived in Buenos Aires, planned to stay for two weeks, and stayed for two months. I was fascinated by the cosmopolitan climate of the city—a city in which you could go to the cinema at any time of the day or night if you wanted! I hadn't found this to be the case in Germany, least of all in Ulm, a very small, provincial town. This initial contact [with the periphery] was purely personal, without any professional intentions.

In 1966, I again traveled to Argentina in order to teach a course in packaging design and packaging technology. The course was organized by the International Labor Organization (ILO), which had contracted me as a consultant. At that time, the United Nations International Development Organization (UNIDO) did not yet exist.[2] So, step-by-step, my encounters with the periphery started to get more intensive.

In 1968, I decided to move to Latin America. My move to Chile coincided with the closure of HfG Ulm. However, it was not motivated by this abortion of one of the most influential experiments in design education in the second half of the last century. I had the chance to go to Milan which, at that time, was already a very attractive place to work in design. But I accepted an alternative offer, again by the ILO, to go to Chile; to work there as a designer on a project for the development of small- and medium-sized industries. In Chile, I entered the "real world."

A decisive influence on this decision had been my Argentinean wife. When we discussed these options, either to go to Milan or Chile, she told me to opt for adventure. At that time, I didn't know Chile. I didn't even speak Spanish. She said simply, "Look, in Europe, everything already has been done in design. Let's go outside, where there are new challenges."

02: In 1973, UNIDO commissioned you to write the report "Development Through Design." How did this come about?

At the beginning of the 1970s, ICSID, our international professional organization, became more and more interested in what was happening in developing countries—we didn't yet have the name "peripheral countries." Josine des Cressionières, the Belgian secretary general of ICSID at that time, approached me to write this report. As far as I remember, there was a draft paper already written by an American colleague, Nathan Shapira; but this paper had certain shortcomings, mainly because he didn't have substantial, firsthand experience in a developing country.

The deadline was six weeks—a very short deadline when you consider that the Internet did not exist at that time. I collected whatever materials I could get hold of, from India, Cuba, Chile, Brazil, and Argentina, and presented this as a working paper at a meeting of experts in Vienna, where, for the first time,

---

²   It was created in November 1966 (NA).

an international organization explicitly dealt with industrial design policy for those countries which were called at that time "developing countries." This draft then was transformed into a guideline paper for the industrial design policy of UNIDO (see Chapter 4.1, p. 325).

Q3: What are your most significant memories of your experiences designing in Chile and Argentina?

This is a very personal question, and I am not particularly keen on getting involved in my own history. But since you asked the question, the most negative memory I still have of my stay in Latin America was of September 11, 1973, when the military coup d'état was implemented with help from outside secret services and covert military support against the democratically elected government of Salvador Allende. As you know, this coup d'état, with its tortures, killings, and "disappearences," was officially legitimized by declaring that the "occidental and Christian values of our culture had to be defended." So much for the values of our culture: this was the negative side.

I then moved to Argentina, for obvious reasons. Fortunately, I had a German passport; otherwise, if I had had a Brazilian or an Argentinean passport, I probably wouldn't be sitting here talking with you. It took me several months to get over the traumatic Chilean experience, and, in nine months, I wrote the book *Theory and Practice of Industrial Design*. Written in German, it was published in Italy in 1975, and later translated into Spanish and Portuguese.

On the positive side, I had the good luck of meeting and getting acquainted with, and later getting to know, a group of very passionate design students who had just finished, or were finishing, their university courses. These courses did not fulfill their promise: to educate industrial designers. Their titles were something like "craftsman in decoration," which was somewhat distant from "industrial design," and still dominated by an interpretation of design as a kind of art or, worse, applied art! Furthermore, I found positive resonance within higher government official circles for the design approach that I practiced. This was, for me, a very fertile environment.

The political experience I had gained in Europe was limited. I was interested, of course, in political issues, which was inevitable in the fervent climate of the 1960s. During my education in Ulm, reading books on critical theory such as Ernst Bloch, Theodor W. Adorno, Walter Benjamin, Herbert Marcuse, and Jürgen Habermas, as part of our seminars, was a must. So, I had some consciousness of what was going on, and what makes economies tick, but I did not have any experience of a direct relationship between professional work and the sociopolitical environment or a sociopolitical program. In Chile, it was possible to map professional practice to a sociopolitical program.

Q4: You are quoted in an article in 1976 by S. Newby[3] as being a "parachutist from Ulm." This phrase often has been used in a negative way to describe Western intervention in a developing country. What

---

[3] Sonia Newby, "Ulm in a Peripheral Landscape," *Design* 332 (1976): 40–1.

steps did you take to limit any negative influence caused by your "landing" in Chile after your experiences at Ulm?

I used the term "parachutist" in the interview with Sonia Newby in 1976 (during the *Design for Need Symposium*), without negative connotations, but with a slight touch of irony to refer to my ignorance about the country in which I would start working: Entering from a blue sky.

Now as to the negative influences, I am not quite sure what these might be. The pragmatic rational "Ulmian" approach that made it possible to draw a profile of the industrial designer, and to consolidate his education, apparently met a latent need. Otherwise, the resonance would not have been as strong as it has been. There seems to exist a hidden romantic notion of the periphery: that it should maintain its status of pristine purity that would be contaminated by any outside contact. It might be advisable to distinguish between different types of influence. I don't see anything negative in the endeavor to contribute to a project of social emancipation. I did not come as a missionary to Latin America. What I did was to provide an operational base for concrete professional design action. I offered some operational tools in order to do product design, from agricultural machinery to wooden toys for children and low-cost furniture, and get rid of the ballast of art tradition and art theory.

This operational know-how was not provided by the universities at that time because the teachers of those courses often did not have firsthand design experience. I wonder how you can teach design if you don't practice design. For this reason, there was a vacuum and a very fertile breeding ground, and thus receptivity for any relevant information and methodological tools which would help to resolve practical design problems.

05: In the Design for Need conference at the Royal College of Art in 1976, you made the statement:

"My summary, 'Design for Dependent Countries,' based on eight years of continuous work in peripheral countries, should read 'Design in Dependent Countries' or 'Design by Dependent Countries.' The center does not possess the universal magic formulae of industrial design which have to be propagated to the inhabitants of the periphery whom the intelligence agencies ideologically conceive as . . . [the] . . . 'underdeveloped'". . . . Do you still hold this view?

To a certain degree, yes. I would not move one millimeter from the position or the statement that, according to my opinion, design should be done in the periphery and not for the periphery as the result of some kind of benevolent paternalistic attitude of the center to these countries. I insist and always have insisted on local design practice. Design problems should be resolved in the local context, and not by outsiders coming in for a stopover visit. This typifies one of the great disadvantages of short-term consultancy jobs, with people flying in from the central countries with very little knowledge about the local context and believing that issues can be resolved by remote control.

06: At this same time, Victor Papanek was writing about similar issues. Design historians have put the two of you together as the key figures in what has become known as the "Design for Need Movement." Did you discuss your theories together or collaborate on any projects? How do you feel your ideas differ from Papanek?

In 1964, when I spent one semester as guest professor in basic design at Carnegie Mellon University, Victor Papanek invited me to go to North Carolina, where he was teaching industrial design, to show me the design approach he had developed. I had high esteem for Victor Papanek because he dared to swim against the stream, and against the complacency in design practice and design education. For this courage, he was heavily punished. For a number of years, he almost was prohibited from speaking publicly at industrial design conferences in the United States. However, my esteem for Papanek did not prevent me from writing a polemical review of his book *Design for the Real World*.[4]

He had attacked a sensitive issue, but his approach and the answers he was ready to give seemed to me not adequate. I would say that he had little understanding of the political economy of design. As is known, he became fascinated by the "do-it-yourself" design approach and did not have a strong interest in industrialization and the development of economies. He opted for design services outside of the business and industrial enterprise context, which I considered of limited effectiveness—like that of a maverick. For this reason, I did not share his views. But this does not mean that I have underestimated his contribution to the field. The receptivity of his book, which was translated into many languages, shows that he had touched real issues. But in answer to your question, we never developed projects together. We occasionally met at conferences. I also wrote a review of his book *The Green Imperative*.[5] I think this was his last book. After that, we lost contact.

07: The "Design for Need" movement seemed to draw on a collective desire in the design profession to do something about social need. In hindsight, can you offer any suggestions as to why this movement appeared to founder?

I wouldn't say that it foundered, because it didn't take off in the first place. It was an attempt to find some answers as a profession to the needs of the majority of the world population, which we felt were left out. This movement, sometimes also called the "Alternative Technology Movement," changed into the "Appropriate Technology Movement," and was promoted particularly in Great Britain, where they had an office with consultants offering services in appropriate technology, especially to African and some Asian countries.

---

4   Gui Bonsiepe, "Bombast aus Pappe," *form* 61 (1973): 13–16. Also published under the title "Piruetas del neo-colonialismo" in the Argentinean journal *summa* 67 (1973): 69–71. The review is included in this volume; see Chapter 4.2. Papanek's reply is also published.
5   Gui Bonsiepe, "Im Grünen," *formdiskurs* 1 (December 1995): 69–71. The review, along with Papanek's response is included is this volume as chapter 4.2.

Later in the decade, this activity lost momentum and went into oblivion. I suppose the reason was that the "Appropriate Technology" and "Design for Need" movements could never quite get away from the prejudice (and it is a prejudice really) that it deals only with second-rate and third-rate technology. It seemed to continue with a class distinction between two types of technology: high-tech for the central countries and low-tech, do-it-yourself technology for the periphery. The appropriate technology movement in the 1980s was influenced by the writings of E. F. Schumacher, who wrote *Small Is Beautiful.*[6] Increasingly, the main protagonists of this movement were coming from the fields of engineering and economics. Designers played a marginal role in these efforts to do something about design in what was, at that time, called developing countries.

08: In a paper in 1993, Pauline Madge quotes correspondence with you in which you reflect on the design movement in the 1970s[7]:

"I consider it a merit of the representatives of the appropriate technology movements to have asked some uneasy questions about industrialization and its effect on the third world; furthermore, of having shifted attention to the rural (poor) population in the Seventies, there still was the hope that a different social organization would give rise to different products and a different mode of consumption. This hope is today shattered."

The statement that hope is today shattered is a very strong one. Can you explain the thoughts that led to this conclusion?

You see in the 1960s and 1970s, and even up into the 1980s, there still was a vague hope called the "third way" between the Eastern block or socialist countries, and the Western block or capitalist countries. With the demise of the former socialist countries of the Eastern Block, at this moment there seems to be no alternative outside the general configuration of capitalism. The only alternative nowadays can be found within the system of globalization, which perhaps we will talk about later.

So, taking up the notion of "shattered hope," I am, by temperament and by decision, not a depressive character. Rather, I would characterize myself as a constructive pessimist and, therefore, I don't agree at all with the well-known "TINA" (There Is No Alternative) dictum by Margaret Thatcher. I would claim there always are alternatives.

09: In recent years, you have not written very much about the issues relating to role of design in a developmental context. What triggered this apparent shift in focus?

---

[6]   Ernst F. Schumacher, *Es geht auch anders – Jenseits des Wachstums* | Technik und Wirtschaft nach Menschenmaß (München: Verlag Kurt Desch, 1974).

[7]   Pauline Madge, "Design, Ecology, Technology: A Historiographical Review," *Journal of Design History* 6, no. 3 (January 1993): 149–66.

I worked in Brazil from 1981 to 1987 as a consultant to the National Council for Scientific and Technological Development, participating as designer in the formulation of an industrial development policy. While there, however, I had only limited access to computer technology. The technological revolution information/computer technology attracted my interest. I perceived that a radical change was approaching, an enormous challenge for designers. One day, I got a letter with an offer to work as a designer for a software firm in Berkeley. I took this job and started to work there in this new field of technology, which I felt was of utter importance similar to the invention of movable type for the printing press in the fifteenth century.

If I had had access to computers and software development from a user's perspective in Brazil, I probably would have remained there. But I didn't, and so I moved to the United States and worked there for three years. The practical work as a designer in a software office permitted me to reinterpret design, getting rid of the traditional topic of form and function, and developing an interpretation of what design is all about, based on language and action theory.

At about the same time, I rediscovered the work of Heidegger. As a German, it was very difficult for me to read Heidegger after the devastating critique by Theodor Adorno, *The Jargon of Authenticity*. However, while in Berkeley, I was fortunate to be able to attend several classes on philosophy, among others, by Hubert Dreyfus. I got a better understanding of Heidegger through the English translation and interpretation. Taking some ideas from Heidegger and computer science, I developed a reinterpretation of design as the domain of the interface where the interaction between users and tools is structured. I consider this not a minor contribution to design theory.

Having said all this, let me just add one thing. My interest in peripheral countries has not diminished. On the contrary, it has increased due to their economic decline and to what I consider to be the symptoms of the end of a one-dimensional socioeconomic model. In my last book available in English,[8] I assess the role of design in the center from the perspective of the periphery and vice versa. In addition to this I have established, created, and coordinated the master's program in information design at the University of the Americas in Cholula, Puebla, Mexico, and continue to work on this program. I live part-time in Brazil, where my main base of operation is located, returning to Latin America whenever my teaching obligations in Germany permit me to do so.[9]

010: It is well known that in the 1970s and 1980s you were a significant influence in the "Design for Need Movement." Despite this prominence, it has been said that you have received little or no recognition as a

8   Gui Bonsiepe, *INTERFACE: An Approach to Design* (Maastricht: Jan van Eyck Akademie, 1999).
9   Since the interview, Gui Bonsiepe has moved to Brazil permanently, where he is teaching at the ESDI in Rio de Janeiro.

designer, and, in fact in the 1980s, you were quoted as saying this is due in part to a political agenda.

Both Er and Langrish and Madge state that, despite Bonsiepe's involvement in the area since 1968, he still is relatively unknown in design circles, and has remained marginal in the design literature. The reasons given are "because the subject itself did not attract any interest within a design world dominated by theoretical underdevelopment and self-centered design discourse"[10] and "because the issue of design in developing countries increasingly has been seen as a political rather than design issue and associated with the political left."[11] Could you expand a little on this?

Recognition is a relative issue. It is not one of my major concerns. I am not particularly inclined to self-branding and self-promotion.

There are universes of language, and if we limit ourselves to the universe of discourse of the English language, by definition, we are cut off from the English-speaking readership when publishing in another language such as Spanish, Italian, German, or Brazilian Portuguese. In Latin America, where I am teaching, living, and writing for a great part of my time, I cannot complain about an absence of recognition.

Q11: What would you say your own contribution has been to the field? What lessons have been learned, and what would you do differently?

These are various questions, so I will take them step-by-step.

I consider my function in Latin America mainly as a catalyst, simply being there at the right place at the right time with the right kind of people, just by chance, mixed with an ingredient of personal decision because of my general interest in the Latin American culture—the great variety of different cultures which I find very stimulating. I feel at home or at ease when I am in Latin America, be it Brazil, Chile, Mexico, Cuba, or Argentina. I don't feel like a foreigner there. On the contrary, I find a receptive climate for what I am teaching and writing and doing as a professional. The hospitality and solidarity of Latin Americans is proverbial.

Now, assessing what I have done so far. I would say that I helped, in a critical moment of industrialization, to define the profile of the industrial designer in Latin America, perhaps even with extrapolation to India and other peripheral countries at that time. Apart from this professional role, I educated or put some students on a track where, on the one hand, they acquired the capacity for critical discourse and, at the same time, became efficient professionals. During our meetings at this conference, the conflictive issue between

---

[10] Alpay Er and John Langrish, "Industrial Design in Developing Countries: A Review of the Design Literature" (Institute of Advanced Studies, Manchester Metropolitan University, 1992).
[11] Pauline Madge, "Design, Ecology, Technology: A Historiographical Review," *Journal of Design History* 6, no. 3 (1993): 149–66.

practitioners versus theoreticians frequently arose. I find this a very damaging tradition. I do not accept this bipolarity that labels you either as theoretician or a practitioner. This either/or proposition has its roots in the origin of our profession, namely vocational training. However, in university courses you are obliged to think about what you're doing, and to reflect on your activity and not just on your own activity but what is going on around you. This is typical of the Ulm approach, of which I would consider myself an exponent—an exponent of "critical operationality."

So, in summing up, my approach was to reorientate young people who did not find answers to their questions in their own context, to provide them with design tools and to propagate industrial design as an autonomous activity separated from art and architecture, and engineering. And not only in Argentina, Chile, and Brazil, but also in other countries such as Mexico and Cuba, where I spent two months in 1984, again under the contract of a United Nations consultancy job, in order to help get their ambitious project of the National Office for Industrial Design (*Oficina Nacional de Diseño Industrial*) into shape.

012: In the field of design for development, what would you think the criteria should be for judging a successful design?

I wouldn't say the criteria have changed, though we cannot talk today anymore about development policies. These have been abandoned. In peripheral countries today, the former development policies have been replaced and dislocated by policies of financing the external debt. Finance-driven policies don't take into account local industrialization, local needs, and local populations. The present imperative is: export or die, and more: pay your debt at whatever price. Whole countries live only to service their debts, debts that grow and grow and grow, provoking social misery and a potential for conflicts. Banks "*Über Alles*" that is the present dogma. In Latin America, we can observe a return to a situation similar to the agrarian feudal economy of the Middle Ages under which the majority of the population lived only to pay tribute to their rulers. Today, whole nations mortgage their future due to the enormous amount of money they have to pay back on international loans, loans of questionable value and outside any democratic control because the local populations that are supposed to "benefit" from these loans are not asked at all. It just happens to them, like a thunderstorm from above. As I said yesterday in my short presentation, the capital flow from the South to the North is bigger than vice versa. So contrary to popular opinion, the North is not "helping" the South at all, but the South is transferring value to the North.

Returning to the question of the criteria, I interpret the role of design professionals as being responsible for the quality of use of artifacts and information. Designers are specialists in the quality of use of artifacts material or immaterial. Let me add that the domain of "quality of use" includes the formal-aesthetic dimension that is intrinsic to design and design work, and not simply an add-on that you can dismiss. It also includes environmental criteria. Designers intervene in helping to assimilate the artifacts into our everyday practice. That is for me the main issue about industrial design and graphic design. So one criteria of success could be paraphrased in the words of Brecht: to make the world more habitable, not a bad aim for a profession! Formulated in more general terms, I claim that the most important criterion for successful design is any attempt to contribute toward autonomy, be it the autonomy of the user on micro-level or the autonomy of the economy on macro-level.

013: "Design for Need" and "Design for Development" are both terms that have been attached to this area in the past. What terms would be most appropriate today to describe design activity in this area?

The design for need and the appropriate technology movements cannot be removed from their historical context; their time has passed. Today, the general settings, particularly the macro-economic settings, have changed drastically into a situation characterized by the anything-but-clear notion of "globalization."

When I was working as a consultant for different governments and private institutions or companies, the focus was on material production, artifacts, machinery, tools, toys, and furniture. Whatever the products, the industrialization process was linked to hardware. Nowadays, I would say, the hot design questions have shifted from a material culture to an information culture that is, the media.

If I were called on today to assist in some program, I would focus on the importance of information technology and communication, which have not been considered as decisive factors in industrialization policies so far. I don't know of any government plan in peripheral countries that takes into account, and tries to do something about, this sector of communication and information technology from a design perspective that puts people in the center. And I would say that design has a vast new field for activity.

014: What message do you have for designers and design educators working in the development context today?

I have always resisted the label of "design-guru," and of having the magic solutions up my sleeve. I don't have any magic solutions. What I do is to go to a particular context and then see what I can do there.

I would divide your question into three parts: professional action, education, and research.

We all know that design is a scandalously underresearched phenomenon, compared with other domains of human life and academic life. As I wrote elsewhere,[12] a profession which does not foster and promote research, and

[12]   Gui Bonsiepe, "Design as a Cognitive Tool: The Role of Design in the Socialization of Knowledge" in the proceedings of the *Design Plus Research* international conference, Politecnico di Milano (May 2000).

incorporate research intensively, building up a proper knowledge base, has no future. We are confronted today with the challenge of constructing a proper body of knowledge about design issues with the help, of course, from many other disciplines such as sociology, computer sciences, philosophy, and history, among others.

Particularly in peripheral countries, design research is necessary and has a legitimate function because, through this research, the design discourse is promoted, and people start to reflect on it. I am, however, aware of a danger related to what we would call esoteric research issues. If we look at some research work, which is very well done of course, obeying all of the rituals of scientific procedures, I sometimes ask myself what is the relevance of the issues that are dealt with? So, my recommendation would be to stick to the local context, this is the rich stuff which cannot be substituted, and which is proper. Start from this local ground without, of course, losing the international perspective. I am definitely not advocating a parochial view of design.

Turning now to education, this is a very thorny question not only in peripheral countries but also in central countries. In all the countries of the periphery, we can observe that design is far more rooted in the academic sector than in professional practice.[13] It is an alarming fact that we register a demographic explosion of design courses, some of questionable quality. For example, consider evening courses which last three semesters, and then you become supposedly a designer. If you tried to do this in medicine or engineering, they would laugh at you! Design has the image, the unjustified image, of being an easy career. It tends to attract the wrong people.

We also face the problem of the "banalization" and "trivialization" of design during the 1990s under the labels "design for fun," "designer jeans," "designer food," "designer drugs," "designer hotels," "designer . . .?" I'm not against fun, but I think it's misleading to put exclusive focus on this aspect of design and the designer's intervention. I am definitely against the notion of design as an ancillary function of marketing.

With regard to design education, I recommend (although I know this recommendation is very difficult to implement) that the people in charge of design courses have professional experience. Otherwise, we get into an academic closed and sterile circuit in which no innovation will occur: the so-called "title factories." Design and design education both live from contact with real-world problems, and in searching for and accepting problems from the outside and bringing them into the learning environment. Design education has to reassess its foundations, that often are taken for granted, and "academicized" and "bureaucratized." Breaking with traditional paradigms, addressing the unresolved relationship between design and the sciences, and getting relevant design research done, these are the issues that are relevant to design in general.

[13]   In appointing a candidate for a teaching position in a design course, more and more weight seems to be assigned to formal academical requisites per se, whereas professional design experience is more and more dismissed.

Now as to the professional issues, I do not feel authorized or legitimized to tell colleagues what they should and should not do. You probably know the very recommendable book *Advice to a Young Scientist* by the British molecular biologist Peter Medawar. I think every designer should read this very clarifying book. He does not talk about design, fortunately enough, but in a typically British ironic manner gives a good X-ray of what a scientist is and should and should not do. Scientists do research and write papers. They produce knowledge, and these papers then are presented at conferences and later published in learned journals or books. He quotes from a manual of the British Society of Electrical Engineering a manual on how to deliver a conference paper and how to prepare a text for a lecture. He states that all persons who are giving a public lecture are under certain amount of stress. The manual recommends that, if you want to feel secure, then you should stand in front of your audience with a 40 centimeter distance between your feet. Note the precision: it must be 40 centimeters and not 38! This, of course, illustrates one of the ridiculous aspects of advice on what to do and not to do.

I would recommend that professional designers working in industry or working as professionals in their own design studios or in public institutions never forget what I consider the basic claim of our profession: "design for autonomy."

I would like to end with a quote from the Argentinean writer Juan Filloy.[14] He opted not to live in Buenos Aires, but in a small provincial town in the Córdoba province. When he was asked why he preferred to live so far away from the fascinating metropolis of Buenos Aires, he answered with a poignant phrase (and I ask you not to misunderstand me if I paraphrase this assessment, transferring it to design):

"The center knows nothing about the periphery, and the periphery does not know anything about itself." This provocative sentence might serve as a breeding ground for reflections about the dialectic relationship between different discourses and practices of design. After all, we live in different places, but in one world.

[14] Juan Filloy, 1894–2000. Argentinian writer.

This chapter provides its own introduction

12 INDUSTRIAL DESIGN IN CHILE 1971-3

Interview with Hugo Palmarola[1]

In 1968, the German designer Gui Bonsiepe began his work in Chile, causing a decisive change in the discourse and practice of Chilean design, incorporating in the discipline a necessary "project drift." Also, he formed the section of Industrial Design of the Institute of Technological Investigation of CORFO that lasts until 1973. After thirty years of his work in Chile, Bonsiepe elaborated a new proposal for design, transferring the project from rationalism to a problem of language. In the interview Palmarola approaches the different aspects of this design proposal.

THE PROJECT IN DESIGN

With the closing of the Hochschule für Gestaltung at Ulm in 1968, along with a diaspora of its proponents, its project-based methodological model extended to the most diverse countries, considerably influencing the condition of design with a rational and critical character. The same year, within the framework of the United Nations for the development of small and medium-sized enterprises, the German designer (HfG ulm) Gui Bonsiepe began his work in Chile, bringing a decisive change in the discourse and practice of Chilean design, incorporating in the discipline a much needed "project mindset."

The technical and pragmatic language of the "ulmian" discourse permitted, at a great scale, the integration of design in Chilean industries and state institutions. In this context, Bonsiepe formed, in 1971, the new section of Industrial Design of the Institute of Technological Investigation of CORFO. Under the direction of Bonsiepe, this group began the design of a large number products to be industrially produced. The primary objective of this project, demanded by the socialist government, centered its efforts in the design of solutions to social problems of massive scale. The variety of products included dosing devices for powdered milk, crockery, furniture for basic housing, portable disc players, calculators, and agricultural machinery, among others, forming the first group of products designed in the country and produced at a massive scale. Although Bonsiepe was never a part of a teaching staff in Chile, with his proposals he motivated a reformulation of the curricular content within leading design schools, linking academic thought and industry.

Thus began, through a concrete incorporation of design to industry and society, the phase of gestation and institutionalization of design, consolidating

[1]   Originally published as Hugo Palmarola, "Entrevista a Gui Bonsiepe," *ARQ* 49 (2001): 54-6. Translated by Estefina Acosta.

in the country a new discourse of what was, until that point, an undefined and diffuse activity.

After September 11, 1973, with the fall of the government led by Salvador Allende, the design processes led by Bonsiepe and his team disappeared. Experiences around these projects would be presented later in his book, *The Theory and Practice of Industrial Design*, published in Italy in 1975.

INTERFACE AND LANGUAGE

Thirty years into Bonsiepe's work in Chile, phenomena such as globalization and computing radically transformed our way of thinking about the world and design. In this context, Bonsiepe, after working in Argentina and Brazil, moved to Berkeley, California, in 1987, where he specialized in the design of "human user interfaces" in a software company, spurring the second phase of his discourse on design.

With the development of computing, new ways of using artifacts would make evident the immaterial character of design in anticipating affordances in software. Thereby, the concept of "interface" gained traction, and the specific materiality of an object, which until then was tied to design, began to disappear.

Moreover, in Berkeley, new ways of thinking and working on language based on the contributions of thinkers such as Martin Heidegger and Humberto Maturana were generating radical changes in different areas unlocking new ways of looking and acting in design as well. One interpretation of humans as linguistic beings who create themselves in language and through it was modifying a long tradition on the concept of language. What was until then a mere descriptive instrument of an internal or external reality was understood as a "generative" domain capable of creating new realities.

From these experiences Bonsiepe developed a new proposal for design, keeping in its core its critical and modern character. The perspective changed. Design from the perspective of rationalism was replaced by design from the perspective of language.

At its core this proposal examined interface as a broad concept upon which Bonsiepe would delineate the entire field of action of the design project. User, artifact, and action would connect in this space, where the concern of design was focused on the affordances of the product and its effective actions, revealing a condition shared by all categories of design.

This reinterpretation proposed, additionally, an origin and foundation for design based on the distinctions that an observer brings from language, being that only on these distinctions are functions created and designed. In this way, a form initially has nothing to do with functions, as these are not found within the product, but in language, and it is within this language where the space of distinctions that render an object available is articulated. This interpretation of language, contained in Bonsiepe's proposal, granted design a primary active role in participating in discussions on possibilities and ways of being in the world, since, as a reorganization of the space of distinctions, it created a new language and a new way of being.

In his latest book, *Dall'oggetto all'interfaccia*, Feltrinelli, Milan, 1995 (Spanish version *Del objeto a la interfase*, Ediciones Infinito, Buenos Aires, 1999), he described this radical conceptual change. To dig into the principal theses of this design proposal, we spoke with Gui Bonsiepe.[2]

What is the need that your new proposal of design is responding to?

Not to a specific need, but to my nonconformity with the issues and interpretative frameworks and their usual vocabulary of form, function, rationalism, functionalism, design brief, product semantics, emotion versus reason. I searched for a basis from which one can refer to design of both physical artifacts (industrial design) and communication artifacts (graphic design and visual language). I was unsatisfied with the postmodernist contributions and attempts at studies of design that grew in design discourse in a purely literal way, without empirical basis, constructing a world closed off, academic, in which theoreticians speak to each other, a revival of a new scholasticism, self-absorbed and preciously defined. From there, the strangest reflections on the reality of design can arise. It is stranger still when they cultivate normative pretensions, particularly in the field of design teaching and research. A theory cannot be created in a purely discursive space, reading texts—although this is indispensable. It would not hurt design students to deepen their reading practice.

From Heidegger in philosophy to Maturana in biology, language is no longer seen as an operational medium, but as a constitutive (organizing) medium. What implications would these ideas have on traditional concepts of function and identity in design?

Language "reveals" and does not reproduce a reality. Analogously, the act of designing forms a space of action. This is why design comes from nothing, particularly not from supposedly objective functions. Design constitutes functions; it does not satisfy them. Functions are also designed. The crux of design, the essence of the identity of design, consists in that, similarly to language, it constitutes a reality. Design and language are two anthropologically constitutive dimensions.

In the general discourse of the design, where do you think the interest for incorporating theories on language was born from? Was it born in postmodernism?

No, postmodern design is largely a mix of the most varied programmatic ingredients. It served, among others, to establish the neoconservative discourse

---

[2]   Cologne, Germany, July 2, 2001.

in architecture and design, of course without ever admitting to it, since as its prefix suggests, it pretends to have overcome a modernism which is largely presented in a cartoonish way.

What do you mean when you pose the concept of "ready-to-hand"(*Zuhandenheit*) in Heidegger as a central category of design?

Heidegger distinguishes between "*Zuhandenheit*" (ready-to-hand) and "*Vorhandenheit*" (present-at-hand) in the sense of being physically there. A simple example can illustrate the difference between these two concepts: walking on the beach a stone is found—it is there, it is present. But only through an act of transformation is this stone revealed as a possible tool, as something available for use. It passes from the domain of mere physical presence to being present for someone. This is where the concept of concern enters, beyond mere need. Through project acts materiality is transformed into an artifact tied to an interest and concern.

What effects do you think the work of Fernando Flores has had in the development of design? What has been your experience?

Fifteen years ago, Fernando Flores with Terry Winograd formed a new approach to design based on language.[3] But I'm afraid that their contribution has not profoundly penetrated the discourse of design professionals and academia—alas, coffee-table books filled with pretty illustrations are easier to absorb than books on reflections about design. They have done work similar to the contribution of Herbert Simon toward the end of the 1960s in his book *The Sciences of the Artificial*.

After the devastating critique by Adorno to the jargon of authenticity (*Jargon der Eigentlichkeit*), Heidegger seemed to me to be irrelevant, and I believed I would not read his texts again. However, in the environment at Berkeley and with an illuminating translator like Dreyfus, I again approached writings by Heidegger, and some of his concepts seemed useful to shed light on certain aspects of design. Although Heidegger himself did not develop a theory on project, perhaps due to his negative view on technology, he did use the terms "*Entwurf*" (project) and "*Geworfenheit*" (thrownness).

How much has the immaterial quality (character) of computing contributed to clarifying and expanding the concept of "interface" to other categories of design such as industrial design?

Design finds its identity in making material and immaterial artifacts amenable to use. This space, in which we find the structure of interaction between artifact

---

[3]   Terry Winograd and Fernando Flores, *Understanding Computers and Cognition – A New Foundation for Design* (Reading: Ablex Publishing Corporation, 1986).

and user, I call "interface," because this concept contains an interpretative potential richer than the old form–function duality. Product design from the perspective of the industrial designer means concentrating on the interface, its quality of use, and its formal-aesthetic quality, obviously without neglecting other factors such as costs, production, and economic feasibility. The essence of an industrial designer's contribution to the configuration of products consists in structuring the interface, and, in this way, its focus differs from engineering.

On this concept, how relevant is the project for cybernetic management, Project Cybersyn that you led in 1972?

The project initiated at the suggestion of Fernando Flores was developed by Stafford Beer, and his team composed of mathematicians, programmers, economists, and engineers. I was heading the design group responsible for the Operation Room or Opsroom with its equipment and graphics.

In 1972, we didn't have the term "interface,"[4] perhaps it did not even exist in the field of computation, though it was used in the theory of fluid mechanics. Looking back, however, we can say that we designed an interface to interact in some neuralgic variables in the economy, its centers of production and of distribution. Today we have another concept for the work done thirty years ago: visualization of complex data. Here, we touch upon one of the most exciting fields for the contemporary designer: using the power of the retinal space for cognitive and operational ends.

It seems like a science fiction story, but it's a fact that in that moment Chile was at the vanguard. Most striking is perhaps the fact that this work was demanded by a public entity—today so denigrated and even demonized by market fundamentalists. With the advancement of computation and the technologies for visualization that came with it, I wonder: Has the teaching of design followed this process? Or have we relegated them (designers) as pixel movers that only use Photoshop for embellishment of screens?

The saturation of "graphic design" has forced a shift toward "information design." What conceptual parallel can be drawn with "industrial design"?

The term "graphic design" today appears to me to have expired. It emphasizes the medium, that is, graphic resources rather than the object of design that is information. The "Manifest Oullim" from Seoul, the genesis of which I participated in as a member of one of two workgroups last year,[5] recognized the loss of relevance of the term "graphic design." Obviously, it isn't about changing

---

[4]   This is the first scientific area in which the term "interface" has been used.

[5]   The Icograda Design Education Manifesto was developed in 2000 by an international group of designers. Participants represented a geographically, politically, economically, culturally, and socially diverse cross-section of the design education community. Prof. Ahn Sang-Soo (South Korea) led the project which was translated into seventeen languages and presented at the Icograda Millennium Congress Oullim 2000 Seoul.

one term for another, but about reflecting on the consequences, particularly for teaching design. And these can be dramatic. The term "industrial design" I believe is not affected by this process of wearing out.

Nowadays there is a design proposal oriented toward responding to very specific habits of small demographics. This proposal suggests that only small and medium enterprises, rather than industries of mass production, would be able to respond to such customized demands (Chilean presentation "ICSID 2001 Seoul") What is your opinion?

This statement can only be confirmed through concrete studies. But we may observe that the dichotomy between small-scale production (made by small industries) and in large numbers (made by large industries) has lost its empirical basis. Customized production can be viable only within the most advanced production technology that happens already in some automotive enterprises.

In relation to this, I extend a question that Alex Kufus[6] poses in the text for the German exhibition "bewusst, einfach—new Design from Germany": "Can there be a style, not from the perspective of the consumer but the producer, oriented towards reasonable and sustainable production?"

I would suggest avoiding in this context the term "style" and rather use the term "corporate identity." A producer, a manager, may implement a company policy incorporating environmental considerations and establishing a lasting relationship between client and business. This is what defines the CI of a business and can imply resistance to the frantic change of one model for another. It implies long-life design.

---

[6]   (NA) Alex Kufus, German furniture designer and representative of what is called New German Design. The quote refers to a note about a design exhibition in 2000 in Wellington, New Zealand.

If you want to characterize the ulm approach—apart from its critical-rational orientation—you can emphasize one characteristic: The reservations against any appeal to a supposedly hidden being in the depths of the past. Regarding the design in peripheral countries, this means questioning attempts to find a local or, even more so, a nationalistic characteristic or essence by digging in the past. In the first instance, attempts of this kind attach themselves to the visual expression, to a supposed style. Confronted with the design of the core countries, the adoption both of a defensive attitude and of a desire for recognition by those countries can take hold. However, this overlooks the fact that the special character of design in any culture is less evident in a configuration of stilemi *than in the specificity of local problems to which the design provides an answer. The programmatic orientation of the ulm program toward industry as the material basis for design exercised a strong attraction for those countries that from the 1960s onward saw industrialization as a path of development. Yet more important than the relation to industry per se was the attempt to provide models for thinking how specific and local problems could be rationally explored through the exercise of design capability—what I called, in the interview in Chapter 2.1, "an operational base for concrete professional design action."*

13 THE ULM MODEL IN THE PERIPHERY

(1987)

It was certainly not coincidental that the composition of the teaching staff and students at the Institute of Design in Ulm (Hochschule für Gestaltung Ulm—HfG) was international. Similarly, the curriculum had features that went beyond the local conditions of the Federal Republic of Germany. That does not mean that the HfG claimed universal significance. It had been conceived in the context of the industrial countries, and it was primarily designed for their needs and scope. But its influence was not limited to the relatively small number of industrialized countries—the center, or metropolis—it also reached those countries that saw industrialization as a tool to reduce their technological dependence and create wealth, and that were also trying to achieve an independent material culture of a contemporary kind.

The Ulm concept was based on the fact that the modern environment is mainly determined by industry, especially manufacturing, communications, and building, and that the traditional university education does not tackle adequately the problems which these industries raise. The HfG filled a gap that the classical universities were not filling and could not fill. The concern was no longer to attach art as a civilizing element to industry from outside—the fundamental error of the applied arts—but to unfold the possibilities for design inherent in industry. That openness to industry did not mean an uncritical and affirmative acceptance, for it were the evident functional, aesthetic, and social shortcomings in industrial production which the HfG wanted to tackle, and which its program was designed to help alleviate.

To the extent that industrialization is now on a global scale, despite occasional voices of industrial oversaturation in the metropolis, it is hardly surprising that the Ulm model was attractive and still is for those countries in the periphery that have a basis for a manufacturing industry in all its many forms, even if only a minimal basis—beginning with the micro-plants in the informal sector, with their often-rudimentary production technologies, and going on through small and medium-sized firms to large companies.

The attraction is not only due to a positive interpretation of the idea of "industrialization," despite all the ecological criticism that is made and justified, it is due to the need to reform the education system, particularly the universities. They often function more as factories for honorary academic titles than as dynamic institutions ready to meet the needs and possibilities of the societies that support them. The curricula for the technological professions, particularly engineering, are more suitable for training administrators of

Article written for the exhibition catalogue on the hochschule für gestaltung ulm: *Die Moral der Gegenstände*, ed. Herbert Lindinger (Berlin: Verlag Wilhelm Ernst & Sohn, 1987). Translated by Anke Grundel.

imported technology than technological innovators. That is an after-effect of colonialization. In order to overcome the design (project) deficit of this phase, the programmes of the engineering faculties—their objectives, structure, and didactics—would probably have to be revised. Only in this way can the technical and cultural intelligence be formed, which is necessary to create a material culture on a modern basis, that is, on an industrial basis.

Whereas initially the policy of import substitution was regarded as a desirable goal by the planning authorities, the inadequacy of this concept of industrialization soon became apparent. Industrialization, if it is to be taken seriously, cannot be achieved simply by producing and reproducing goods; it has to include the design dimension as well. For without innovation as a factor to accelerate industrial dynamic, the peripheral countries will not move beyond a passive reflex form of industrialization.[1]

Already by the early 1960s the ulm model was meeting with response in some peripheral countries, among others in the Escola Superior de Desenho Industrial (ESDI) founded in Rio de Janeiro, Brazil, in 1962.[2] The National Institute of Design in Ahmedabad (NID, founded 1961), India, also took up some of the ideas in the HfG. These institutions based their programmatic approach, their equipment, curriculum, and didactics (problem-oriented learning) on the HfG.

While the ESDI and the NID were new institutions incorporating experiences of the HfG, that is, they were outside the traditional university framework, the Industrial Design Centre (IDC, created in 1969) in Bombay was incorporated right from the start in a technical university. It was intended for postgraduate students and given a high degree of autonomy. Both the NID and the IDC combine teaching with consultancy work for industry and public institutions; that was also part of the HfG programme. It is particularly important for the peripheral countries, where generally there is a social and cultural gap between the insular academic world on the one hand and society/industry on the other.

In Latin America, too, design was seen as an instrument for industrialization. Owing to the strong influence of government institutions on industrial development, industrial design was introduced into multilateral and bilateral technical cooperation programmes and promoted, as in Chile at the end of the 1960s, and especially under the Salvador Allende government. Here the attempt was made to free design of its association with sophisticated and expensive luxury goods for the 5 percent economically privileged in the population and extend it to capital goods.

---

[1]   In economic theory, a difference is made between growth, development, progress, and evolution. Julio H. G. Olivera, "Crecimiento, desarrollo, progreso, evolución: Nota sobre relaciones entre conceptos," *El Trimestre Económico* 26, no. 103 (3) (1959): 410–21.
    Design and design policy can be related to these four terms. Furthermore, the criterion of distributive justice as a form of social inclusion would have to be added.
[2]   The influence of the ulm school in Latin America is documented in the publication *Diseño / 5 documentos* (La Plata, 2002), by Javier de Ponti, Heiner Jacob, Silvia Fernández, Alejandra Gaudio, and Valentina Mangioni.

That would certainly hardly have been possible without the ulm concept, which was distinguished by a view of design as an activity on a technical foundation—an activity that was to be practiced with a basically rational approach. To turn one's back on rational discourse would have closed access to industry and the promotional institutions right from the start, for they had to decide whether design should be allowed a role in the industrialization process. In that process the political dimension of design revealed itself as an instrument for the reduction of ruinous technological dependence. The design rationalism of the HfG proved operable in the periphery. It acts as an antidote to the playful games that make the design of a door handle or a living room lamp into a problem of profound existential significance. The rationalism of ulm also shuts the door on the romanticization of poverty and the exotic, and it prevents the paternalist gestures of "assistantialism." Ultimately, the design problems of the periphery can only be solved in the periphery itself.

A glance at the range of design programs in the metropolis will show that the ulm model is still relevant; for, so far, no viable alternative has been created which could help the peripheral countries, even if only as a critical point of reference, to develop their own product culture.[3] The education centers in the metropolis are even less prepared to consider the specific design problems of the periphery and differentiate their teaching accordingly. Recommendations to "tropicalize" design in the periphery are rather to be interpreted as tourism clichés than as a serious alternative. And the reduction of design questions to questions of style also distorts the view of the problems. But with the mass of short-lived formal innovation that is now being documented in many of the trade publications in the metropolis, the ulm morphology will certainly prove to be an achievement that cannot be bagatellized.

With its broad theoretical approach, which opened the way to critical reflection for the students, the teaching program of the HfG has frequently served as orientation in the periphery. We still do not have documentation of the diaspora of the Ulm Institute of Design. But the continued influence of its model and experience should prove so strong that they can form a chapter in a future history of design in the periphery.

---

[3] This does not mean to assign an exclusive role of the HfG ulm program regarding issues relevant for peripheral countries. In the meantime, a variety of programs with valid approaches have been created, among others—to name a few—in the Netherlands (Eindhoven Technical University) and in Finland (Aalto University).

The comparison of the development of design in different Latin American countries allows one to identify structural similarities that can be explained by the historical development as former colonies. For example, in the 1960s, thanks to the option for industrialization and the associated reduction of the primary role as exporter of raw materials and agricultural products, a consolidation of industrial design can be observed, which was then thwarted by globalization and the policy of deindustrialization that accompanied it, especially since the 1990s. It seems paradoxical that as the labor market for professional design activities shrank, the number of design courses expanded, thus offering some of the graduates of the design disciplines an opportunity to work as teachers in the field for which they had been trained.

14 INDUSTRIALIZATION WITHOUT DESIGN
(1991)

The history of industrial design is part of the history of industry. It goes beyond the traditional history of technology, which does not generally focus attention on the social protagonists of industrial production, that is, on the firms where technology and design occur. A comprehensive and well-founded history of industrial design in the periphery can only be made when historians have written an account of industrial development, including the history of trade, science, technology, and, most importantly, everyday life in these societies. You cannot make a comprehensive history of technology without including these other domains.[1] A project of that size is beyond the scope of this first and less ambitious attempt to collate a range of material from the periphery and present it in a brief overview. The concern here is to identify common structural features, without excluding local particularities. The common elements can be structured into a pattern that will show that the history of design in the periphery has its proper characteristics, without suggesting a polar opposition between the periphery and the center. The use of these two expressions, in reference to the *condition périphérique*, is supported by facts. There are countries with the power to invent new games and fix the rules of play, with winners and losers in politics, trade, finance and technology, and also design. Then there is a second, numerically larger group of countries that have to accept these rules, whether they want to or not. Attempts to invent their own games have been blocked— as history has repeatedly taught us—in order to force their participation in the games that are established by the powerful players.[2]

The only chance for peripheral countries to find ways for inventing their own history and shaping their own future is a process requiring steadfastness, patience, and caution. The asymmetry of international trade has moved the

Originally published in extended form: Gui Bonsiepe, "Paesi in via di sviluppo: la coscienza del design e la condizione periferica," in *Storia del disegno industriale—1919-1990 Il dominio del design*, ed. R. Ausenda (Milano: Electa, 1991), 232–69. Published in English in Dawn Oxenaar Barrett (ed.), *Interface—An Approach to Design* (Maastricht: Jan van Eyck Akadamie, 1999), 83–96. Translator Dawn Oxenaar Barrett .

[1] In this argument, a history of design in the periphery would not primarily be a history of individual designers.
[2] In the nineteenth century, a small country like Paraguay had an industrial infrastructure to make armaments. The war of the Triple Alliance between Argentina, Brazil, and Uruguay on the one side—supported by the British with their financial and trading power—and Paraguay on the other ended in 1869 with the population of Paraguay reduced to 24 or 15 percent of its original size. The male population was almost completely wiped out: only 3.5 or 2.1 percent survived the war. The different numbers result from the different sources on population figures at the start of the war, some giving 800,000 and others 1,300,000. Julio José Chiavenatto, *Genocidio Americano: A Guerra do Paraguay* (São Paulo: Editora Brasiliense, 1979), 151.

role of technology, technological innovation, and design into the center of hermeneutic enterprises. They are seen as the new ways of making history and determining its course, going beyond the traditional forms of the exercise of power.[3] With the dependence theory, attempts have been made to explain the difficulties Latin America faces in breaking out of its peripheral position.[4] It was therefore appropriate to use the dependence theory as a framework to explain the importance of design for development. That has had hermeneutic value, but it was less appropriate.

INDICATORS OF DESIGN DEVELOPMENT

A number of indicators will help to arrange the unstructured mass of data covering the state of design development in the periphery. The following six criteria are used:

— Management in private firms and public institutions
— Professional practice
— Public design policy
— Design education
— Design research
— Design discourse (publications)

Design management is visible in the range of products on the market. It is displayed in the following three, closely related offered areas:

— Product concept (what services does the product offer the user?)
— Quality control in production
— Aesthetic qualities (not regarded as an extra, but as a fully integrated category)

The maturity of professional practice can be seen from the extent to which designers are integrated in the manufacturing industry, in the practice of the acquisitions offices (public tenders) and in promotional programs.

---

[3]   From the innumerable examples of the disadvantages of exporting raw materials, one case may be cited: Brazil has large reserves of high-quality quartz. It exports this raw material at a price below \$1/kilogram and imports the finished products made from the material at more than \$2,000/kilogram. J. W. Bautista Vial, "A guerra silenciosa que vamos perdendo," *Terceiro Mundo* 30 (1990): 47–52. This chapter attacks the damaging approach of some economists who treat technology and technology packages like any other commodity, for example coconuts.

[4]   For a discussion of the dependency theory, see Dudley Seers (ed.), *Dependency Theory: A Critical Reassessment* (London: Frances Pinter, 1981). Of particular interest is the article by Luc Soete, "Technological Dependency: A Critical Review," 181–206. The author points out the errors of technological dependency theory, which fails to recognize the important dynamic aspects of technology, which concentrate most of the discussion on the specific trade value of technology and the way it is used for the "international redistribution of economic surpluses," p. 182.

Design education is reflected in the independent planning of curricula and their implementation in universities.

Design research shows up in the production of innovative knowledge relevant and specific to design.

"Publications" here is an abbreviation for all actions intended to create and establish design discourse.

With the help of these six criteria the development of design in the periphery can be classified in a five-stage arrangement that begins with the proto-design phase and ends with the full integration of design in industry. If one looks at the most highly industrialized countries in Latin America, one sees that during the period from 1960 to 1990 they advanced to the third stage.

That process can be described as follows:

1. The phase of proto-design (from Independence to the end of the Second World War, though there is an example of corporate design for public institutions already starting in the 1920, for example YPF Argentina—the National Petrol Company). This phase is hardly researched. It would be wrong to assume that industrial design only existed from the moment when the special term was introduced. Although consumer and capital goods were largely imported, there already was a local manufacturing industry. It expanded particularly during the Second World War, when imports from the industrial countries declined.
2. The embryonic period of industrial design during the 1950s. This is when avant-garde movements emerged in art and architecture, confronting the phenomenon of industry and regarding industrial production as the central issue in modernization and modernity.
3. The period of emerging institutionalization in the 1960s and 1970s, when design offices and education centers were established.
4. The period, during the 1980s, when consolidation began despite endemic instability and the often-abrupt changes in monetary policy (especially the manipulation of exchange rates, through either overvaluation or undervaluation of the local currency).
5. A period of maturity that may be reached in the future.

THE MODEL FOR DEVELOPMENT

During the 1960s, industrial design was integrated into bilateral and multilateral technical cooperation programs, under the influence of development doctrine. In 1972, a conference of experts was held in Vienna by United Nations Industrial Development Organization (UNIDO), in order to discuss the role of industrial design in peripheral countries.[5] The arguments in support of industrial design

[5]   Gui Bonsiepe, "Development through Design—Working Paper for a Workgroup Meeting" (Vienna: UNIDO, 1972). A working paper was written for internal use in UNIDO, entitled "Design

put forward then were later taken up again in the Ahmedabad Declaration in 1979.[6]

It is well known that the attraction of the development doctrine lay in the promise that developing industrial production would increase the gross domestic product. As we have seen, increasing the gross domestic income can provide the necessary condition for development, but not the sufficient conditions. Islands of abundance can exist in an ocean of poverty. The dynamic of social change, which is supposed to be promoted by technical development, is more complex than was first imagined.

The development doctrine remained on the agenda throughout the 1970s and the 1980s. The aim was import substitution, with the explicit objective of building up a national product culture. In the 1970s, the development projects were revised. In the process it was discovered that technical cooperation programs had not always fulfilled their promises. The mistakes were christened "development pyramids," another name for the dumping of scarce resources. The relatively high costs of creating a job induced Kurt Schumacher to propose a different approach, which he called "intermediate technology."[7] Schumacher wanted the suitability of a technology to be measured by the number of jobs created per unit of investment. Moreover, he thought that this type of technology should be suitable for the local needs of the peripheral countries.

Fed by the student revolt, social criticism of modern technology took concrete form in alternative technology and alternative design. Intermediate, appropriate, and alternative technology had a good response from international organizations and a number of development institutions (which paradoxically were not in peripheral countries). The projects were mainly influential as demonstrations and showcases rather than as projects with more far-reaching consequences. Although these approaches were certainly not new for the peripheral countries, those who supported them did succeed in raising some unorthodox questions on technology and society, and beyond this, in drawing attention to the rural population. A critical reappraisal of the results was presented by Witold Rybczynski. Judged by the declared intentions of "Appropriate Technology," it was revealing.[8] Appropriate Technology aimed to design "tools that would take account of poverty as a frame of reference and that would be suitable for genuine development."[9] The list of requirements included the following: low costs, using local materials; creating jobs; having a small scale; being understandable for the rural population; able to be produced in small blacksmith workshops; using decentralized energy sources; and avoiding patents, payment of license fees, and dependence on sophisticated

for Industrialisation"; later it was revised as "Industrial Design: Basic Guidelines for a Policy of UNIDO." This document contained examples of industrial design in Latin America, especially Chile, and India. An edited version of this paper is published as Chapter 4.1 of this book.

6   *Ahmedabad Declaration on Design and Development* (Ahmedabad: National Institute of Design, 1979).

7   E. F. Schumacher, *Small Is Beautiful—A Study of Economics as If People Mattered* (London and Tiptree: Blond & Briggs, 1973).

8   Witold Rybczynski, *Paper Heroes* (Garden City: Anchor Books/Doubleday, 1980).

9   K. Darrow and R. Pam, Appropriate Technology Project, Volunteers in Asia 1975.

financing arrangements. The products included low-cost solar steam boilers, bicycle rickshaws, hand grain threshers, ox carts, and multifunctional tool bars to work the soil.[10]

KEY FACTORS IN INDUSTRIAL DESIGN IN LATIN AMERICA AND INDIA

The rich history of design differs in the various countries of the region. In some countries, craft production accounts for a considerable share of the product culture, and it can be taken as the starting point for industrial design. However, in this context a range of craft products, like textiles, ceramics, utensils made of wood or basketwork, *papier mache* products, and feathers, are excluded since they form a separate history. The products of the informal sector of the economy are also beyond the scope of this brief outline.

In 1949, the Students Organization of the Faculty of Architecture of the National University Buenos Aires at the University of Buenos Aires published an article by a representative of Concrete Art. To judge this text by available sources, it is the first document in Latin America that considers industrial production as a subject for design. The author wrote: "The shape of the spoon is also a cultural problem. . . . Functional design begins with the principle that all the shapes created by human beings deserve to be regarded as cultural manifestations."[11] He appealed to artists to "turn their attention to the socially pulsating area of mass production, everyday products for everyone, for they are ultimately the direct reality of modern man."[12]

The BKF or Batwing Armchair developed shortly before the outbreak of the Second World War is probably the best internationally known example of Argentinian design. It employs iron struts used in building and a seat of sailcloth or leather. In an innovative way it continues the design of the folding Tripoli chair that was invented in 1887. A commentator in the local design scene criticized the rigidity and lack of mobility in local industry during the years of protectionism. He said: "In many branches of industry, particularly in sanitary installations, not a single model has been changed for more than twenty years."[13]

On the other hand, the policy of protectionism also enabled growth, particularly in light industry. The wide range of product classes—a useful indicator for the state of industrial development—can be surmised from the list of awards for good design found in the exhibitions. In 1970, for example, a total of fifty-six companies showed 270 products in the annual design competition in Argentina. The number of examples show the advanced state of design there at the end of the 1960s. Design was more advanced there than in some

[10] It should be evident that the excursions into design *povero* in the center have nothing to do with design *povero* in the periphery. Design *povero* in the periphery results from a situation of poverty, while in the center it emerged in a context of abundance.

[11] Tomás Maldonado, "El diseño y la vida social," *CEA—Boletín del Centro de Estudiantes de Arquitectura*, no. 2 (1949): 63–5.

[12] Ibid.

[13] Rodolfo Möller, "Crónica del diseño industrial en La Argentina," *summa* 15 (1969): 22–5.

European countries. Official design policy in Argentina began in 1962 when the Centro de Investigación de Diseño Industrial (CIDI) was founded. Progressive entrepreneurs began to play an influential role. The firm of Siam di Tella was the first in Latin America to raise industrial design to the status of a development division for corporate identity—at a time when the term "CI" was still alien to many managers in the advanced industrial countries of Europe. Most people who identified themselves as designers had been trained as architects. In 1962, the first department of industrial design was set up at the University of La Plata. Design discourse was encouraged when the magazine *summa* was launched in 1963. The climate of intellectual curiosity can also be seen from the translations of specialist literature. The influence of the ulm school was spread in the 1960s by Tomás Maldonado, who had been invited to join the teaching staff in ulm in 1954 by Max Bill.[14]

Under the military government from 1976 to 1983, a policy of deindustrialization was pursued following a strict neoliberal program. With the argument that Argentinian companies must be able to compete with foreign products, the local currency was driven up against the dollar, while the markets were opened to imports. In sardonic humor it was said that the only reason canned air was not being imported was because this product was not available on international markets.

In October 1968, a group of Chilean design students took part in an international seminar in Buenos Aires on Education of Industrial Design in Developing Countries. A number of well-known design educators had also been invited. That seminar opened new perspectives for the students, since at that time in Chile design education was regarded as a craft-oriented subsection of interior decoration. From 1971 the group was accepted into the newly founded design group at the Technological Research Institute (Comité de Investigaciones Tecnológicas), to develop designs that corresponded to the social and economic policy program of the Salvador Allende government. For the first time, it was possible to liberate design from its marginal role and to design a broad range of products, from wooden children's toys to agricultural machines. The military putsch in September 1973 put an end to that promising beginning.

After the initial attempts in São Paulo in the 1950s, intensive development of design began in Brazil in the 1960s. In 1962, the College of Industrial Design, Escola Superior de Desenho Industrial (ESDI), was founded in Rio de Janeiro. It was strongly influenced by ulm, and it in turn influenced many design courses that were set up later in Brazil. In 1981, industrial design was included in the official science and technology policy.[15] Design studios were established in

---

[14] In a homage to the "Great Tomás" Reyner Banham called the new approach in design education "the cool breeze blowing from the Ulm hills in the early sixties, disturbing the climate of self-satisfaction in design." Reyner Banham, "El testamento del Doctor Maldonado," *summa* 24 (1970): 54–5.

[15] This decisive step was taken by Lynaldo Cavalcanti de Albuquerque (an engineer with focus on industry, whereas formerly scientists with focus on research had headed the institution) as president of the National Council for Scientific and Technological Development (CNPq).

industrial centers, particularly in São Paulo. Designers succeeded in designing unconventional types of products as well as public transport and equipment for the steel industry. The influential São Paulo Federation of Industries set up its own design promotion department in 1978.[16]

The development of industrial design in Mexico was similar to that in Brazil: a broad, almost hypertrophic academic field, with strong emphasis on national identity and design promotion by the government, but not to the same extent as in Brazil.

Although it has only a small manufacturing industry, Cuba's industrial designers have been trained since the 1960s in a loosely structured design institute. When the National Industrial Design Office (Oficina Nacional de Diseño Industrial) was founded in 1989, industrial design spread into six areas: product development, training education, design assessment, standardization, promotion, and information.

In India, two institutions have played an outstanding part in developing design: the National Institute of Design in Ahmedabad and the Industrial Design Centre in Bombay. In 1958, Charles and Ray Eames were invited by the Indian government to study the possibilities for design in small firms. The founders of the National Institute of Design drew substantially from their report. The Industrial Design Centre was founded in 1969 to train students at a postgraduate level. It has always had a strong interest in theory. A recurrent theme is the creation of an identity that would differ from European and American design models. About the contrast between the center and the periphery, a critical observer of the design scene said that " to judge from the predominance of furniture and lamps in the design periodicals in the industrial countries, there appear to be two main concerns for designers working there, namely, to find new ways of sitting and lighting rooms. . . . The basic needs in India are quite different. In a rapidly industrializing economy like India, industrial designers can make a meaningful contribution to the development process. Design faces huge challenges in key areas like agriculture, education, transport, communications and housing construction."[17]

BIBLIOGRAPHY

Blanco, Ricardo. "Los diseñadores de Buenos Aires." *Nuestra Arquitectura* 511/12 (1980).

Bonsiepe, Gui. *Diseño industrial, tecnología y subdesarrollo.* Vol. Cuadernos Summa - Nueva Vision. Buenos Aires: Ediciones summa, 1975.

Bonsiepe, Gui. *Diseño industrial, Tecnología y Dependencia.* Mexico: Edicol, 1978.

Bonsiepe, Gui. *Diseño industrial en América Latina.* Buenos Aires: Ediciones summa, 1979.

---

[16] NDI—Núcleo de Desenho Industrial.
[17] Kirti Trivedi, "Industrial Design in India: Problems and Potential," in *IDC Faculty on Design* (Bombay: Indian Institute of Technology, 1989).

Bonsiepe, Gui. *A 'Tecnologia' da Tecnologia*. São Paulo: Edgard Blucher, 1983.
Bonsiepe, Gui. "O *futuro* do design na América Latina." *Design & Interiores* 16 (1989): 139–41.
Bonsiepe, Gui. "Diseñando el futuro - Perspectivas del diseño industrial y gráfico en América Latina." *tipográfica* IX, no. 8 (1989).
Congon, R. J. (ed.). *Lectures on Socially Appropriate Technology*. Eindhoven: Committee for International Cooperation Activities, Technische Hogeschool, 1975.
Möller, Rodolfo. "*Una década* de activa promoción del diseño industrial." *summa* 50 (1972): 104–7.
Neumeister, Alexander and Gudrun. Low-cost Furniture for the Philippines, Rattan Furniture for Export. Reports on two design workshops 1982 and 1984 on the Philippines, 1984.

(b) Design in the "Periphery"

Design history keeps the past in memory in order to understand the present and prepare the future. Therefore, it constitutes an essential component of a design study program. One of the roots of design history with its branches in industrial design, communication design, exhibition design packaging and design, textile/fashion design is embedded in art history, sometimes with links to the history of architecture or general cultural history. Confronted with the lack of specific courses in design history in Latin American universities, classes in design history were—and are—given by representatives of different academic disciplines, not necessarily linked to design. Furthermore, representatives in charge of design history obviously can—and do—have their own agenda of what they consider to be design and consequently object of design history.

One of the criticizable and weak points becomes visible when design history does not take into consideration the broader context of industry, companies, and industrial policies. In the face of this deficit, a small group of design professionals not necessarily historians decided to prepare an overview of what in the future could become a full-fledged history of design in Latin American countries. The project began in 2004 outside of any academic structures. The result of this work was presented four years later in 2008 in form of an illustrated book with country reports and a set of essays about specific issues of the design discourse—Silvia Fernández and Gui Bonsiepe, eds., Historia del Diseño y Comunicación Visual en América Latina y el Caribe—Industrialización y comunicación visual para la autonomía *(São Paulo: Blucher, 2008). Sadly, the book remains untranslated.*

15 HISTORY OF DESIGN IN LATIN AMERICA

(With Silvia Fernández and Lujan Cambariere)

(2016)[1]

At the "3rd International Conference on Design History and Design Studies—Local Histories of Design," held from July 9 to 12, 2002, in Istanbul research was presented on the influence of the HfG , on design education in Latin America.[2] Among others, this particular research with focus on design education revealed that though local histories of graphic design and industrial design in Latin America existed, there was no general history of design in this area in which the full story is told.

In the beginning of the twenty-first century, design history in Latin America was a poorly researched field. Moreover, the study programs of design history were determined within the framework of history of art or history of architecture, and occasionally as an issue of social sciences, so that design history tended to be presented according to extrinsic criteria. In the conceptual framework of these disciplines, especially art history, essential categories for design such as industry, technology, economics, and industrial and social development hardly appear, if at all.

Though the publication of local histories with sound and rich primary documentation has begun, mainly in Brazil, Chile, and Mexico, research projects developed in academic institutions predominantly approach design as a closed entity without links to the general context, and often don't get distributed. While there have been tentative institutional initiatives to collect and organize systematically historical pieces, industrial products and graphic materials are mostly dispersed in second-hand stores or auction houses of material without commercial value, or abandoned in a private garage or repair shop. The graphic history of industry or commerce can be read solely in the

---

[1]   This chapter is based on an article by Silvia Fernández, "El libro Historia del Diseño en América Latina y el Caribe: industrialización y comunicación visual para la autonomía," *Ecos de la Academia* 3, no. Junio (2016): 170–3, and also on an interview by Lujan Cambariere, "Una historia americana," published in *Página 12, suplemento*, August 16, 2008, 1–3. An early version of the text was translated by Niberca Lluberes Rinicon.

[2]   Silvia Fernández, "The hfg ulm: On the Origins of Design Education in Latin America," *Design Issues* 22, no. 1 (2005): 3–19.

frontside of local stores. Some documentation is kept in the archives of public institutions and companies. Libraries and newspaper archives are the refuge for editorial graphic memory, as are a few second-hand bookstores. There exist private archives, journals, publications, catalogues, and conferences with varied material. And above all, there are the protagonists of the history themselves.

Confronted with this panorama, the idea to elaborate a broader more comprehensive version of design history by putting Latin America on the map came from Silvia Fernández and a team—Javier De Ponti, Alejandra Gaudio, and Valentina Mangioni—with the name *Nodal*[3] starting with small book editions. The network of existing contacts was enlarged, and in 2004 a group meeting with fourteen participants was held place in La Plata (Argentina) in order to discuss objectives, approaches, and procedures for developing this ambitious research project. The outcome of the discussions listed in the following paragraphs was summarized in a programmatic document with a series of recommendations. The guidance did not claim to be universally valid but to provide a frame of reference that would ensure a more coherent result than just a compilation of individual papers. A second meeting with twelve participants took place in October 2005 in the Center for Advanced Studies (CEAD) at the University of the Americas in Cholula, México, to discuss the first version of the contributions. In March 2006, the authors sent their final version.

— The central issue of the research is the history of graphic design and
   industrial design in Latin America, touching also the areas of fashion and
   textile design, graphic design in film and TV. The work is part of a much
   broader project than the one planned back then: the history of material
   and symbolic culture in Latin America and the Caribbean.
— The common denominator of the research project was the rescue of
   the political and public dimension of design, which has fallen almost
   into oblivion during the first decade of the twenty-first century. In other
   words, to maintain awareness of the sociocultural and socioeconomic
   context in which industrial and graphic design manifest themselves.
— It is recommendable to place design in the context of the communication
   industry and manufacturing industry, because companies and institutions
   form to considerable degree the link between project and society.
— A difficult point to resolve will be (and is) the interaction between
   political and societal facts of the local history and design facts. The key is
   to find the links between these two domains.
— The world of popular and anonymous design should be included.[4]
   Several countries of the region—particularly México, Colombia, Perú,

---

[3]   NODAL is a network with members ad hoc, starting from a common project. The aims are
       presented in a manifesto that can be accessed: http://www.nodalatina.net/ (accessed August
       20, 2020).
[4]   Excerpts from the programmatic document formulated at the First meeting of the Network
       History of Design in Latin America and the Caribbean, La Plata, October 12, 2004.

Bolivia—count with a rich craft culture whose results can be taken as token of local cultural heritage. When an industrial context is either lacking or weak, this local material crafts culture can be interpreted as an expression of a supposed Latin American identity—an issue that regularly turns up during design congresses in Latin America. But it is difficult to avoid the trap of essentialism with an idealization of a preindustrial past as if that past can provide a reference for design that by definition is oriented toward the future.

Although the historical process cannot be reduced to contextual circumstances (trying to explain it from one mono-causal and simplistic perspective), it is necessary to provide contextual information in order to discover the links with macro-economics, politics, and social studies of an activity that has its origin in the social, economic, and productive development of each country. It may suffice to analyze the British design since the industrial revolution, the political origin of the Bauhaus and the Swiss School, or the growth of the Industrial Design Center (IDC) at the Indian Institute of Technology, Bombay (1965), the education project of the HfG Ulm (1953), or the political origin of the ESDI in Brazil (1962), or the insertion of design into the program of the socialist government of Salvador Allende (1971), and the eagerness that China includes design in its educational programs and in the industry, to understand the need to relate these processes. Therefore, background information in the form of a list of relevant data was provided at the beginning of each country report.

The whole project rested on the hypothesis that there exist certain similarities and structural affinities across Latin American countries that allow their integration into one shared history of design.[5] In the background of the collective work stood the intention not to use design as an object for theoretical undertakings; the goal was considerably more modest: to document and understand design—the material and semiotic artifacts—as an embodiment of technical-industrial civilization. The idea of this research project grew in the context of a deep financial, economic, and social crisis in Argentina in 2001, in which no project seemed possible, and the dictum "No Future" seemed the only reasonable answer. Counterintuitively, the lack of resources stimulated the desire to start the project against all odds and against all prospects of success.

Researchers from Europe suggested to apply for a research grant within a collaborative program between European and Latin American institutions, with challenging conditions, demanding signatures of the Deans of each of the participating universities, six Latin American and six European. Passing the administrative and bureaucratic hurdles to get the signatures of the authorities was time consuming and took one year. The research proposal was presented but—competing with other projects from computer science and hard sciences—Design did not make it—as might be imagined. Notwithstanding this

---

[5]   Emir Sader, Ivana Jinkings, AA.VV, *Enciclopedia Contemporánea da América Latina e do Caribe*, Laboratório de Políticas Públicas (São Paulo: Boitempo, 2006).

breakdown, the work continued without official support and independent of academic institutions.

The research revealed, among other things, that in the 1960's—a period characterized by the promotion of national development policies—design was part of these policies, in which industrialization and communications were organically assisted by design, and that this process was opened but also remained truncated in several countries. In other words, the development of design in that decade was determined to a considerable degree by the public sector, with the creation of associated agencies and programs developed by the Ministries of Economy, Industry, Trade, and Commerce. It was a period in which design was part of the political discourse in terms of establishing autonomy or reducing the state of heteronomy. It also showed that in the 1960s until the mid-1970s, design as a profession and teaching subject was more consolidated in several Latin American countries like Mexico, Chile, Brazil, and Argentina than in several European countries at that time under authoritarian regimes such as Spain and Portugal.

In 2008, the collective work, translated from six languages (English, German, Italian, French, Portuguese, and Dutch) into Spanish, was published in Brazil.[6] (*History of Design in Latin America and the Caribbean: Industrialization and Visual Communication for Autonomy*). The preparation of the edition of the book with contributions from twenty-four authors counted on financial contributions from several participants and volunteer work of professional services. The book was designed by Carlos Venancio and Fabian Goya (Argentina) using the font Borges designed and ceded by Alejandro lo Celso (Argentina).

The main part of the publication offers an illustrated overview of the development of industrial and graphic design of eight countries of the region from the beginning of design activities that differs in the countries until the start of digitalization about 2000 in the context of economic, social, and industrial policies. The second part with the title "Influences and Prospectives" consists of practical and theoretical influences received mainly from European sources such as Swiss graphic design, public graphic design in the Netherlands, the semiotics of products (product language from the HfG Offenbach), and the migrations of professionals between Spain and Latin America. In addition, there are contributions to understand the context in which design operated, organized around themes such as the role of public design, the theory of objects, the difference between design and art, and design and crafts, and design and sustainability.

An annex contains the programmatic document with an outline of the general approach and the specific guiding lines for the research, the 1964 English manifesto "First things First"[7]—Ken Garland authorized the reprint of this

---

[6]   Silvia Fernández and Gui Bonsiepe (eds.), *Historia del Diseño y Comunicación Visual en América Latina y el Caribe—Industrialización y comunicación visual para la autonomía* (São Paulo: Blucher, 2008).
[7]   http://www.designishistory.com/1960/first-things-first (accessed August 20, 2020).

important document that has maintained its relevance—and three synchronous tables:

1. Listing the policies, programs, and creation of public organizations that had an impact on design activities
2. The most important Latin American designs with broad impact, social relevance, influence in the public space, and direct or indirect result of government policies
3. The different automobiles designed and produced in Latin America

The foreword addresses general questions approaching history and the fundamental importance of design history for teaching design.

One characteristic has been present—to a greater or lesser degree—in the works gathered in this book: not treating design as an isolated cultural phenomenon, closed in on itself, but as a variable of socioeconomic and sociopolitical processes.

According to the vision of the hegemonic countries, Latin American economies would be predestined to limit themselves to the export of commodities. The economist Aldo Ferrer writes: "The growing gap in the technological and value-added content between imports and exports revealed that the Argentine economy was returning to a productive structure based essentially on the exploitation of its natural resources and increasingly removed from a diversified and complex structure inherent in the dynamics of development and the capacity to participate in the segments of international trade."[8]

This characterization is not limited to Argentina, but is valid to a greater or lesser extent for all Latin American economies subjected to a process of subordination to interests that inhibits a development of a less polarized, less antagonistic, social organization with less structural violence.

When writing about the history of design in Latin America, it is inevitable to insist on the project understood as an action for autonomy or, more modestly, an action for the reduction of heteronomy. A society that abdicates the right to participate in the project of contemporary material and communicational culture neglects its possibilities for the future. Without design there is no future, and without new teaching there is no design for this future.

A characteristic of the approach to design history documented in this book is its contextual framing of economic, social, and industrial policies. In Latin American countries, in general, it existed—and still exists—a pronounced policy resting on the role of exporters of *commodities*—primary products— and importers of industrial products. This predominant role of being suppliers of raw material exists since the European "discovery" of Latin America that survives basically by exporting nature and its resources—with devastating consequences—copper, core iron, lead, oil, silicon, silver, wood, coffee, bananas, soya, biogas, and so on.

---

[8] Aldo Ferrer, *La economía argentina* (Buenos Aires: Fondo de Cultura Económica, 2004), 324.

In contrast to this policy, there exists an enormous design vacuum that offers a chance and opportunity for industrial design and graphic design. But to counteract the obnoxious policy of commodity exporters, the appropriate policy decisions needed to be made. These became known under the term "Import Substitution Policies" putting emphasis on a local industrialization process oriented toward the internal market. These ISP policies continuing a neoliberal program by the military government that started in the second half of the 1970s were gradually abandoned in the 1990s.

Graphing the organizations and programs that directly and indirectly influenced the design activity in the different countries, the conclusion history of discontinuity in Latin American countries is amazing. Even Che Guevara[9] is the one who carries the project of incorporating design into the development of the Ministry of Light Industry in Cuba.[10]

Obviously, the project of sketching a map of the history of design in Latin America could not claim to be more than fragmentary. The entire work makes visible the intricate entanglement of design issues with social issues. In Latin America, the societal contradictions become visible in design with undeniable force.

[9]   Looking back, Bonsiepe considers highly likely that Guevara knew about the Concrete Art Movement of the 1950s in Argentina and that he was aware of what design was about. Bonsiepe believes that Guevara might have had knowledge about the concrete art movement, and its importance for getting industrial design into discussion due to political affinity as the core members of the concrete art movement were or had been affiliated with the communist party or at least shared similar views. Bonsiepe notes that when Tomás Maldonado, Claude Schnaidt, and himself were invited to give conferences in Cuba in 1969 for the first time, they were indeed invited by the Ministry of Light Industry. More evidence of this hypothesis is corroborated by an interview by Roberto Segre published in [the Cuban magazine] *Casas de Las Américas* (noviembre-diciembre 1969 año X, n.57) that suggests that Che Guevara worked together with [Cuban designer] Clara Porset even though she was not a party member.

[10] The Ministry of Light Industry in Cuba (Ministerio de la Industria Ligera), now part of the Ministry of Industries, encompassed consumer goods, such as earthenware, plastics products, textiles, leather, graphic, cultural, and domestic and personal goods.

This contribution reflects the influence of the exploration of language-theoretical texts (particularly by John Searle)—an opportunity that opened up in 1987 while working at Fernando Flores' software house, Logonet/Action Technologies in Berkeley/Emeryville.

Practical work as a designer in a software office permitted me to develop a much more differentiated understanding of design, getting rid of the traditional topics and divisions of form and function, and developing an interpretation of design based on language and action theory. Partly this came about through being in the Berkeley environment and the sudden exposure to a range of literature I was previously unaware of. I remember the shock when I entered for the first time the Computer Literary Bookstore in Sunnyvale. I was looking for literature about interface and did not expect to find a single publication. Instead, I was confronted with a one-meter shelf filled with publications about computers and interface. Furthermore, living on the Berkeley University area and having the chance to listen to conferences, you "breathe" an atmosphere where something is happening. At the home of Fernando Flores, I had the privilege to listen to the conversations of Dreyfus and Maturana. A comparable, even more mind-opening atmosphere I found only in Milan and the environment of Maldonado and the Feltrinelli publishing house.

Among other things that became clear for me at this time was the central role of language for the assessment of designs—especially when formulating judgments about candidates for prizes in competitions, why a design wins a prize. Inevitably, automatic assessments come into play, as it were, which are then grounded as rationally comprehensible decisions in order to avoid the accusation that they hardly present more than personal preferences and whims. For this reason, the "Likes" and "Dislikes" used in social media don't go beyond been bean counting and signal a state of regression of a state of mind. Because what counts is only the positive or negative reaction, then, even if they were taken in to consideration and played a role, the reasons or arguments provided for the judgement remain hidden or suppressed.

Where this plays into looking at design in general is in a much sharper analysis of the words through which design practices are not only described but organized.

16 ASPECTS OF DESIGN IN THE PERIPHERY

(1989)

Gradually, during the 1970s, industrial design and graphic design each took on their own profiles. One indicator for the consolidation of a new profession is when education programs are established. At that point an informal activity is being changed into a structured activity. Another indicator is the activity in the economics ministries of peripheral countries, institutions have been set up in the form of design centers and project groups.

In the programs for international technical cooperation, which are particularly intended to strengthen the SME sector, industrial design has been included in order to raise the quality of products. But despite all these efforts, it has not been possible to incorporate design fully, in either the consumer or capital goods industry. The gap between entrepreneurial activity and design activity suggests that we should look for the reasons. A number of explanations can be identified.

One widely accepted explanation is the dependency theory developed by Latin American social scientists in the 1960s. It is concerned with the trade imbalance between the industrial countries and the peripheral countries. Despite its critical value, the discourse of dependency suffers from weakness on a pragmatic level. It looks primarily abroad for those responsible. More specifically, it looks in the small group of industrialized countries. Multinational corporations come under fire, and are particularly accused of exerting excessive influence on local economies. But the presence of multinational firms does not explain why firms have not made more intensive use of the possibilities of design locally.

According to another explanation, the local firms are too weak financially to invest in design. Ultimately, profit on the investment would only become available over the medium term. By comparison, operations on the financial market, in a highly inflationary context, and the transfer of financial resources to accounts abroad (flight of capital) bring short-term and secure gains.

According to a third explanation, the lack of integration of design in the production system can be attributed to a management decision. It is cheaper to copy a design developed abroad than to undertake design development oneself.

The fourth explanation introduces the argument of the gap between university education programs and the needs of industry.

Paper originally given at the meeting *Diseñando el Futuro*, Mexico, April 1989. First published in English in *Design Issues* 7, no. 2 (Spring 1991): 17–24. Also, in Dawn Oxenaar Barrett (ed.), *Interface – An Approach to Design* (Maastricht: Jan van Eyck Akadamie, 1999), 105–9. Translated by Gui Bonsiepe.

All four explanations provide a greater understanding of the problem. But which explanation will encourage the integration of design as a strategic element in industry? In order to estimate the scope for design in Latin America, we need to cast a glance at design discourse there.

The traditional link between design and drawing induces people to imagine a designer as a person who makes sketches. Although the link between design and drawing enjoys a long tradition, it has no future. Anyone who sees drawing as the outstanding feature of design is making it impossible for design to become a basic discipline in a new paradigm for education.

This is not to underestimate drawing as a skill. For some design areas, it is certainly important; it may even be the necessary or the only way for a design to be experienced or communicated visually. The strength of the drawing paradigm lies in the fact that both industrial design and graphic design belong to retinal perception. It is mainly through the visual channel that users approach products. Consequently, an ability to draw makes the designer appear as a specialist in forms, even more, as a cultural representative of industrial production. This idea is reflected in the profession's occasional claim to humanize technology. In this way, design becomes a process in order to lend the products of industrial civilization a certain visual quality a posteriori. That approach signalizes weakness, and it fixes design as something at the margins dealing with surface and appearance, without conceding how necessary it is. Users automatically assess products as soon as they leave the factory, and the designs only take on their identity in this social field where such assessments are continuously made.

People say: the design of a product is good, it is practical, or it is out of date; it is attractive, boring, well made, good fun, or useful; it will last a long time; it is playful, ephemeral, expensive, pleasant, cold, typically British, and so on. We talk about products in terms of assessments that exert a strong influence on the way we use them.

Accepting this premise, the designer can be seen as a specialist dealing with assessments. That is quicksand. However, assessments do express the essence of design, which is diffuse, indeterminate, and fragile on the one hand, but creates tangible realities on the other.

The problem of design as a search for identity—particularly a stylistic identity—has been discussed repeatedly in Latin America. It has been possible to create an identity in the field of literature, through the work of writers and poets like Julio Cortázar, José Lezama Lima, Octavio Paz, Mario Vargas Llosa, Garbriel García Márquez, Carlos Fuentes, Mario Benedetti, Juan Ruffo, Jorge Luis Borges, and Pablo Neruda[1]. These writers are part of world literature. They probably did not set out to create a Latin American identity. They wrote their works, and that was enough. If there ever was a Latin American "soul" (*latinoamericanidad*), it revealed itself in participation in the broader literary discourse. As long as design is not practiced comprehensively and to a greater

---

[1]   These are some key writers and poets who have defined Latin American modernity. It is relevant to note that it is a gender and ethnically homogeneous group.

extent in Latin America, the search for an identity through design will be like the search for a chimera. Identity is not the realization of a potential that lies hidden somewhere in the lower depths of what is believed to be a Latin American character.

It has been suggested that identity should be sought in the past and that formal regional codes should be revitalized or that stylistic elements dug out of pre-Colombian epoch should be applied to design. That seems to me a questionable undertaking. Identity is not in the past; rather, it needs to be created. How could a pocket electronic calculator or a water tap be designed with formal elements of the Mayan culture? Nostalgia is not an effective way to prepare for the future.[2]

It has to be said that in Latin America there is no deeply rooted design discourse.[3] This is one cause of the difficulties that emerge when design is introduced to companies. The existence of a design discourse, in Italy, for example, can be regarded as one of the reasons for the success of Italian design. From this it may be concluded that the future of design in Latin America depends not so much on the relatively small professional group of designers but on other players in industrial production.

A brief story will help to underpin that statement. Douglas Davis is not talking about Latin America or other peripheral countries, although his statement applies to them as well. He wrote:

Democrats and Republicans are engaged in closing the gap in foreign trade. . . . They blame our commercial tactics and strategies for our difficulties instead of blaming our unattractive coffee pots and—until very recently—ugly cars. The cruel truth is that American consumer products are, with a few exceptions, stylistically meagre, not very attractive, and poorly put together. In a decade of design, when investigations show that sophisticated buyers, both foreign and domestic, rate the quality of design on the same level as the price, such a state of affairs constitutes a costly error.

Japan, Italy, West Germany, and Great Britain pay great attention to design quality. The governments of these nations spend substantial sums of money on design education, establishing competitions and contests, prizes, design centers. In Japan, design competitions are analyzed by the media with the same fervor that we dedicate to baseball games and movie stars. On this point, our news media are frugal in disseminating information on design; we usually function at the level of gossip or fashion, and therefore reveal the

---

[2] That statement should not be understood as turning away from history. On the contrary, greater familiarity with Latin American and pre-Hispanic civilization will enable us to understand the present better, and from there look into the future.

[3] There is a difference between design discourse and design. A discourse as an ever-present background makes design an everyday transparent reality in which one lives. During the past three decades, the production of academic papers about design has grown considerably as a result of the master's and doctoral programs, as well as the bundle of measures and activities that have settled and crystallized around design. Design has become an object of events.

same superficial notion to our business leaders relating to the significance of inventing and perfecting a product.

The period of mass production has ended. The new frontier, with its unprecedented technologies, will permit, and even require, exuberant design strategies. If we do not wake up, our debt to the rest of the world will surely increase with profound social, political, and personal consequences for all Americans.[4]

That is a cool dismissal of the United States that once—in the 1950s—had a leading position in design. During the 1973 oil price crisis, the United States lost that lead and has been trying to regain lost ground since the mid-1980s.

Design will play a dominant role in the economy in the next century. A country that claims to be an actor and not a mere spectator will have to make design a pillar of its technological and commercial activities. Entrepreneurs and politicians in some Asian countries have understood that. There is nothing to prevent the lesson from also being learned elsewhere. Resignation is the state of anti-design.

[4]   Douglas Davis, "Design Gap - Not a Trade Gap," *The New York Times*, Col. 1 (August 12, 1988):
        27.

Design discourse in Latin America at the end of the 1940s/beginning of the 1950s was as up-to-date and on par with the design discourse as in European countries devastated by the Second World War and dedicated to reconstruction after twelve dark years of fascism.

It would be revealing to research and make more evident this discourse and at the same time to show how, counter to historical expectations, the understanding of the project of modernity was reimported from Latin America into Europe after the Second World War.

From European or American perspective, one can observe a tendency of considering modernity in Latin America as something second-rate or, at best, as a kind of "compensatory alternative" to the negative phenomena of industrialization: alienation, loss of communitarian life, destruction and devastation of the natural environment, and increase of social cleavage between rich and poor.

In this picture of the "peripheral" countries, post–Second World War crafts and craft design exerted a particular attraction because it could be taken as the characteristic of these countries in respect of design. However, this is a reductivist vision. Worse, imposed on these countries—who were allegedly at this time supposedly not compatible with the rationality characteristic for the modern development process—it produced externally a fetishism for "craft" and, in the countries themselves, defensive reactions alluding to a "hidden soul." This reaction is as old as the beginning of the industrialization process and expansion of Enlightenment, confronted by romanticism with its defense of emotions and disconformity with rationalism.[1]

Instead of looking for local stilemi as outstanding feature of the identity of Latin American design, project activity should be interpreted as activity with much broader aspiration. Design in dependent countries poses a fundamental, perhaps uncomfortable background question: How should one contribute to a reduction of dependence?

This paper, written originally as a contribution to the catalogue for an exhibit held in Munich in 2017, and focusing on the asymmetrical power relations involved, tries to take a much more differentiated view of the relation between Latin American design and design in Europe and the United States since the 1930s. It pushes toward a position that sees design as a possible moment of resistance and therefore to be a step, even if a very small one, to reduce heteronomy and to increase genuine autonomy.

---

[1] I quote only several examples of the rich literature on postmodernism and its historical origins: Peter Bürger, *Ursprung des postmodernen Denkens* (Weilerswist: Velbrück Wissenschaft, 2008); Perry Anderson, *The Origins of Postmodernity* (London; New York: Verso, 2006 [First edition 1998]); Heinz Paetzold, *The Discourse of the Postmodern and the Discourse of the Avant-Garde: A Series of ten Lectures concerning the Link between Social Philosophy and Aesthetics under the Conditions of Postmodernity* (Maastricht: Jan van Eyck Akademie, 1994).

17 BETWEEN FAVELA CHIC AND AUTONOMY

Design in Latin America

(2018)

Theories do have political implications.[1]

From a Western perspective, it often seemed obvious to read the development of design in so-called periphery countries[2]—once condescendingly dubbed "developing countries"—as the same process as the development of "the centre," just staggered by several decades.

Such a simplifying view, however, fails to do justice to reality. The reflections below are intended to offer a differentiated picture of the relationships, ranging from complex to antagonistic. Rather than seeking to present an encyclopedic overview of design in all Latin American countries, the aim here is to elaborate a characteristic profile by examining selected but significant data.

Although this sketch primarily relies on material dealing with the development of design in Latin America since the 1930s,[3] possible parallels can also be identified, given the structural similarities, with Asia and Africa and their former colonial countries. The region offers revealing material illustrating the decisive influence of the Western economic—and political—decisions on the ups and downs, advances and setbacks, in design's development in the periphery. These countries, each in their own context, are characterized by a dependence created by asymmetrical power relations, if not crude dominance

Published in English in *Flow of Forms—Forms of Flow—Design Histories between Africa and Europe*, eds. Kerstin Pinter and Alexandra Weigand (Biefield: transcript, 2018), 124–33.

[1]   David Graeber, *Toward an Anthropological Theory of Value—The False Coin of Our Own Dreams* (New York: Palgrave, 2001), 88.
[2]   The term "periphery" is used in the sense of Immanuel Wallerstein's world-system theory referring to the asymmetrical value transfer from dependent countries to core countries. According to this mode of analysis, Latin American countries belong to those countries with semi-periphery status. As early as the 1960s, it was above all social scientists from Latin America who analyzed the relationship between core countries and the periphery as part of their work on dependency theory. In contrast to advocates of dependency theory, Andre Gunder Frank (1929–2005) supported the radical idea that underdevelopment was not a state, but was a condition produced on a daily basis. The topic of development, and, with it, the political role of periphery countries, was first officially voiced in a speech by President Harry Truman in early 1949. At that time, he talked of "underdeveloped" countries, a disparaging term rightly avoided today. His speech marked the official start of the Cold War.
[3]   Fernandez and Bonsiepe (eds.), *Historia del diseño en América Latina y el Caribe*.

relations, which forges the framework for practicing the complex of activities involved in design in Latin America.[4]

A set of criteria comprising the following six categories can be used to sketch the present state of design:

— Consolidation of the professional practice of design
— Training available for design (provision of design courses)
— Integration of design in business policies on the local level
— Funding for design within government programs for industrial development (establishing design centers, integrating design into institutes for innovation research)
— Design discourse (publishing design journals and books, media presence)
— Design research (first and foremost in the form of design history)[5]

Against the background of this structure, the following provides an outline of the development of designing material artifacts in Latin America and the Caribbean, including the role of craft production, for example the design approach known as "nativism," though without the political and restorative connotations recently informing this term.[6]

In 1949, the *Boletin del Centro de Estudiantes de Arquitectura*, the student journal of the architecture department at the University of Buenos Aires, published a short piece by Tomás Maldonado entitled *El diseño y la vida social*. In this article, he put the case from the visual arts perspective for revising the cultural importance of designing industrial products. Rather than design being a kind of "second rank" applied art, it was an independent field in a technical and industrial civilization and formed an integral element

---

[4]   Two sources are mentioned here from the wealth of economic statistics available: the e-magazine QUETZAL–Politik und Kultur in Lateinamerika (www.quetzal-leipzig.de/themen/globalisierung-und-regionalisierung/die-stellung-lateinamerikas-in-der-weltwirtschaft-1909 3.html) with its article on the position of Latin America in the global economy (in German) by Peter Gartner, and the mOXLAD (htt p://moxlad-staging.he rokuapp.com) database and blog with figures, for example on economic indicators in Latin American countries, presented visually as charts and graphs (accessed June 26, 2017). The data is largely drawn from studies conducted by the United Nations Economic Commission for Latin America and the Caribbean. These figures show, just to give two examples here, that over a period of 110 years—from 1900 to 2010—the unit price for imports in Argentina rose twelvefold (from 50 to 596), while the unit price for exports only increased sevenfold (from 51 to 366). The figures for Brazil show the same tendency: the unit price for exports increased by a factor of just under nineteen (22 to 425), while the unit price for imports rose by a factor thirty-six (25 to 899). This widening gap is not only caused by economic factors, but these figures nonetheless are very revealing.

[5]   Gui Bonsiepe, "Paesi in via di sviluppo: la coscienza del design e la condizione periferica," in *Storia def disegno industriale 1919-1990—Il dominio del design*, ed. Raffaella Ausenda (Milano, 1991), 252–69.

[6]   During the US presidential election in 2106, the term "nativism" was given a definite right-wing spin—a direction the term had not had before. In the context of the resistance movement of native peoples, nativism had previously meant a turn toward preserving and reviving own indigenous cultures. As a radical stance, "nativism" can be read as an intensified reaction to immigration which, in its most extreme form, calls for a halt to all immigration.

in the process of modernization.[7] Maldonado's article is regarded as the very first in Latin America to explicitly address the subject of industrial design and clearly elaborate the distinction between art and design. He also touched on the controversial topic of whether design can be viewed as an artistic activity, an issue often surfacing in a perennial debate, and unequivocally took a stand against the tendency to read design as applied art—a questionable, if not outdated, concept derived from middle-class notions of art removed from strategic utility benefit and an industrial sector driven by profit and utility.

One may wonder how a country in Latin America came to call for endowing industrially produced materials and everyday semiotic artifacts with a practical and theoretical dignity at a time when Europe was still suffering from the consequences of devastation under fascism. The reason, though, can be found in the emigration/forced migration of artists, writers, composers, philosophers, and scholars from diverse European countries, including those from Spain fleeing the Franco regime as well as Italians, French, Germans, Austrians, Hungarians, Poles, and Czechs, who were victims of a prevailing political and racial persecution in their own countries.[8] These migrants brought with them writings and documents facilitating the access of Argentina's young artistic avant-garde to previously unknown material, in particular documents on the Russian Constructivists who explicitly addressed and explored the connection between politics and the design process. The young avant-garde, in turn, wanted to overcome their isolation and expand their sphere of activity by focusing on industrial products and their role in everyday life in the belief that it was feasible to link a technical and industrial revolution with a social revolution.

The birth of industrialization in Latin America was shaped by two major global events: the worldwide economic crisis in 1929 and the Second World War. As the main combatant nations increasingly focused industrial production on weapons, they cut back or halted the export of industrial products; in some cases, exports were also banned for political reasons. In the area of furniture production, the architects in Argentina,[9] who had been developing designs for this manufacturing sector since the 1930s, laid the foundation for the term "industrial design."[10] Yet the industrial production of complex goods, for

---

[7]   Maldonado, "El diseño y la vida social," 7–8. Republished in Carlos Méndez Mosquera and Nelly Perazzo (eds.), *Tomás Maldonado—Escritos Preulmianos* (Buenos Aires, 1997).

[8]   Tomás Maldonado 1989. Entrevista. Flash Art, 151. Spanish version in Maldonado 1997, 117–27.

[9]   These were architects born and trained in Argentina as well as those coming to live in the country. Two of three designers of the iconic BKF "butterfly chair" were Argentinian (Hardey and Kurchan), while Bonet was from Catalonia. They intensely followed developments in modern architecture in Europe—primarily Le Corbusier and Gropius as representatives of modernism. The first edition of the influential Argentinian cultural journal SUR (South), founded by Victoria Ocampo in 1931, published writings by Gropius on the Total Theatre. When Victoria Ocampo visited Gropius in Berlin in 1929, he showed her works of modern architecture in the city. See: Silvia Fernández, *Señal Bauhaus: Mujeres en el diseño en Argentina* (La Plata: Ediciones Nodal, 2019).

[10] Martha Levisman, *Diseño y producción del mobiliario argentino 1930-1970* (Buenos Aires: Archivos de Arquitectura Contemporánea Argentina, 2015). In terms of the reception of modern furniture, the new tubular steel furniture was initially praised for its comfort and functionality,

example agricultural machinery, which integrated industrial design, did not start until the 1950s.[11] Given that the profession of industrial designer did not yet exist in Latin America and there were no official programs for design, the industrial designs were developed by engineers, technical draughtspersons, and people who were self-taught.[12] The first generation of designers—in this case, working with the entire complex of activities involved in the design process[13]—comprised those from the fields of architecture, mechanical engineering, artisan crafts, and art. They faced the task of transforming design skills into an activity relevant socially, industrially and economically, and hence delineated the profile of a new profession.

The early 1960s saw countries such as Brazil and Argentina launching training programs for design influenced by institutions in Europe and the United States—principally by the Ulm College of Design, London's Royal College of Art, the IIT Chicago, and New York's Pratt Institute.[14] At this time,

    as well as for its hygiene and cleanliness. But it was also viewed as being an element alien to the popular tradition, if not condemned from the outset as "Bolshevist and Jewish." Carlos Mazza, "Tradicional y moderno en la producción de muebles en Argentina: 1930-1950—Equipamientos para hoteles de turismo y oficinas administrativas," *Registros* 8 (2012): 52–68.

11   The company policy—today known as the corporate identity program—of the Argentinian state oil company YPF, founded in the early 1920s, was inspired by the model of corporate identity created by Peter Behrens for AEG. Javier de Ponti (ed.), *Diseño, identidad y sentido— Objetos y signos de YPF 1920-1940* (La Plata: Dicere, 2012).

12   From 1951 to 1953, a design course was offered at the Instituto de Arte Moderna at the Museu de Arte Moderna São Paulo (MASP). The director of the institute was the Italian art historian and critic Pietro Maria Bardi, who emigrated to Brazil in 1946. However, it was not possible to gain support from the industrial sector for this precursor institution. See Ethel Leon, *IAC primeira escola de design do Brasil* (São Paulo, 2014). In 1963, the Escola Superior de Desenho Industrial (ESDI) in Rio de Janeiro officially launched a complete, fully developed university program for industrial design and visual communication which explicitly referenced the program at the Ulm School of Design (HfG Ulm), and had a considerable influence on design training in Brazil. Similarly, a program for industrial and communication design was also launched in Argentina in 1963 at the National University of La Plata, which also drew heavily on the ulm program. The crucial role played by the Olivetti company, represented by Adriano Olivetti (1901–60) and Roberto Olivetti (1928–85), has as yet only been partially documented and researched. The company had branch offices in São Paulo and Buenos Aires. Hence, three cultural spheres provided important influences on the development of design and design education in Latin America: English-speaking (the United Kingdom and the United States), German-speaking (West Germany and Switzerland), and Italian.

13   Spanish use the term "diseño industrial" for product design and "comunicación visual" for graphic design. Since the 1990s, with the expansion of the term "design," the general term "designer" is sometimes used. Under the influence of the multitalented artist and activist Aloísio de Magalhães (1927–82), Brazilian Portuguese has adopted "industrial design" as an umbrella term covering both product design and—contrary to international usage—visual communication (programação visual). This is because the Portuguese word "desenho" means sketch/drawing and so does not reference the design components evoked by the English word "design." For many years, there was an ongoing disagreement over whether desenho industrial, meaning "technical drawing," should not be expanded to include industrial design. To avoid confusion with "technical draftsperson" as a profession, the industrial designers call themselves by the abbreviated term "designer" from "designer de produtos."

14   The publications by Tomás Maldonado had and still have the greatest influence. The meticulous and painstaking work by Laura Escot, *Tomás Maldonado: itinerario de un intellectual técnico—The itinerary of a technical intellectual* (bilingual edition) (Buenos Aires, 2007), presents a detailed account of Maldonado's development. In his work as an artist, Maldonado, cofounder of the Asociacion Arte Concreto-Invención in 1944, is regarded as belonging to the Argentinian avant-garde movement. He took a practical and theoretical view of the culture

design centers were also founded, for example the Industrial Design Research Centre (CIDI) in Buenos Aires in 1962. These fulfilled a similar function to the design centers in Europe[15] with the aim of educating society at large, and, as such, were comparable to the *Gute Form* (good design) movement at that time in West Germany and Switzerland. These design centers undertook such tasks as organizing exhibitions of industrial products, holding competitions, presenting lectures, publishing materials, and establishing a library for the field where the current leading design journals could be consulted. Industrial engineers in leading positions played a crucial role in integrating industrial design into development programs, though in general they have received little acknowledgment for their work.[16]

The term "industrialisation" certainly had positive connotations at that time, since the social and ecological impact of industrialization processes on the environment either had not been registered or were disregarded as collateral damage of relatively little importance. In general, development was viewed, beyond any shadow of a doubt, as a thoroughly worthwhile goal, with industrialization rightly anticipated as producing an increase in aggregate income. However, the question of who would ultimately benefit from the fruits of industrial development—apart from the undeniable advantages of a modernized infrastructure and public services—evaporated in an atmosphere of generalities and vagueness. Even today, in comparison to other regions of the world, Latin America still has the most pronounced income disparity between its various social groups.[17] In popular imagination, high-ranking politicians liked to be seen with their hands grasping a control wheel on some pipeline in an oil refinery or chemical plant or inspecting a sheet metal factory. The smoking factory chimneys and intermeshing cog

of technical industrial objects and design and regarded the training of designers as a new challenge. In 1954, the same year he started his tenure at the Ulm School of Design, he painted his last work in the twentieth century but, after a long break, returned to artistic work in 2000 in Milan. His new works have been shown at venues including galleries in Berlin and Lugano as well as the Museo Nacional de Bellas Artes in Buenos Aires. Although Maldonado's crucial role in cultural development has been widely acknowledged in Italy and Latin America, such appreciation has not yet been forthcoming in Germany, despite his leading role at the Ulm School of Design and his thirteen years there as a teacher, vice chancellor, designer, and author.

15 For example, the Design Centre in London opened in 1956, based on the Council of Industrial Design founded in 1944 during the Second World War.

16 The three examples of engineers here all played a key role in promoting and consolidating industrial design. In Chile, Fernando Flores (1943–) introduced industrial design as a subject in a technological research institute in 1970; in Argentina, Basilio Uribe (1916–97) founded the Design Centre in Buenos Aires; and in Brazil, Lynaldo Cavalcanti de Albuquerque (1932–2011), in his role as president of the National Council for Scientific and Technological Development (CNPq, 1980–1985), called for integrating industrial design into policies on industrial development. As a rule, designers are not admitted to these institutions—a situation which can be attributed, on the one hand, to a lack of information and, on the other, to the widespread public image of designers as specialists for creating attractive product shells and differentiating side issues trivialities, or as mavericks not used to working in teams (designers in love with creativity and outsiders unwilling and incapable to work in teams).

17 Goran Therborn, "Dynamics of inequality," *New Left Review* 103 (2017): 67–85. He references the Human Development Report 2015, which also contains an inequality-adjusted index compiled using the Gini coefficient.

wheels symbolized industrialization's potential. Even though individual voices critiquing consumer society could already be heard in the second half of the 1960s, the doubts expressed about design and the need for a critical approach increased significantly in volume after the 1990s. This development came in the wake of the concept of design unraveling, its absorption into marketing, as well as its one-dimensional alignment to the glorified market. In 1968, at a design exhibition in Rio de Janeiro, for example, students at the Escola Superior de Desenho Industrial (ESDI) presented an installation critical of design described by a Brazilian design historian as follows: "Packaging from industrial products were displayed on a long table surrounded by 10 of the Series 7 chairs designed by Arne Jacobsen in 1952. The installation was crowned by a vacuum cleaner whose suction tube was replaced by a broomstick."[18] This installation can also be read as part of the wave of student protests in 1968, inasfar as it exposed the antinomies between modern design and industrial and social realities, but it did not advocate overcoming these contradictions by rejecting industrial production and returning to vernacular design.

While development policy in the 1970s focused on the manufacturing industries, exemplary designs were created, above all, for the public sectors— the industrial design of street furniture and visual communication for urban guidance and orientation systems (in São Paulo and in Buenos Aires)—which can certainly stand international comparison. The industrial development policy promoted by various governments was flanked by multilateral projects especially the United Nations Industrial Development Organization (UNIDO).[19] Here, the aim was to use import substitution to improve the foreign trade balance of periphery countries, support local manufacturing industries, satisfy demand on the domestic market, and develop those products meeting local requirements which were unavailable on the international market. This policy was directed against the traditional role aligned to the periphery as an exporter of commodities in the form of wheat, soya, meat, wood, copper, iron ore, crude oil, rare earth elements, coffee, spices, palm oil, and cotton, that is, "designless" products. The economic policy of extractivism or neoextractivism[20] fitted perfectly with the periphery's role imposed since colonialization as a supplier of raw materials—a farce of a type of development that left—and leaves—wastelands behind. The programs integrating industrial design were directed against this constellation of international trade derived

[18] Ethel Leon, "Design em exposição," PhD thesis, School of Architecture and Urbanism (University of São Paulo, 2013).

[19] In 1973, the author of this chapter published a UNIDO report in Vienna instigated at the initiative of the International Council of Societies of Industrial Design (ICSID). The report then served as the guidelines for programs supported by the UNIDO (Gui Bonsiepe 1973, Design for industrialization, Vienna). At that time, he had been working for five years in Chile, and had extensive experience in this area. Just a few weeks before the putsch, he received the request to write the report and had just six weeks to complete it. This report served as a point of reference for the Ahmedabad Declaration on Industrial Design for Development, published in 1979.

[20] The dominant media rarely report on resistance from affected local populations against the consequences of such policies, and especially against strip mining, which robs them of the basis of the livelihood.

from colonial dependencies. Bilateral programs were inevitably permeated by political interests. For example, the Alliance for Progress (1961–70) funded and designed by the United States also sought to block the spread of the Cuban revolution in Latin America.[21]

At this point, one should mention in passing the appropriate technology movement, promoted especially in the United Kingdom in the 1970s, and the approach propagated under the rallying cry of "Small is beautiful"[22] calling for technical demands on manufactured products to be so simple that they could be made in a village smithy. This movement resonated particularly with students who, through their attraction for the supposed simplicity of rural life, had strong reservations about urban life and industry, if not capitalism in general. Similarly, California's New Age movement should be also mentioned here with its attempts at developing alternative lifestyles—virtual enclaves in an industrialized world. This movement was characterized by its appeal to the American way of life's strong tradition on self-reliance (e.g., "Do-it-yourself"). Here, the *Whole Earth Catalogue*, published in several editions since 1968, provided nothing short of an encyclopedia of self-made designs.

In 1976, the theme of design in the so-called developing countries—until then only a footnote at best—was given an international profile by the *Design for Need* conference held in London at the Royal College of Art.[23] The decision to address this sociopolitical complex of themes was necessitated, to an extent, by the wave of student protests around the world in 1968. The debate revealed the discontent over the role of designers in a consumer society as accomplices in accelerating the circulation of commodities. The question was raised of how it would be possible to satisfy needs beyond the markets, a point specifically addressing the situation of periphery countries. At that time, these were bundled together under the collective name "Third World," today a concept entirely lacking in substance. Under neoliberalism, as the present phase of capitalism is known, there is little left of the tangible unease at the congress and the underlying mood of concern. The critique of the capitalist mode of production only lives on, at most, in the idea of sustainable design, even if the term "sustainability" cannot be saved from the danger of being instrumentalized as a mere label leaving everything just as it was.

---

[21]   This also included using the Peace Corps, even though these mostly young volunteers need not necessarily have been aware of the fact since, officially, it was promoting an agenda for peace. However, as has been noted, "The main reason that John F. Kennedy proposed this idea was to ultimately halt communism." Lindsay Boshak (after 2008), *The Peace Corps*, http://www.cold war.org/articles/60s/PeaceCorps1960.asp (accessed June 26, 2017).

[22]   Ernst Schuhmacher, *Es geht auch anders—Jenseits des Wachstums* / Technik und Wirtschaft nach Menschenmaß (Munich, 1974).

[23]   The lectures were published in Julian Bicknell and Liz McQuiston (eds.), *Design for Need* (London: Pergamon, 1976). The two UNIDO reports mentioned earlier deal with the role of design for industrial development, but do not address craft production in any greater detail. The UN agencies promoting design education are organized in specializations by their main focus, including technological and industrial development, where engineers and business economists are primarily responsible, and cultural affairs, involving the human sciences. In this categorization, craft production falls under culture (assigned to the UNESCO). In line with this division of labor, the approaches of the design programs are also different.

A topic often addressed in the discourse of design in Latin America revolves around the identity of local design—a question which can be explained as a reaction against the influence of hegemonic[24] designs of the core countries as well as the attempt to counter this influence with something of their "own," indigenous and familiar. The turn to the "own" signalizes the desire to challenge the occidental patterns of thinking and axiologies internalized in the course of the history of colonialization. Since the late 1990s, the body of critique linked to this has been termed "decoloniality."[25] A similar interest in the autochthone was already evident in the 1960s in the visual arts, identified under such terms as "indigenism," "campesinismo" (rural and peasant motifs in art), and "obrerismo" (subjects in art rooted in the world of work).[26]

Countries such as, for example, Mexico, Ecuador, Peru, and Colombia are renowned for their highly differentiated tradition of craft products, each with its own individual material, formal, and chromatic expression. In these products, as is well known, the functions of design and production form a single unit. When industrial designers adopt craft designs and deliver products then made by craft workers—primarily by women—this brings with it the danger of using these workers purely as labor force rather than fostering their innovative abilities. Moreover, the rich stock of forms for craft products is linked to a traditionally rather narrow range of products. By romanticizing the notion of "design," it then becomes possible to present these products as authentic design informed by a hypostatized Latin American essence. The enthusiasm with which this option of ontological essentialism is sometimes pursued can be linked to an anti-technological Romanticism with its aura of the supposedly genuine and exotic, intact and unspoilt. Through the appropriate marketing, this creates resonances—not least as "favela chic" offered at art trade prices— in the design boutiques of core countries glutted with industrial products.

The turn to designing less complex products that are simple to manufacture can be explained as a result of Latin America's deindustrialisation and the break with a policy of import substitution and domestic market production which, from the 1960s to the late 1970s, provided industrial design with the foundation and framework for development. The movement to promote industrialization, especially for small and medium enterprises, gradually lost momentum as, among other things, programs specifically developed for periphery countries in the euphemistically entitled Washington Consensus

[24] On the difference between hegemony and dominance, see Vivek Chibber, *Postcolonial Theory and the Spectre of Capital* (London: Verso, 2013).

[25] Pablo Quijano, "Notas sobre la teoria de la colonialidad del poder y la estructuracion de la sociedad en America Latina, Papeles de Trabajo," *Centro de Estudios Interdisciplinarios en Etnolinguística y Antropologfa Socio-Cultural* 19 (2010): 1–15.

[26] For more information on this, see the detailed research by Argentinian art historian Andrea Giunta on the fine arts in Latin America: Andrea Giunta 2008, Vanguardia, internacionalismo y política, Buenos Aires. Chapter 7 on the Aporias of Internationalism, in particular, allows parallels to be established to the development of design. Since Latin America's deindustrialization process was intensified in the 1990s, the spectrum of the design problems tackled has narrowed. In the context of the de-semantization of the designer concept, product design has largely limited itself to small design objects made by craft processes and using autochthone or recycled materials—lamps, fruit bowls, stools, and utensils for writing and desks.

(1989) fueled a wave of privatizations and a concomitant plundering of public resources—a development particularly pronounced in Latin America. In the wake of privatisation and opening up the economy to foreign capital (buying out local companies, first and foremost banks and phone companies) design was increasingly in demand as an instrument of branding and "corporate identity."[27] At the same time, with design seen as a soft option compared to such subjects as engineering or medicine, there was a glut of design courses—with nearly 600 courses covering all manner of areas offered in Brazil alone at state and private colleges. Although the ministries of education are responsible for drafting and monitoring the conditions for setting up design courses, these requirements are lax and far less stringent than those imposed on other study programs. The introduction of postgraduate degrees and doctoral programs in design led to a growth of publications as well as largely academic and traditional research in the field, whereby it is still open as to whether this will further deepen an already-profound gap between design practice and the academic ivory tower.[28]

As is evident from two volumes edited by Nadir Lahiji, the question of design's utopian potential alludes to a problem which, over the last twenty years, has almost entirely disappeared from design discourse, except in architecture.[29] This may be due to the discourse in the field of architecture being able to draw on a long tradition, in contrast to other areas of design. Certainly, I am not aware of anything comparable in the fields of industrial design and visual communication. The colonial past of Latin American countries does enable design or—in a larger sense—the complex of activities involved in the design process to be taken as an example the multilayered and as-yet suppressed attempts at emancipation in such areas as politics, the financial, economic, and industrial sectors, and everyday culture. In the present decade, it is no coincidence that the 200th anniversary celebrations of political independence in most countries in Latin America have brought into debate the topos of a "second independence," that is, redeeming the promise of modernity. American philosopher Susan Buck-Morss has characterized

---

[27] There is no updated and comprehensive data available on the designer labor market. Based on personal estimates, weakening product design in Brazil is partially balanced by, for instance, game design.

[28] In a study of twenty-five countries, the following four indices were compared and listed in a ranking: publications in leading scholarly journals; number of registered patents; extent of investment in R&D (research and development); and the number of doctorates in the sciences and engineering. The Latin American countries in the study included Brazil and Mexico. Brazil, for example, ranked seventh for doctoral degrees, did not generate any sizeable number of patents, came twenty-fourth for publications in high-profile scholarly journals, and was not investing any sum worth mentioning in R&D. Source: Stephan Theil, "Por que a Alemanha ainda produz tanto?," *Scientific American*—Brasil 11, no. 126 (2013): 39–43. Whatever one's reservations against benchmarking, these findings are revealing and an indication of how universities are isolated from the rest of society.

[29] Nadir Lahiji (ed.), *Architecture Against the Post-Political—Essays in Reclaiming the Critical Project* (London: Routledge, 2014); Nadir Lahiji (ed.), *Can Architecture be an Emancipatory Project?* (Winchester: Zero Books, 2016). The publication by Alberto Toscano and Jeff Kinkle (eds.), *Cartographies of the Absolute* (Winchester: Zero Books, 2015), follows a similarly critical artistic activism.

the modernity project as: "the dreamworlds of modernity—political, cultural, and economic—are expressions of a utopian desire for social arrangements that transcend existing forms."[30] As is well known, the mere mention of the word "utopia" and the possibility of changing existing social structures is an anathema for postmodern ideas and its two political variants of neoliberal conservatism and post-structuralism.[31] When it comes to the sociopolitical role of design in the periphery, that role can be read from the questions which prove so difficult to answer: Does design—however understood—contribute to weakening hegemonic relations between the center and the periphery? Does it harbor the potential of emancipation, of dismantling conflict-laden ruling conditions, of limiting heteronomy? One step here, for example, would be to focus on solving local problems—since the only things that investment funds looking for worthwhile objects in the periphery ignore and leave untouched are those countries' problems. Starting from local design problems is far from cloistering oneself away—and given today's networks, any such plan would certainly be an illusion. Instead, it facilitates the elaboration of design solutions that could sensibly be expanded from the local to other contexts as well. If design is to be able to play a role as an export factor for periphery economies at all, then it would have to be aligned with existing standards. And there, the trade agreements and ISO standards, as they stand, already act as a filter in the economies of core countries to neutralize undesirable competition.

It is no secret that design activities in the periphery, as far as they are directed to counter a role limited to the export of commodities and are unwilling to adapt to the dominant media, can hardly be harmonized with geopolitical hegemonic interests. In this sense, far from the potential of design being at the mercy of exaggerated or obstructive practical necessities, it is explicitly exposed to a conflict situation.[32] Here the design process can contain—excluding any heroic idealization—a moment of resistance and be a step, even if a very small one, to reduce heteronomy. The quote at the start of this chapter can thus be amplified to read: Practice also has political implications.

[30] Susan Buck-Morss, *Dreamworld and Catastrophe—The Passing of Mass Utopia in East and West* (Cambridge, MA: MIT Press, 2002), xi.

[31] Hal Foster, *Recodings—Art, Spectacle, Cultural Politics* (Port Townsend: Bay Press, 1985), 121.

[32] As yet, no attempts have been made to pursue the option of opening up the constrained sphere of design by linking design activities to technical and scientific innovation.

(c) The Question of Difference

*This text was presented at the first International Congress of Design held in
a Latin America, in Mexico, in 1979.*

*One could expect that professionals from the United States would be
present at this occasion. It should cause no surprise that an approach to
design from a political perspective could cause lifting of eyebrows indicating
negative assessments. Against a reductionist interpretation of design as
an instrument to make the cash register ring—and not more than that—it
became apparent just how difficult it was to make different (i.e., not only
commercial) approaches to design activity comprehensible.*

*An interpretation of design as a service to the client who finances a project
fits into the dominating pattern. But the term "client" can be understood
covering a broader domain. "Client" can be understood as the user of the
product and accepted as dominating objective and concern for business (for
designers anyway).*

*The expansion of User experience (UX design) is a symptom for the
reorientation of business thinking (and design) refocusing on the domain of
the user pushed into the background when there prevailed a one-dimensional
interpretation of design as tool for one objective fixed by the (financing)
client—design as profit-maximizing tool, as tool for making the cash register
ring. However, there is not ONE design, there is not ONE client who finances
a design; there are contexts, and thus there are different clients. Design is
in many ways a struggle to give weight to these "other contexts" and other
clients.*

18 Between Marasmus and Hope

(1979)

Searching a title for my contribution to this conference, I looked for a combination of concepts which might catch the essence of the present situation of the "Dependent Underworld." Thus, I came up with the title: "Between Marasmus and Hope." I have not selected these concepts with the deliberate intention of shocking, because I am not an addict of verbal radicalism—that good fellow traveler of conservativism. It is not the concepts, but the reality which is shocking. Weakened in their material base of subsistence, subjected to an apparently unlimited drain of resources, suffering a process of dependent accumulation and value transfer of astronomical dimensions, entangled in a multinodal web of control, these countries, which, for lack of a better term, have been named the "Third World," shelter a dialectical complement: the hope to eradicate the stigma of poverty.

Pondering on the future of our profession, I wish to concentrate on its role in dependent countries. I use the word "dependent" without polemical intention, understanding this concept as the situation of certain countries whose reality reflects the movements of the dynamical center of the world economy.

I start with the hypothesis that there is a difference between industrial design in central countries and in peripheral countries. There is not one, and only one, industrial design; there are at least two industrial designs. The reality of the periphery can be described under many aspects as alarming and indeed as catastrophic. It is sufficient to recall the food, housing, health and education shortage, suffered by a great part of the population of this planet— an emergency that industrial design as a technological activity must help to overcome by coming up with concrete proposals.

Doubtlessly, the question of material culture, the question of the "Physique of Culture," is a crucial question for every society. On account of this, the designer as a "physicist of culture" is situated in a strategical point of the system of objects. This system nowadays, in its technical, functional, esthetical, and symbolical attributes, is almost exclusively predetermined in the center.

In response to this reality, industrial design in peripheral countries is sometimes interpreted as an extension of something created in the center and projected toward the Periphery. This expansionist model may occasionally be wrapped up in a rhetorical respect for local traditions and conditions. According to this view, industrial design constitutes a monolithic reality,

A paper presented at the XI Congress and Assembly of the International Council of Societies of Industrial Design, ICSID, Mexico City, October 14–19, 1979. Published in abbreviated form by Excerpta Medica 1980, Elsevier North-Holland. Revised version, translated by Gui Bonsipe.

admitting minor modifications within the conceptual and practical frame of reference pre-established and forged in and by the center.

Historically, the vision is based on an anthropological division between "civilized" and "primitive," on a political division between rulers and ruled, or empire and colonies (in spite of what has been considered as the decease of colonialism). The underdeveloped countries, which with diplomatic courtesy are named "developing" or "less developed" countries, constitute those groups, which the Enlightenment in the eighteenth century called "primitives." But, there is a difference: the "primitive" of times past had an ambivalent character. For more sensitive and critical minds, the concept concealed the attraction of evoking recollections of a distant and lost state of happiness, as suggested by the etymological connotation of the term "primitive", that is, "near to the origins." The "*bon sauvage*" has been a creation, a dialectical product of civilization in a similar way as the present underdevelopment of the Periphery is a dialectical complement of the development of the center. This irritating component has been surgically extirpated from the binominal concept "development / underdevelopment," which indicates a one-dimensional realm of thinking. Memories of a different reality have been lost, though perhaps only temporarily, because the greening of countermovements, which do not accept development and industrialism as unquestioned panacea, would seem to indicate that those memories are recurrent.

I do not refer to the discussions about the limits of growth, but to the wide spectrum of alternative movements in which appropriate technology can be included, although with certain reservations. When transnational corporations begin to appropriate Appropriate Technology, it should be clear that Appropriate Technology, in spite of its attraction, is no guarantee for improving the living and survival conditions of those in need.

Opposed to the hegemonical interpretation of the center, we can identify the inverse snobbism of those who, fascinated by the exotic character of the Periphery, perceive in the same salvation, purity, authentic reality of design, alternative design, and the possibility of working for real basic needs. This tendency is rooted in the unease and even nausea felt by critical designers when faced with the type of projects offered by the center.

In order to understand the structural difference between peripheral and central industrial design, we must compare the origins of industrial design in those countries which at the time were industrialized, with the origins of design in peripheral countries during the past decade.

The good design movement ("*Gute Form*") at the beginning of twentieth century tried to correct the deformations and aberrations of the production potential by recurring either to technical rationality or the artist's subjectivity or inwardness. The first group appealed to the imperatives of standardization, of raise in productivity, and of "honesty in the use of materials" (*Materialgerechtigkeit*); the second group tried to cure the barbarisms of a ruthless industrialism by esthetical treatment. Both groups were occupied with product quality, which is verified according to three indicators: functional quality; esthetical or formal quality, and finishing quality.[1]

Nevertheless, the two opposed groups were the expression of a therapeutical movement for correcting a certain type of industrial development; they acted on an already-existing industrial base. Concerning their economic objectives, they were first and foremost organized to take over international markets. But—and it is important to emphasize this—the production problem had already been solved. This take-off of industrial design became still more obvious in the United States during the 1920s, when industrial design was explicitly assigned the role of an instrument in competition. Bottlenecks did not occur in the area of production, but in the area of distribution and consumption. This is the main cause for the difference between central and peripheral industrial design. In the Periphery, diversified industrial manufacturing industry is lacking.

Starting from a weak industrial infrastructure, the accent should be put on fostering productive capacity without losing sight of the objectives, that is, the *for what*? Development without precise attributes, without political imagination, that is, without clarity about the "matrix of use values" (*Matrix der Gebrauchswerte*), and without analysis of necessities, which should be satisfied by the artifacts, probably wouldn't help to raise the living standard of the majority of the local population.

Evidently, in a context of outer-directed industrialization, industrial design becomes superfluous. This strategy fosters a polarization of a world society in which the dynamical center becomes an insular pensioners' state resting on the Third World.

A policy aiming at the internal market renders industrial design anything but superfluous. But for the existing difference in take-off conditions, the role of the industrial designer and his qualifications and education must be different too. This is the main reason for the limitations to transfer experiences directly from the center to the Periphery. They must be evaluated for their relevance and be adapted.

What are the options for industrial design policy in peripheral countries? If only two centuries of industrial development have caused ecological havoc and the predicable exhaustion of certain nonrenewable, specially energy resources, the expansion of this model encounters physical limits. Taking average values, it is plausible to deduce that within one century the period of the quantitative growth model will come to an end. Consequently, the concepts of "development" and "industrialization" show symptoms of erosion.

We face here a dilemma: on the one side, the idea of industrial development must be perhaps radically reconsidered. On the other side, as Tomás Maldonado observed recently, one must avoid to get caught in the trap of the new apologists of poverty who recommend abstention from an industrialization

---

[1]   Obviously, after four decades later this short list of indicators needs to be extended among others by environmental quality, climate impact, and social relevance.

which benefits local majorities, allegedly because the resources are finite and would not be sufficient for all. This is the voice of the "haves" which is not convincing to the "have-nots." According to historical experiences, only on the material base created by an industrial system there will be a guarantee for a life on a higher-than-subsistence level for the majority of humankind. It is hypocritical to recommend a draconian curtailing of aspirations to satisfy majoritarian needs. On the other hand, copying the center's form of satisfying needs does not get us any further ahead. Therefore, it is not surprising that this state of things causes perplexity.

As far as I know, nobody has found a convincing and feasible answer to this dilemma, neither has alternativism. Generally, these countermovements flourish in niches provided by industrial society. Their viability depends on an industrial support structure, and, thus, their value as guide or reference point for design in the Periphery is limited.

Instead of "alternative technology" or "appropriate technology," I prefer the terms "local technology" and "local design." This term emphasizes the crucial point: that technology and design are created in the Periphery. This is the only way to break the vicious circle of technological dependence, that is, to end a situation in which the Periphery has neither vote nor voice in fundamental decisions affecting the material base of their societies.

A well-known utopist of the nineteenth century wrote that the economical periods differ less in *what* they produce than *how* they produce. From a designer's point of view, this statement must be revised. Without wanting to neglect the importance of the mode of production, it would seem that *what* is produced is equally important. This leads us to the idea of restructuring the material culture.

An even indulgent view at our environment is sufficient to reveal that this system of objects suffers from pathological deficiencies in functional, ecological, and esthetical terms. The culture of the shopping center is not OK, it is not even "almost OK." If industrial design didn't exist, this would be the best moment to invent it. What matters is to fill the design vacuum in the Periphery. Here design problems are obviously not problems of style and form. These at best have secondary importance, if any importance at all. Everyday life can dispense with these niceties.

How can this design vacuum be filled? Among others there are two possibilities: the informal and the institutional approach. When working for rural communities, that is, when making grassroots work with direct interaction of the users and producers, the noninstitutional way may be the best. The other way consists in availing ourselves of public bureaucracy.

Given the weak local industry, the support provided by public and semipublic institutions is of crucial importance for the implementation of a design policy. Governments could work as promoters and producers of technology in form of product designs. These might belong to those areas where the government later on is a buyer or credit source, for example equipment for schools, hospitals, urban areas, rural infrastructure, agricultural machinery, and so on.

The other no-less-important area to anchor industrial design are educational institutions. Sometimes they are accused of elitism and of being separated by a gulf from their social and industrial context. The first reproach loses validity if the characteristics were marked by excellent technical qualifications combined with social concern, and were not understood in the sense of a closed and exclusivist caste. The second danger can be reduced by linking thematically, didactically, and institutionally the education program of the designer with the social and technological context. For this reason, the acceptance of educational models and experiences of the center can be counterproductive. A critical reading of programmatical and didactical experiences of the center will lead perhaps to a farewell of cherished values.

In synthesis, in my opinion, the crucial issue of design profession, education, and research in the Periphery consists in the contribution to cultural and technological liberation. It's absurd to believe that the dependent countries can be emancipated by others. That task cannot be delegated to others nor can it be assumed by others. The liberation in the field of technological culture, of the culture of objects on which industrial design has influence, will be self-liberation or will be no liberation at all.

Though thirty years ago the issue of climate change was far away from occupying public attention in the way it does today, concerns for the ecological implications of design activity have steadily grown since the beginning of the 1970s.[1] *Today, in front of the undeniable climate change and the drastic consequences of this process for the planet, the issue of sustainability has moved into the center of policy decisions—or at least it should do so.*

Concerning design education, it might be recommendable to reassess the appropriate institutional setting for its location—not any longer as independent program, nor embedded in traditional departmental divisions like Architecture, Design, or Art, but as part of a university for sustainable development.

---

[1] An early example for reassessing the role of design in relation to the environment is the publication by Tomás Maldonado, *La Speranza progettuale* (Torino: Einaudi, 1970). Translations published two years later *Design, Nature and Revolution* (New York: Harper & Row, 1972), and in French, Spanish, and German.

19 THE ENVIRONMENT IN THE NORTH–SOUTH CONFLICT

(1991)

The issue of the environment is acquiring increasing political virulence in the peripheral countries. The aim of this chapter is to reveal the antagonisms inherent in the issue; hence, no attempt is made to propose solutions. Instead, a number of reflections on the environmental crisis are presented in the context of the North–South confrontation.

In no profession today can one evade environmental problems like the following: pollution; diminishing natural resources; population growth; threats to biological diversity; disruption of the ecological balance; and the effects that one's own professional activities inevitably have on the environment. The design professions, particularly those specializing in designing the material substratum of modern societies, face the challenge of creating a new professional standard for what could be called ecological design. By this, I do not mean following a trend or indulging in an opportunistic positioning, to latch on to movements that have media presence.

It would be unjust not to give credit to the attempts to redefine industrial design with an ecological perspective. Industrial designers are increasingly concerned with problems like waste disposal, solar hot water storage, soil preparation, and the use of recycled materials. They have begun to work on energy consumption and eco-balances. It took more than ten years and the incentive of the 1973 oil crisis to redesign a washing machine that consumes 50 percent less energy and water. This can hardly be seen as a sign of advanced environmental awareness, or a radically new approach to product development. Is there an ecological washing machine, with ecological standards comparable to electrical safety standards? The answer depends on the standard one applies. The washing machine is an example of redesign that neither questions nor invalidates the predominant norm that relies exclusively on economic criteria as an incentive for change. It can be summarized in the principle: "more output with less input." In the meantime, there has perhaps been some shift toward more ecological motivations from the purely economic tendency to minimize costs and maximize the input–output relation. But despite all the

Lecture originally given at the *2éme Quadriennale Internationale de Design*, *Caravelle*, Lyon, July 1991, and first published in the Brazilian periodical *Design & Interiores* 5, no. 29 (1992): 83–6. Published in English in Dawn Oxenaar Barrett (ed.), *Interface – An Approach to Design* (Maastricht: Jan van Eyck Akadamie, 1999), 97–104. Translated by Dawn Oxenaar Barrett.

improvements, the prevailing attitude is still "more of the same, only better," or "more of the same, but less resource-intensive." It appears to be very difficult to break with the paradigm of all industrialization programs that are based on the Ford doctrine. It is Fordism itself that is in crisis, as it can provide no answers for the current disorientation.

Taking the example of the washing machine a step further, one has to ask whether this indicates a new perspective opening for design, technology, and industry. I have my doubts. The development of the washing machine is based on a conceptual frame of reference that is blind to "the domain of taking-care-of." To prevent misunderstanding, I think it is inappropriate to dismiss this concern as idle chatter from a financially secure middle class that wants to have clean air, clear water, and less traffic noise and have fewer pesticides remnants in their food.

The subject of the environment is not a problem exclusive to the rich countries. The remark by a representative of a peripheral country that poverty is the Third World's environmental problem is hardly more than a clumsy tactic to stifle discussion on ecology in the periphery and turn the discussion back to the status quo.[1] The irrefutable evidence of environmental destruction calls into question the paradigm of industrialization as an instrument to create worldwide prosperity and happiness. But the poor suspect that they are now being called to pay for a feast that they have helped to prepare but to which they were never invited.

With good reason, the poorer countries are asking: Which nations are responsible for the greatest pollution? Comparing energy consumption, an average citizen in the United States consumes twenty times more commercially available energy than an inhabitant of the Third World. It is understandable why peripheral countries refuse to serve as scapegoats and accept the accusation that their population growth is the primary cause of environmental problems.

Is their suspicion about being forced into a discussion on ecology a sign of persecution mania, or is it justified? It stands to reason that peripheral countries are unwilling to agree with a proposal of an international division of labor that would give them the role of exporting cheap bananas and coffee, or other goods that are much more poorly positioned in world trade than industrial products.

If I correctly interpret the approach of the ecology movement and the role of the green parties in both the industrial and the peripheral countries, their concern is to shift the main focus of interest from a view of the environment as a supplier of raw materials to one of the environment as an area needing to be taken care of. That implies new environmental ethics and consequently new design ethics.

Design can no longer mean designing more and more objects. A new paradigm needs to be created for the practice of design and the possibilities

---

[1]   It is counterproductive to play the environment and poverty off against each other.

of industrial production. The unquestionable achievement of eco design is an articulation of an attitude that calls into question earlier paradigms of design and industrial production that we took for granted. We are witnesses to a transition period in which the traditional system of values can no longer cope with a new reality. The situation can be characterized by an interest in ecology and concern with the malaise of the modern movement.

If we look at the history of industrial design, we can identify three fundamentally different interpretations of what constitutes good design, each with its own criteria and standards. In the first, design is placed in the context of increasing efficiency and productivity. In the second, design is set in relation to marketing and business strategy, in order to create meaningful differentiation between products in competition. The third interpretation can be called design as the domain of cultural responsibility.

A new perspective is now opening up in the form of eco design. Its objectives are stable growth with environmental compatibility. These terms are new names for what was placed under the umbrella of "appropriate technology," a term from the 1970s that was explicitly intended to take into account the reality of the periphery.

Here, a brief critical excursion is needed on the point in time when two categories of countries were invented: developed or industrialized countries on the one hand and underdeveloped countries on the other. These two terms reached their peak of acceptance at the end of the 1940s, when the president of the United States, Harry S. Truman, offered a strategy for economic development as a doctrine for those nations that later came to be called the Third World. Similar terms are: "developing countries"; "under-developed countries," and "peripheral countries"—the term that seems more suitable because it identifies the core cause of underdevelopment, which is the lack of autonomy to act. The name Third World suggests an other-directed, not a self-directed world.

A vision of development emerged from the vacuum of power left by the Second World War, at a time of ideological confrontation between two opposing blocs. By inventing the doctrine of development, Western powers were trying to counter the attempts of their former colonies to look for an alternative economic and social model. The development doctrine and the expression "Third World" have their ideological roots in the Cold War. Now that the Cold War is over (at least in declarations of politicians), the term "Third World" has lost its basis.

The development doctrine was based on a plausible argument: more effective production is the key to prosperity and wealth. Higher production is achieved through industrialization. Modern technologies, including modern management techniques, have always guaranteed higher standards of living for societies that are usually considered poor. This model has as its objective the democratization of consumption.

Industrialization was seen as attractive, not only for promising greater prosperity but also as a factor in world trade. The terms of trade are generally unfavorable to economies who depend on the export of raw materials. So two strong motives can be identified for industrialization in peripheral countries:

— An internal reason—to create a material culture with its own identity
— An external reason—to improve the trade balance by exporting more value-added industrial products

That may explain why criticism of industrialization for ecological motives appears less plausible in peripheral countries.

If the whole of humankind wants the same standard of living as that enjoyed by the small group of highly industrialized countries, six planets of the size of the earth would be needed, simply to provide the raw materials and dispose of the waste. This shows that there is something wrong. The answer will depend on the interest of the person posing the question. From the standpoint of the poor Third World, there is only one chance of overcoming the state of disadvantage and the precarious nature of their material existence, and that is by industrialization. So far, a more convincing strategy has not been established.

Concern for the future and the need to participate in shaping the future lead us to the question of design in the peripheral countries. Design can be seen as an attempt to create the conditions for modernity, not only in the industrial infrastructure but also, indeed above all, in social organization. Such an interpretation of design and technology is based on the philosophy of enlightenment: overcoming self-inflicted tutelage (*selbstverschuldete Unmündigkeit*). However, "self-inflicted" needs to be modified here. One cannot hold the peripheral countries solely responsible for their lack of emancipation. Other powers are partly to blame. Asking who is responsible is not intended to provoke a defensive reaction. A debate on the North–South divide can easily become bogged down or too discursive if one side acts as accuser and the other as defendant. However, we must ask whether a North–South dialogue has actually ever taken place. The answer must be "no." The central countries usually adopt a paternalist attitude. The peripheral countries fall easily into an attitude of dependency and resignation, and then assume the role of the victim.

There is an alarming inequity in relations between North and South. That cannot be denied. Divergences emerge when we tackle the question of why that is so, and what can be done to lessen those tensions. Environmental issues cannot be separated from political ones. In recent years, technology is back in focus, drawing attention to the detrimental effects on the natural and social environment. Also, under consideration is the role of technology in world trade and the level of prosperity. Technology and innovation are depicted as a new form of dominance that is replacing the more traditional forms. However, if a framework for a strategy of effective action is needed, it is inadequate to view technology or design as a "thing." They should instead be seen as a social practice.

These remarks should not be misinterpreted as confirming the status quo of the world today. The decisive aspect of industrial design in the periphery is not lack of technological backwardness; it is the unequal context. In the periphery, a design approach is necessary that applies standards other than "good form" or "design for fun."

Firstly, the crass income differential, especially in Latin America, is a constant strain on social structures; so the ethical consequences of design practice clearly emerge. Walter Benjamin's remark that there is no cultural document that is not also a document of barbarism casts the mildew of doubt on any example of "good design" in the periphery.

Secondly, the overwhelming burden of foreign debts forces a constant drain of capital out of the peripheral countries. This raises the question whether and how design can contribute to reducing that progressive inequality in the interests of regional and global coexistence.

Latin America, a region used to importing capital, has become a heavy exporter of capital over the last decade. Brazil alone has paid more than US$100 billion in interest incurred on debt in the last ten years. The business principles of some international finance organizations do not appear to admit that it is technically impossible to ever reduce such a burden of debt. The new rounds of negotiations that are regularly held to reschedule the loans bear witness to this. The orthodox principles that recommend a local renunciation of consumption and an increase in exports not only lead to income concentration and a transfer of assets from the periphery to the center, they also have destabilizing effects on the economy. These efforts could get out of control in the future and prove counterproductive even for the central countries.

Why is design in general and industrial design in particular so important for the periphery? Would it not be much simpler to accept an international division of labor in which the periphery throws cheap raw materials and food on the world market and meets its need for industrial goods with imports? The answer is very definitely "No." Technology is a dynamic economic factor. The same cannot be said for bananas. Today, technology is characterized by permanent innovation, and design has acquired strategic importance. That is precisely what is lacking in the periphery.

In a broader view of design, the periphery can be described as a situation without a plan, or perspective. The periphery is where there is no design discourse that could constitute the basis for daily life in a culture. Design is concerned with the permanent concerns of being human, like the body, work, and play (including aesthetics). The way a culture tackles the permanent human concerns manifests itself in the manner in which we produce and consume the flood of artifacts that are made possible by industrialization and technical progress.

Environmental awareness and design are interlinked. There cannot be a meaningful discussion on ecology without discussing design at the same time. The question of ecological values has not been answered. Presumably, these values will emerge as a reaction to the environmental crisis. Among the new ecological values, one could quote a serene or detached (*gelassen*) attitude to products. In the future, perhaps a new discourse will gain momentum with a resulting serene design, as the expression of new consumer behavior. Continuing with "more of the same" shall easily burst the bounds of the ecosystem. To avoid that danger, a change in attitudes in design will be essential, but also in relation to the environment and in relations between the countries of the center and the countries of the periphery.

Until recently, it can be difficult for readers in central countries to understand the attraction of the topic of "identity" in peripheral countries. As far as design is concerned, the attraction is reflected in the search for an identity as a set of characteristics that allow "Mexican" design, or "Brazilian" design, or "Chilean" design to be set apart as something typical and special. As noted already, there are weaknesses in these kinds of search for identity. Written in 2007, this paper was an attempt to confront these problems in the context of the vastly enlarged fields of concern for "identity" provoked by globalization. Using the question of identity in literature and paralleling but contrasting this to "branding," the article tries to tease out the difficulties—but also the possibilities—of establishing deeper and more genuine notions of "identity" in design.

20 IDENTITY AND COUNTER-IDENTITY OF DESIGN

(2007)

The question about the concept of identity has arisen in design disciplines, especially in visual communication, industrial design, handicrafts, and even architecture. This is important because, on the one hand, the design disciplines form practices of daily life, and, on the other hand, they reflect different cultural practices of daily life.

Faced with the growing amount of research on identity, Zygmunt Bauman writes:

> Lately, the concept of identity has unleashed a veritable discursive explosion, as noted by Stuart Hall, in 1996, in an introduction to a collection of articles. Since then, several years have passed during which this explosion has triggered an avalanche. No other aspect of contemporary life seems to enjoy the same degree of attention from philosophers, social scientists and psychologists. This means that 'research about identity' quickly became a thriving activity. Furthermore, it can be said that 'identity' today has become a prism through which other current aspects of contemporary life are discovered, captured and researched.[1]

We can observe a similar process in the sphere of design: identity and globalization occupy a central position in the current design discourse.

POLITICS AND IDENTITY

Behind the concept of cultural identity in general and cultural identity in design disciplines in particular (mainly industrial design, visual communication, and fashion) lay questions that could disturb the friendly atmosphere in the debate because of controversial political factors that, in principle, may appear innocent. These are questions of the following:

— Domination and submission
— Antinomies and asymmetries

This text is an extended version of a lecture given in a colloquium at University of the Arts in Zurich, on March 22, 2007. Translated from Portuese by Lara Penin. Text edited for length and some content/language update.

[1] Zygmunt Bauman, "Identity in the Globalizing World," in *The Individualized Society* (Cambridge: Polity Press, 2004), 140.

— Autonomy and heteronomy
— Colonialism and postcolonialism
— Globalization and counter-globalization
— Universal standards and local particularities
— Differences and (despite everything) things in common
— Conflicts between center and periphery
— Exclusion and inclusion

Someone who analyzes the theme of identity, asking pertinent questions and inquiring about the literature in the area, would not be free, at first, to feel disoriented. She, who hopes to deal with well-defined concepts, will lose that hope when she finds the concept of "multiculturalism" with the observation that it is a "word riddled with misunderstandings."[2] The questions escalate, and most of them remain open because the answers do not suffice.

DESIGN IN RELATION TO LITERATURE

I will present some literary examples to illustrate the relationship between design disciplines and literature. The questions of identity found in literature are reflected in the design disciplines, rising questions that would not arise solely within design sphere.

There is no doubt that there is such thing as identity within design, and, therefore, it makes sense to speak of "design identity" drawing inspiration from literature. The complexity of the concept of identity in literature can serve to clarify it in the field of design disciplines. This occurs despite the differences between literary creation, that is, the production of discursive artifacts in the form of texts, and the design disciplines whose results, as is known, are manifested in material artifacts, such as objects of use, packaging, textiles and semiotic artifacts, such as movie trailers, weather charts on television, visualization of scientific data, and websites.

When extrapolating literary concepts into design, there is no way to suggest that, for example, architecture is a text. This is a fundamental misunderstanding, albeit widespread, especially in the discussion of architectural theorems.

The following question arises in design conferences in peripheral countries with surprising persistence: Is there a typically Brazilian, Argentine, or Mexican design and how does it differ from Japanese, Italian, or Swedish design? In other words: What is its identity? This question is not limited to the periphery, but is also found in central countries, perhaps reflecting the desire for design's own relevance in the market, and a certain nostalgia for an exclusive field. Before going into that question in more detail, I will analyze the meaning of the term "identity."

---

[2] Francesca Rigotti, "Las bases filosóficas del multiculturalismo," in *Multiculturalismo— ideologías y desafíos*, ed. Carlo Galli (Buenos Aires: Ediciones Nueva Visión, 2006), 31–82.

THEMATIC OF IDENTITY IN LITERATURE

In literature, the theme of identity is treated, among others, through the figure of the double, against the idea that there is only one identity for each person. In the short story entitled El Otro (The Other), Jorge Luis Borges describes a dialogue between a twenty-year-old man and a seventy-year-old man—both being facets of the same person. The older man makes a list of the central events in world history (between the two world wars) as well as his own country: "With each passing day, our country gets more provincial and more full of itself, as if closing its eyes. I would not be surprised if the teaching of Latin was replaced by that of Guarani."[3] We touch here on the topic of the canon, that is, the norms and cultural heritage that can hide a hegemonic claim: the teaching of Latin matters, the teaching of Guarani does not. Borges does not give a scientific answer to the question "What is identity?," but rather a literary answer: identity is a dream that the Other has of the Self.

In contrast to Borges' resigned formulation (fear of cultural decay), an American writer expresses himself aggressively. Reacting to the requirement to organize multicultural study programs and incorporate marginalized literary works from other cultures, he maliciously asks: "Where is the African Proust?" Obviously, this is a question that has no answer. He asks this question only as a provocation. It puts African literary production in confrontation with Marcel Proust's literary production, that is, with the canon of Western culture. One critic could argue that this question is unfair. A defender of the Western canon could answer: it is not about justice, but about literary quality.

Behind this controversy, two questions are hidden: first, the question of the existence of universal standards; second, the question of the legitimacy of these standards. Are the dominant standards always those imposed by the dominators? It would not be then a matter of quality, but a matter of power.

An Italian researcher writes on this theme: "Through this perspective (of quality), even school curricula will have to propose reading the works of the great masters; therefore, the works of Plato and not those of Rigoberta Menchú—the works of the canon of classical European authors: Shakespeare, Dante, Tolstoy, Dostoievski, Stendhal, John Donne and T.S. Eliot. In short, the group of 'pale patriarchal penis people' decorated with their laurels."[4]

ARGUMENTS AGAINST THE CONCEPT OF IDENTITY

Unleashing against the closed character of a culture, the Spanish writer Juan Goytisolo criticizes mercilessly the provincialism of his country: "Thus, in the most fruitful and rich periods of a literature, there are no unequivocal influences,

[3]   Jorge Luis Borges, "El Otro," in *El libro de arena*, 1st ed., Obras Completas (Buenos Aires: Emecé Editores, [1975] 2007), 13–20.
[4]   Francesca Rigotti, "Las bases filosóficas del multiculturalismo," quoting the characterization of authors formulated by Robert Huges, *The Culture of Complaint* (Oxford: Oxford University Press, 1993).

national essences or exclusive traditions: only polygenesis, miscegenation, promiscuity."[5] He quotes the Syrian-Lebanese poet Ali Ahmad Said: "Identity cannot be accepted as something finished, nor definitive, on the contrary, it is an always open possibility" and continues: "True identity is a continuous current that is nourished by an infinite number of streams and creeks."[6] This is a clear rejection of the dreams of a fixed identity or a nationalist being.

I want to quote a representative of the social sciences who is also against the idea of identity as something fixed, lasting, closed, proper, essential. Without mentioning it explicitly. Zygmunt Bauman criticizes the slogan "Think globally, act locally." He writes: "There are no local solutions to problems created globally. . . . The global forces, overwhelming and indomitable, thrive in the fragility of the political scene and in the decision of potentially global policies, always fighting for a larger portion of the crumbs that fall from the party table of the barons of the global assault. Anything that supports 'local identities', as an apparent antidote against the wrongdoing of globalizers, is actually submitting to their game."[7]

Equally, he critiques any oversimple notion that we live in a "multicultural era": "The announcement of a 'multicultural era' reflects, in my opinion, the experience of a new global elite that, when traveling to other countries, meets members of the same global elite who speaks the same language and who cares about the same things. . . . However, the announcement of the multicultural era is a declaration of incompetence: of the refusal to formulate a judgment, to assume a posture; a declaration of indifference, of washing your hands in the face of petty fights over preferred lifestyles and values."[8] In the same direction, Goytisolo said: "The 'identity' is revealed to us as something that needs to be invented and not just discovered." In other words: identities are not entities hidden in some secret and deep place, but something that needs to be created (or, in design terminology, something that needs to be designed). Naturally, this concept goes far beyond branding or corporate design.

In the publication with the revealing title *The Illusion of Identity*, the French scientist, Jean-François Bayart, criticizes substantialism, that is, the belief in the existence of permanent cultural characteristics, and this generates the concept of identity with dangerous political potential. He writes: "These wars [in the former Yugoslavia, the Caucasus, Algeria and the Great Lakes, in Africa] and the revolts revolved around the concept of identity and extracted their deadly potential from the assumption that a 'cultural identity' necessarily associates with a 'political identity'. However, each of these identities is, at best, a cultural construct, a political and ideological construct, that is to say, in the end, a historical construct."[9]

---

[5]   Juan Goytisolo, *Contracorrientes* (Barcelona: Montesinos, 1985), 168.
[6]   Ibid.
[7]   Zygmunt Bauman, *Identidad—Conversaciones con Benedetto Vecchi* (Madrid: Losada, 2005), 187.
[8]   Ibid.
[9]   Jean-François Bayart, *The Illusion of Identity* (Chicago: The University of Chicago Press, [1996] 2005), ix.

DEPENDENCY THEORY

The quotes presented are contrary to the good intentions of developing an identity of design, which are recurrently discussed in conferences in periphery countries, whether Mexican, Brazilian, or Chilean. In the context of existential dependence and the rebellion against this dependency, it is sometimes proposed to recover the native tradition proper of objects of use and ornamentation.

The dependency theory[10] that was developed as a genuine contribution of the social sciences in Latin America, in the 1960s, makes it possible to understand the political face of the issue of design identity in the periphery, an often thorny subject. This theoretical contribution appeared in several contexts characterizing the end of colonialism: the Cuban revolution; the Second Vatican Council, with the option for the poor and the liberation theology; the 1964 coup d'état in Brazil; the invasion of the Dominican Republic in 1965 by the United States; and the 1966 coup in Argentina.

Dependency theory aimed to find an explanation for the failure of Latin American development, despite its large territorial extensions, its cheap labor, the variety of natural resources, the cultural homogeneity and good communication infrastructure. The causes of the delayed development were attributed to the dominant model, making it responsible for the subcontinent's economic failure to take off due to the supposedly feudal social structures. The central thesis said: Latin America is not underdeveloped by the lack of social capitalist structures, but, on the contrary, by the predominance of these structures. Underdevelopment was not considered a historical state, but a result of the capitalist development process. These countries *were* not underdeveloped, but they *have been* and *still are* underdeveloped.

For industrial design in Latin America, the dependency theory played an important practical role. Technological and industrial policies, conceived within this theory, and oriented to the policy of import substitution, opened a space for local design activities. This policy of search for identity was based on industrialization and, decidedly, did not turn to the past in search of the supposed roots of Latin American design in pre-Columbian cultures. This search is a chimera, not suitable as a starting point for the development of an autonomous and valid design for the future. Instead of looking for design identity in a romantically idealized past, it would be more appropriate to change the direction and look toward the future.

The opposite of the dependency theory is the ominous Washington Consensus, formulated in the late 1980s. During the 1990s, it guided the policy of the World Bank, the International Monetary Fund, and the World Trade Organization—with disastrous social and economic consequences for the affected countries.

---

[10] Atilio Borón, "Teoría(s) de la dependencia," *Realidad Económica* 238 (2008). See also: http://www.iade.org.ar/modules/noticias/article.php?storyid=2661 (accessed December 12, 2008). This summary of dependency theory is largely based on this article.

IDENTITY DESIGN|BRANDING

The *l'imaginaire* of the other (target audience) can be built intentionally, through an identity policy in the form of branding. In the professional discourse of branding, identity is defined as "the sum of all the characteristics that make a brand or company unmistakable and unique."[11] This interpretation of identity as a sum of individual attributes has two characteristics—a fixed one, of static constancy; and another that is changeable, of flexibility and the exchange of identity: "Alongside the aspect of fixed identity and continuity, the aspect of permanent change coexists. Nothing can remain immutable. Everything is changeable."[12] Comparing these two characterizations, one can see the scope of the concept of identity that extends from the pole of constancy (static) to the pole of change (dynamic). In the presence of the *l'imaginaire* of the other, there is a self-image that does not necessarily coincide with the image on the *l'imaginaire* of the other. Divergences and dissonances between these two images are inevitable. The designer must be aware of this potential divergence between the company's reality and its image. Thus, the designer can avoid the danger of performing a mere face-lifting, aiming to make a company more attractive for sale on the stock exchange, by means of a simple visual improvement.

At the time of the European maritime explorations, the movement left Europe to the periphery aiming at the occupation of large tracts of land, in the centrifugal sense. At that time, the stranger, the exotic, was sought. And that was the object of an exploratory gaze. Today, the movement has reversed direction. The stranger (the foreigner) arrives in the metropolis in a centripetal movement. Through the processes of migration from the periphery toward the metropolises, their culture is confronted with the "stranger" (another culture) in their own country. Thus, the center experiences, in its own territory, the "harassment" of the foreign immigrant.

In relation to tourism, the traditional flow has been directed from the center to the periphery,[13] in search of landscapes, fauna, peoples, and exotic cultures. However, with the strong migratory current from peripheral countries to Europe, reverse tourism emerged. Now, Europeans are faced, in their own countries, with another reality, which they did not seek. The "stranger" reached them in an uncomfortable way. The cultural confrontation that has taken place

---

[11] Robert Paulmann, *Double Loop—Basiswissen Corporate Identity* (Mainz: Herrmann Schmidt, 2005), 125.

[12] Eberle Gramberg, Gerda e Jürgen Gramberg, "Stadtidentität—Stadtentwicklung ist Identitätsentwicklung," in *Stadtidentität—Der richtige Weg zum Stadtmarketing*, coordenado por Maria Luise Hilber e Ayda Ergez (Zürich: Orell Füssli, 2004), 27–35.

[13] The issue of the presence of "the other" in Europe goes of course far beyond tourism and in fact touches upon immigration from non-European countries and regions, related to work, study, war, and refugees. While many forces behind immigration in Europe have been planned, for example when countries have invited workers from specific countries at certain times or higher education attracting students globally, the author is pointing to the unplanned aspect of immigration, something "they did not seek," that perhaps, in fact, escaped Europe's original colonial design, and forced a cultural confrontation.

forces them to review the supposedly universal values of European culture. In this cultural shock, there is a potential for conflict, with demonstrations of hostility and even aggression.

NATIONAL BRANDING AS IDENTITY DESIGN

After these reflections on the issue of identity, I dedicate myself to some concrete examples of identity design, firstly to the national branding of peripheral countries. Opportunities have arisen for international branding consultancies that operate globally in the market constituted by states and countries that intend to develop a rebranding process, in a way that applies to companies. The reasons for these new communication efforts, with an emphasis on the visual aspect, are to present themselves in a more attractive way at the international level. They aim to promote tourism, attract international events to the country and, above all, create an attractive climate, especially for international investors, radiating a positive identity, and aiming at possible financial returns.[14] The universal character of these promotions now also covers cities, regions, and countries. However, the authorities responsible for branding contracts have little understanding of the matter or are unaware that "a renewal of identity goes far beyond flags and logos."[15]

It is symptomatic that relatively small countries in Latin America (Uruguay, Nicaragua, Guatemala, Chile, Ecuador) strive to position themselves internationally, promoting the renewal of their identities through such marketing operations. Perhaps they believed in the idea that a visual refashioning or visual enhancement of a logo, with the corresponding slogan, would be positive to promote identity. Possibly they hope that, through a branding program, a country can be included in the group of dominant countries—that is, national branding serving as an entrance ticket for the international identity club.

To create attractive identities in the international market, *vision programs* are applied. As a result, this whole creation procedure takes on, whether intentionally or not, almost mystical-religious traits of enlightened visionaries. To take full advantage, it is recommended to fit not only companies, but also cities, regions, and countries. By applying branding techniques, competitive advantages are generated—a process in which the creation of symbolic identity has a determining role. The ethnic feel or ethnic look is used with the use of local particularities to apply mainly to goods in the textile and fashion sector. The momentum for expanding branding finds its limit on the planet Earth for now. However, it would correspond to the branding logic to soon submit the Terra satellite to the branding strategy as well.

---

[14]   Simon Anholt, *Brand New Justice—How Branding Places and Products can Help the Developing World* (Oxford: Elsevier, [2003] 2006).
[15]   Mark Leonard, *BritainTM—Renewing Our Identity* (London: Demos, 1997), 10.

The indefatigable expert, Wally Olins, recommends a seven-stage program for branding a country.[16] The branding of countries presents itself almost as a historical fatality—the naturalization of social processes camouflages the interest in maintaining the status quo and serves to defend itself from uncomfortable questions. In this context, the appearance of a devastating criticism would not be surprising.[17]

A clear distinction must be made between long-term visual identity and that limited to a period of government. Mexico applies a systematic identity policy to mark export products. The use of the logo for premium products (certain subtropical fruits) is subject to quality control. Only when a product meets certain quality criteria is the use of the new logo for transport and consumer packaging allowed. In Brazil, the registration of the company that wants to use the logo (such as Made in Brazil) is simply requested, without any subsequent quality control of the products in which this brand will be applied.

THE SYMBOLIC DIMENSION OF PRODUCTS

Branding deals with communication problems, essentially aiming at creating a predisposition for positive valuations. Thus, it attaches great importance to the symbolic aspects of design, especially industrial design. Branding is linked to the phenomenon called "the theological whims of merchandise."[18] Today, the two concepts of classical economics, "use value" and "exchange value," are complemented by the third concept, that of "symbolic value": "Initially, the symbolic value of the commodity was attributed to the systemic effect of relations of production. Later, it was consciously instrumented through the design of logos and branding, acquiring its own dynamic to transform illusion into reality. For example, 'designer mineral water' was created, turning water into a luxury branded good. Symbolic value really becomes exchange value."[19] These words prove the nefarious connection of the designer's profession with expensive, elaborate, eccentric, and cute things. The question remains as to the ability of design to break free from this embrace of branding.

Branding reached its climax with the "theological whims" and the "metaphysical subtleties" of the merchandise, including the "sensorial-trans-sensorial" attributes. Techniques for creating symbolic aspects of products and companies have reached a degree of maturity that, 150 years ago, would have been difficult to predict. In view of the importance acquired by the symbolic dimension of goods and companies, it seems necessary that product design and

[16]   Wally Olins, *Trading Identities—Why Countries and Companies Are Taking on Each Others' Role* (London: The Foreign Policy Centre, 1999), 23–6.
[17]   Terry Eagleton, "A Fresh Look at Wally Olins's Highly Regarded Branding Manual, now in Paperback," *eye* 14, no. 53 (2004), http://www.eyemagazine.com/feature.php?id=116&fid=508 (last accessed August 21, 2020).
[18]   Karl Marx, "Der Fetischcharakter der Ware und sein Geheimnis," in *Das Kapital* (Berlin: Dietz, 1947).
[19]   Sven Lütticken, "Attending to Abstract Things," *New Left Review* 54 (November/December 2008): 101–22.

the teaching of design incorporate the study of emotions (emotional design). The designer should be concerned with these aspects of emotional design instead of dealing with trivial things like usage, practicality, and technical details. This process is facilitated by the provision of software for rendering. However, its use, as is known, does not replace design activity.

The symbolic aspect of an anonymous technical product, such as a screw, at most, is present in secondary features. On the other hand, this aspect can be inflated to the extreme in the scope of consumer products, assuming superlative dimensions and reaching absurdity, as shown in the example of a fine porcelain kettle in the shape of an animal skull covered with beaver skin. In this category of symbolic products, you can also include a lemon squeezer that has become a design icon, in which the primary characteristics of use are completely subordinated to a formal concept. This may be one of the reasons why this juicer is considered a sculpture to decorate the desks of CEOs. The enthronement of the symbolic dimension corresponds to the arrogant contempt for the flatness of practical functions.

In a schematic opposition Good/Evil of the ten commandments of emotional branding, you can read: "From the product to the experience: products fulfill needs—experiences fulfill desires" and then, "From function to emotion: the functionality of the product deals with its superficial qualities—emotional design is about experiences."[20] It doesn't matter how the consumer reacts to the sensory experience in handling a bread knife when she cuts her finger—that would be the simple result of a secondary, superficial, practical function that would have little importance. According to the commandment of emotional design, the consumer would be interested only in sensory experiences, not caring about the finger cut. At the apex of this class of products, there would be objects in which any characteristic of a practical function is eliminated. There the consumer would live in the apotheotic sky of high trans-sensorial experiences over any materiality.

The priority of formal-aesthetic factors also explains the interest of the *marchands de tableaux* who transform their galleries into antiques of modernity and now exhibit and sell design objects as well. Above all, products such as furniture and lamps, in which price does not play a relevant role.[21] In an exemplary way, this creation of identity is manifested in author design. The signature of a famous designer guarantees the identity of the unique, unmistakable, authenticity, highlighting the product from the mass of everyday products without identity, elevating it to the sphere of exclusive objects, and bringing it closer to the status of art objects.

The effort to flaunt an unmistakable identity is manifested in the New Design Cities initiative—presumably different from established design cities such as Milan and London. These New Design Cities showcase the so-called

---

[20]   Marc Gobé, *Citizen Brand—10 Commandments for Transforming Brands in a Consumer Democracy* (New York: Allworth Press, 2002).
[21]   Alain Badiou, *Dritter Entwurf eines Manifests für den Affirmationismus* (Berlin: Merve, 2007), 10.

creative industries to prove their exceptional character to which the sectors of the film industry, television, marketing, advertising, trend research, fashion, publishing, architecture, exhibition and event design, graphic design, industrial design and new media—that is to say, the creative people—in addition to the traditional cultural industries in the form of museums, theaters, concert halls, and galleries—all this accompanied by a myriad of culinary offerings and shopping possibilities. To measure the degree of attractiveness of these cities, you can consult the Bohème Index[22].

Basically, it is about city marketing aimed at a social group that is able to enjoy the benefits of a certain lifestyle.[23] The loss of the unmistakable profile of the concept "design" and its subordination to regressive trends are manifested in the double use of the word: on the one hand, as a generic term in the distinctive "design cities," and, on the other hand, as a specific term for "creative" activities. By placing design alongside gastronomy services, it follows that, in public opinion, design is associated, in good measure, with a party. The degree of "self-love" of the "creatives" has reached a dimension in which there is no lack of attempts to declare them as a new class—the creatives class. The degradation of the term "design" has already been observed by several authors: "In the late 1980s, the term 'design'—almost became an abuse term. At least he suggested vacuity and superficiality, or the useless repackaging—to generate profit—of the most common things 'mineral water designers'. In the worst case 'designer drugs', 'designer violence', the term fantasized a world of glamorous surfaces that hid an underlying amorality, lack of affection and even corruption."[24]

"CULTURAL INDUSTRIES" IN THE AFFIRMATIVE

After the profound economic crisis that hit Argentina (2001–2), design was promoted to the rubric of cultural industries by the federal government (Ministry of Culture) and, above all, by the administration of the city of Buenos Aires. Apparently, the critical dimension of this concept coming from the Frankfurt School was unknown to the leaders of the program. This process was accompanied by the distinction granted by UNESCO to the city of Buenos

[22] Marie-Josée Lacroix (ed.), *New Design Cities/Nouvelles Villes de Design*, Ville de Montréal, Ville de Saint-Etienne (Montréal: Les éditions Infopresse, 2005). The first initiative to establish the new design cities was attended by Antwerp, Glasgow, Lisbon, Saint-Etienne, Stockholm, and the Time Square business association in New York. The minting of cities' identities and their promotion are motivated by local commercial and political interests.

[23] In a critical analysis of the initiative to propose the city of São Paulo as a candidate for the select group of "design cities," the author writes: "It is . . . to question the appropriation of concepts and phenomena of public interest and property by small private groups, who speak in the name of design, cities and culture under the command of cultural legitimacy, which is enshrined by economic power." Ana Claudia Berwanger, "O design e a cidade: considerações e perspectivas de análise," *agitprop—revista brasileira de design* 28 (2010).

[24] Rick Poynor, "on 'some virtues of design'," in *Design beyond Design—Critical Reflections and the Practice of Visual Communication*, ed. Jan van Toorn (Maastricht: Jan van Eyck Akademie, 1998), 111–13.

Aires as the "City of Design"—a result of the city's marketing. The support granted by the government focused on the symbolic and formal-aesthetic aspects of consumer products of low technical complexity, produced by hand, for which the term "object design" is used: necklaces, accessories, fashion objects. The emphasis on symbolic aspects suggests the search for sources of inspiration for the creation of identity. For example, the resort to the world of the Mapuche symbols in Patagonia.

PERSPECTIVES OF HANDICRAFTS

The use of local resources (graphic motifs, color combinations, materials, and labor-intensive production processes) in relation to the design and creation of identity can be seen, in an exemplary way, in peripheral countries. To a large extent, these activities belong to the informal sector of the economy and generally apply simple, noncapital intensive processes. The theme of crafts and design can be studied by the following postures, which can appear in pure or mixed form[25]:

1. Conservative approach. It seeks to protect the artisans against any influence of design from outside. This attitude is occasionally found among anthropologists who reject any approximation between design and artisans, as they want to keep the craftsperson in a pure state, immaculate and immune to contemporary influences. Without wishing to question the anthropologists' good (or not so good) intentions,[26] the impression is that they want to preserve the exclusivity of the research field, pretending to be the only legitimate experts to give their opinion on the artisans and their products. Basically, it is a territorial dispute: Who can touch what?

2. Aesthetic approach. In this approach, artisans are considered representatives of popular culture, and their work is raised to the status of art, using the term "popular art" in reference to "classical art." The repertoire of popular art forms (ornaments, color combinations) is used as a starting point or source of inspiration for creations. Those outside the communities approach its formal-aesthetic language to produce

---

[25] Lately, these programs and initiatives are characterized with the attributes "sustainable" and "socially responsible," with which an ethical stance is highlighted. Nothing is said about the capacity of these initiatives to promote the autonomy of artisans and artisans, thus avoiding a relapse to welfare (government social assistance programs).

[26] In the last ten to fifteen years, there has been a great approximation between design and anthropology; ethnography is now very much incorporated in design practices. Beyond the disciplinary discussion, it is important to make it clear that artisans are agents of their own practices, rather than passive subjects who need to be "protected" from outside influences. They are the ones who should consider their own material and cultural production, and make decisions about their own future, rather than let it be decided by either anthropologists and designers. Bonsiepe is of course here being provocative to make explicit what should *not* be the approach toward crafts.

design objects. Concretely, this focus is manifested in the so-called ethnodesign.

3. Productivist approach. It considers artisans as a qualified and inexpensive workforce, using its capabilities to produce objects developed and signed by designers and artists A certain disingenuous mindset is needed here to accept this approach, presented as a "help" for handicrafts in the periphery. Humanitarian interests are alleged to produce "inspired" designs in local popular culture or designs brought directly from the center to take advantage of the cheap labor of these communities. Such design practice tends to perpetuate dependency relationships, instead of contributing to overcoming them.

4. Culturalist or essentialist approach. It considers local artisans' projects as the basis or starting point for true Latin American or Indo-American design. Sometimes, this approach is accompanied by a romantic attitude that idealizes the supposed "bucolic" past.

5. Paternalistic approach. It considers artisans, in the first place, as political clientele of welfare programs and plays a mediating role between production and marketing (marketing), in general, with high profit margins for sellers.

6. Focus on promoting innovation. Advocates the autonomy of artisans to improve their livelihood conditions, which are often precarious. In that case, an active participation by the artisans producers is required.

THE SEMANTICS OF WEAVING

Often, artisanal production is understood in a reductionist or limited way, due to a purely aesthetic-formal vision. This can be explained using the example of diamond patterns applied to textiles, ceramics, gourds, and wood products in Mexico. "In Q'ero (form of weaving) the diamond is divided into four parts, being one of the elements most used to represent cosmological concepts. The graphic elements of this diamond are: rays, a vertical dividing line and arrows that refer to concepts of space and time. . . . The vertical dividing line structures the diamond in hatún inti (big sun) which, according to the Q'eros and Kaulis informants, means the sun at noon and a dualistic social order. . . . Other graphic elements are used by them, representing the time of day, the period of the year and the four-part division of the Earth."[27]

In the design research project quoted earlier, the visible difficulties in interpreting these patterns were found. An example shows the misinterpretations that a person from outside and unfamiliar with the local culture may have:

[27] Fernando Shultz, "Diseño y artesanía," in *Historia del diseño en América Latina y el Caribe— Diseño industrial y comunicación para la autonomía*, eds. Silvia Fernández and Gui Bonsiepe (São Paulo: Blucher, 2008), 308–22.

"The colonization process has suppressed these semantic dimensions or cosmological visions. In research on the Amuzga language in the Guerrero State, Mexico, it was discovered that the translation of the Amuzga language into Spanish was done by volunteers from the US Summer Linguistic Institute, which led to a loss of the correct meaning of all historical-conceptual content. These translations are so wrong that a geometric figure, in which the participants of the summer course visualize the shape of a 'shoe', that sign is literally translated as 'shoe', despite not wearing shoes in this region."[28]

MANIFESTATION OF DESIGN IDENTITY

To understand the meaning of an identity, it is recommended to make a list of its different manifestations. The design identity is materialized as follows:

1. In the form of a group of formal or chromatic features (*stilemi*)
2. In the structure of the taxonomy of products, that is, the types of products characteristic of a culture, for example a gourd that was created in the Guarani culture
3. Using local materials and corresponding manufacturing methods
4. In the application of a specific design method (empathy for a tradition and use of these attributes rooted in a given region)

In Santa Clara, in the state of Michoacán (Mexico), the Purépecha ethnic group has been working with copper since pre-Columbian times. First, copper disks are forged in collective work. Then, each artisan works the raw material in their own way, applying hammer blows. The tools are manufactured by the artisans themselves using truck parts, preferably springs and other suspension components.

When it comes to improving living conditions through cooperation projects, it is not enough to work with design alone. Other support measures, such as the granting of microcredits and marketing, should be included. Often, artisans live on the edge of extreme poverty, not even being able to buy the raw material for production. For this reason, they depend on traders who supply them with the material, forcing them into debt and taking most of the profits.

The Casa de las Artesanías in Michoacán prevents this danger of exploitation by giving full support to artisans. Thus, it does research for the discovery and supports the preservation of local cultures. The works of artisans and groups are disseminated through competitions and exhibitions. It offers improvement courses about forms of organization and marketing. In addition, it adopts protective measures against unfair international competition, which

---

[28] Ibid., Shultz quotes a research report here.

reproduces these designs outside the country, applying industrial methods of serial manufacturing.[29]

In addition to subsistence problems, the relationship with the environment and nature plays an important role for handicrafts. The motifs of flora, fauna, and astronomy serve as a starting point for legends and traditions that later materialize in the products. The destruction of the environment and the elimination of animals and plants are a threat to the continuity of this culture. "Therefore, ecology is not only a 'bio-physical-chemical' issue, but a cultural issue of fundamental importance for the permanence and development of artisans."[30]

BIBLIOGRAPHY

Anholt, Simon. *Brand New Justice—How Branding Places and Products Can Help the Developing World*. Oxford: Elsevier, 2006 (1ª edição 2003).

Badiou, Alain. *Dritter Entwurf eines Manifests für den Affirmationismus*. Berlin: editora Merve, 2007.

Baltes, Martin (coord.). *Marken—Labels—Brands*. Freiburg: Orange Press, 2004.

Bauman, Zygmunt. *Flüchtige Moderne*. Frankfurt am Main: Suhrkamp, 2003. [*A modernidade líquida*, Zahar, Rio de Janeiro 2001.]

Bauman, Zygmunt. "Identity in the Globalizing World," in *The Individualized Society*, 140–52. Cambridge: Polity Press, 2004.

Bauman, Zygmunt. *Identidad—Conversaciones con Benedetto Vecchi*. Madrid: Losada, 2005.

Bayart, Jean-François. *The Illusion of Identity*. Chicago: The University of Chicago Press, 2005.

Borges, Jorge Luis. "El Otro," in El libro de arena, in *Obras Completas*, Buenos Aires: Emecé Editores, 2007.

Borón, Atilio. "Teoría(s) de la dependencia," in *Realidad Económica* 238, 2008. http://www.iade.org.ar/modules/noticias/article.php?storyid=2661

Bredekamp, Horst. "Bildbeschreibungen. Eine Stilgeschichte technischer Bilder?," in *Das Technische Bild—Kompendium zu einer Stilgeschichte wissenschaftlicher Bilder*, por Horst Bredekamp, Birgit Schneider e Vera Dünkel, 37–47. Berlin: editora Akademie, 2008.

Eagleton, Terry. "A Fresh look at Wally Olins's Highly Regarded Branding Manual, now in Paperback," in *eye* 53, Herbst, 2004. http://www.eyemagazine.com/feature.php?id=116&fid=508

Galli, Carlo (ed.). *Multiculturalismo—ideologías y desafíos*. Buenos Aires: Ediciones Nueva Visión, 2006.

---

[29] With reference to handcrafted products that exceed local demand, the question arises for the authenticity and separation of the so-called "airport art" or "tourist art," that is to say, products explicitly oriented to tourists. To guarantee authenticity, especially of the most expensive products, a quality seal is placed on which the name of the craftswoman or craftsman also appears. The identification of the author guarantees the identity of the product.

[30] Shultz, "Diseño y artesanía."

Genovese, Alfredo. *Manual del filete porteño*, Buenos Aires: Comisión para la Preservación del Patrimonio Cultural de la Ciudad Autónoma de Buenos Aires, 2008.

Gobé, Marc. *Citizen Brand—10 Commandments for Transforming Brands in a Consumer Democracy*. New York: Allworth Press, 2002.

Goytisolo, Juan. *Contracorrientes*. Barcelona: Montesinos, 1985.

Goytisolo, Juan. "Abandonemos de una vez el amoroso cultivo de nuestras señas de identidad," in *Pájaro que ensucia su propio nido*, 86–90. Barcelona: Random House Mondadori, 2001.

Hilber, Maria Luise e Ayda Ergez. *Stadtidentität—Der richtige Weg zum Stadtmarketing*. Zürich: Orell Füssli, 2004.

Huntington, Samuel P. *The Clash of Civilizations and the Remaking of World Order*. New York: Simon & Schuster, 2002.

Kozak, Claudia. *Contra la pared—sobre graffitis, pintadas y otras intervenciones urbanas*. Buenos Aires: Libros del Rojas, Universidad de Buenos Aires, 2004.

Kafka, Franz. "Die Verwandlung," in Kafka, Franz, *Die Erzählungen und andere ausgewählte Prosa*, eds. Roger Hermes. Frankfurt: S. Fischer, 1999.

Lacroix, Marie-Josée (ed.). *New Design Cities/Nouvelles Villes de Design*. Montréal: Ville de Montréal, Ville de Saint-Etienne, Les éditions Infopresse, 2005.

Leonard, Mark. *Britain™—Renewing Our Identity*. London: Demos, 1997.

Lütticken, Sven. "Attending to Abstract Things," *New Left Review*, n. 54 (2008): 101–22.

Maldonado, Tomás. "Neue Entwicklungen in der Industrie und die Ausbildung des Produktgestalters," *ulm, Zeitschrift der Hochschule für Gestaltung*, n. 2, 1958.

Maldonado, Tomás. *Disegno industriale: un riesame—Definizione Storia Bibliografia*. Mailand: Giangiacomo Feltrinelli, 1976.

Marx, Karl. "Der Fetischcharakter der Ware und sein Geheimnis," in *Das Kapital*, 76. Berlin: Dietz, 1947.

Olins, Wally. "Trading Identities—Why Countries and Companies Are taking on Each Others' Role," The Foreign Policy Centre, London, 1999.

Paulmann, Robert. *Double Loop—Basiswissen Corporate Identity*. Mainz: Herrmann Schmidt Mainz, 2005.

Pratt, Mary Louise. *Imperial Eyes—Travel Writing and Transculturation*. London, New York: Routledge, 1997.

Rigotti, Francesca. "Las bases filosóficas del multiculturalismo," in *Multiculturalismo—ideologías y desafíos*, coordenado por Carlo Galli. Buenos Aires: Ediciones Nueva Visión, 2006.

Said, Edward W. *Orientalism*. New York: Vintage Books, 1979.

Said, Edward W. *Culture and Imperialism*. New York: Vintage Books, 1994.

Sandall, Roger. *The Culture Cult—Designer Tribalism and Other Essays*. Boulder: Westview Press, 2001.

Shultz, Fernando. "Diseño y artesanía," in *Historia del diseño en América Latina— Diseño industrial y comunicación para la autonomía*, eds. Silvia Fernández e Gui Bonsiepe. São Paulo: Blucher, 2008.

Ulanovsky, Julieta e Valeria Dulitzky. *El libro de los colectivos*. Buenos Aires: La Marca Editora, 2005.

PART THREE
Design, Visuality, Cognition

Gui Bonsiepe: Framing Design as Interface

Hugh Dubberly

Gui Bonsiepe has had a remarkable career—distinguished by its length (sixty years and counting) and by its variety—a career that's difficult to classify, because it has many dimensions and requires many "keywords" to describe in a way that approaches completeness. And yet, Bonsiepe's career may serve as a signal of where design is heading or even as a model for a new generation of designers—a model of how designers may explore the "space" of design and also expand that space as they adapt to a continuously changing world.

Bonsiepe described his alma mater, the Hochschule für Gestaltung (HfG) Ulm (1959), as "a school in Germany but not a German school." As art historian Pamela Lee has noted, the HfG had a "cosmopolitan" population with faculty from a range of disciplines (including cybernetics, information theory, operations research, physics, semiotics, systems theory, as well as more formal design disciplines) and students "from some 49 countries" filling 40–50 percent of enrollment in any given year.[1]

Similarly, one might describe Bonsiepe as born and trained in Germany but not a German designer. He is much more: Polyglot; Polymath; Polydesigner. At home (literally) in many countries, having lived and worked in Argentina, Brazil, Chile, Italy, and the United States, as well as Germany.

Bonsiepe's work bridges boundaries:

— Twentieth century—Twenty-first century
— Old world—New world
— Global north—Global south
— Center—Periphery
— Visual—Verbal
— Form—Structure
— Object—Interface
— Material—Digital
— Technology—Use
— Production—Presentation
— Theory—Praxis
— Teaching—Writing—Designing

---

[1] Pamela Lee, *Think Tank Aesthetics: Midcentury Modernism, the Cold War, and the Neoliberal Present*, (Cambridge, MA: The MIT Press, 2020), 193.

For me (of course, this is an idiosyncratic view), three aspects of Bonsiepe's career stand out:

— his association with the HfG Ulm
— his work on Project Cybersyn
— his book *Interface: An Approach to Design*

HfG ULM

Much has been written (including by Bonsiepe in this volume) about this small design school that lived for a scant fifteen years—a gray concrete modernist cloister, on the edge of a cow farm, high on a hill ("Der Kuhberg") in the countryside between Stuttgart and Munich, in Southern Germany. And yet, the HfG remains unknown to most practicing designers. Some who do know it describe the HfG as "the new Bauhaus" (though that term was perhaps more accurately appropriated by Moholy-Nagy for his school in Chicago). And in any case, when Gropius offered to bestow this "blessing" (or brand extension) on the HfG, the faculty didn't exactly embrace the idea—and rightly so (see Chapter 3.9 in this volume).

Yes, the Bauhaus (with its several incarnations) had ties with the HfG (most notably Max Bill and Josef Albers, though by the mid-1950s both had moved beyond their Bauhaus roots, inventing their own paths forward). Yet the HfG was something new and original: its own reformation.

At core, the Bauhaus and the HfG shared the modernist credo that design could and should make the world better for everyone. But the Bauhaus was a school of architecture and art, which later added design as something of an afterthought, whereas the HfG articulated itself as a school for design, stated as such, right from the start. (Hochschule für Gestaltung translates in English literally as "high-school for design.") Most importantly, though, in so doing the HfG became a school for a new kind of design, modernism of the second generation, postwar, high modernism—the modernism of Helvetica, Lufthansa, Braun, and ultimately Apple under Jony Ive.

Yet, in the long run, "the perfect radius" will not be the HfG's most important contribution. What was transformative about the HfG is that the curriculum embraced "environmental design" (not to be confused with designing for the environment, an issue the HfG largely missed but rather designing the entire "built" environment), what the Dutch team of Wim Crouwel, Friso Kramer, and Benno Wissing called "Total Design," echoed by the approaches of British counterpart Pentagram and US counterpart Unimark, multidisciplinary collaboratives founded in the early 1960s to tackle complex systems design projects for large organizations, for example the Schiphol Airport (Amsterdam) signage system—design firms as a collegial faculty rather than a hierarchy dominated by a "pater familias" starchitect.

Their concern for systems—what West Churchman called "the systems approach"—was in the air (or in the Zeitgeist for the Germans). The allies had defeated the existential threat of fascism, in large measure through better technology (i.e., the radar, the bomb, and the computer) and better planning (i.e., operations research, a forerunner of "systems thinking"). The Bauhaus' industrial-machine optimism was replaced by a more advanced techno-optimism. Planning methods, which had won the war, would surely improve the peace. And increasingly, computers would aid management.

In a sense, improving the peace was the whole point of the HfG (the "raison d´être" for its original funding by the US Marshall Plan). To that end, the HfG embraced information theory, operations research, cybernetics, and the ethos of the systems approach. (For example, Norbert Wiener, who named the field of cybernetics, lectured at the HfG in 1955.)[2] And out of that embrace grew the design methods movement. (Two of the key founders of the design methods movement, Horst Rittel and Bruce Archer, taught at the HfG.) Design methods comes down to us today, rebranded most recently as "design thinking". This environment, in which designers read and discussed ideas and tried to work out what they meant for practice, was formative for design and for Bonsiepe.

CYBERSYN

Around the world, 1968 was a horrible year. In Vietnam, the Tet offensive and My Lai massacre. In Prague, the Soviet invasion. In many countries, student protests. In Paris and other parts of France, occupation of schools and factories, battles with police, and strikes. In Tlatelolco, the Mexican army massacre of as many as 400 students. In Brazil and South Carolina, protesting students were killed by the police. In Germany, the head of the Socialist Students Union was seriously wounded. In the United States, Martin Luther King and Robert Kennedy were assassinated.

Amid this broad turmoil, the closing of the HfG is a footnote. Yet, the events are not unrelated. Politics played a role in ulm, too. The school ran out of money. The Marshall Plan funds were gone, and local government support evaporated amid claims that the HfG faculty were communists. Bonsiepe reported "only one nominal communist," though he didn't say who.[3] He added that an element of the defunding may also have been that local burghers were tired of HfG students corrupting their daughters.[4]

Three years after the HfG closed, Bonsiepe was in Chile working for the democratically elected socialist government of Salvador Allende. As the account by Eden Medina in Chapter 4.10 details, Allende had made Fernando Flores finance minister. Flores (then just twenty-eight years old) engaged

---

[2]   A more surprising visitor, at least at first sight, was Martin Heidegger. See René Spitz, *The Ulm School of Design: A View Beyond the Foreground* (Stuttgart: Edition Axel Menges, 2002), 235.
[3]   Ibid., 111.
[4]   Personal communications between the author and Gui Bonsiepe, 2007.

British cybernetician Stafford Beer in an audacious plan to build a networked computer system for managing Chile's economy. (Keep in mind this was 1971. The IBM PC was ten years away. The public internet was more than twenty years away; at the time, the internet's precursor, ARPANET, had only thirteen research nodes.) Flores' plan was called Project Cybersyn—cybernetics + synergy. More colloquially, it was known as El Sistema Synco (Sistema de INformación y COntrol), "system of information and control."[5]

Earlier, Flores had worked on an operations research team of the Chilean State Railways.[6] From that project, he knew of Beer's work and of his book *Decision and Control*,[7] which describes how frameworks from cybernetics might be applied to business management—the sort of work Beer had done for United Steel and International Publishing Corporation in the United Kingdom, forming the consulting firm SIGMA (Science in General Management). Flores was impressed by *Decision and Control*, and, on a visit, saw the book in Bonsiepe's library. Later, Flores told Bonsiepe that he imagined Bonsiepe was the only other person in Chile to have the book and that sealed Flores' decision to hire Bonsiepe to work on the Cybersyn project. That Bonsiepe had the book is an effect of the ethos of the HfG and its interest in systems theory.

The Cybersyn operations room (the network's central node, which was to be housed in the Chilean presidential palace, Palacio de La Moneda, in Santiago, before the 1973 coup intervened) may be among the most well-known artifacts Bonsiepe designed with others (he has been very careful to note that the project was a team effort). The room and its command chairs look a bit like Eero Saarinen's Tulip Chair (Knoll, 1955) meets the StarTrek Bridge (Matt Jefferies, 1966) in Winston Churchill's cabinet war rooms bunker (1939). (See Chapter 40 of this volume for Bonsiepe's description of the design of the Cybersyn Ops Room.)

In addition to the room and furnishings, Bonsiepe also managed both the interface and information design for Cybersyn. And information was the whole point. The system was designed to tell the government what was happening in factories and also how citizens were feeling. Some 500 remote telex machines were to feed information to two mainframes in Santiago. A parallel effort, Project Cyberfolk, was to follow soon after. The project was designed to use "algedonic" meters to get information on the pain ("algos") or pleasure ("hedone") created by government policies through polling devices. "Beer built a device that would enable the country's citizens, from their living rooms, to move a pointer on a voltmeter-like dial that indicated moods ranging from extreme unhappiness to complete bliss. The plan was to connect these devices to a network—it would ride on the existing TV networks—so that the total national happiness at any moment in time could be determined."[8]

---

[5]   Lee, *Think Tank Aesthetics*, 133.
[6]   Ibid., 157.
[7]   (Chichester: Wiley, 1966). The book is still in print.
[8]   Evgeny Morozov, "The Planning Machine: Project Cybersyn and the Origins of Big Data Nation," *New Yorker*, October 13, 2014. For more extensive description of the project, see, in this volume, Chapters 4.10 (a), Bonsiepe's own description of the Opsroom, and 4.10 (b), an

WHY ALL THIS EFFORT?

The primary complaint about centralized economic systems is that they allocate resources poorly, in large part because they lack the required information. (We put up with the inequities of market economies because they are supposed to allocate resources more efficiently.) Yet, living beings (like you or me) are centralized (at least to a degree) and fairly competent at allocating our internal resources. Beer had reasoned that organizations, too, can allocate resources efficiently—thus his "viable systems model". Unfortunately for Beer, Allende, and Chile, the technology available to them in 1971 was not up to the task. (For example, the Cybersyn information displays required a great deal of labor to keep up pace, as they had to be changed by hand.)

Fast-forward fifty years though, and today's technology is more than ready. Technology critic Evgeny Morozov, quoted earlier, has suggested that "Big Data and distributed sensors"—coupled with the internet, cloud computing, machine learning, and mobile devices—"insure [sic] that the market reaches a homeostatic equilibrium by monitoring supply and demand."[9] Morozov used Uber as his prime example, but the leading US tech companies—Amazon, Apple, Facebook, Google, Microsoft, and so on (and their Chinese counterparts) — are all involved in "Cybersyn capitalism." Social psychologist Shoshana Zuboff later called it "surveillance capitalism." Recently, art historian Pamela Lee updated the label to "algorithmic capitalism."

Measurement (or continuous surveillance, if you like) is just a part of the whole operation, which includes monitoring (comparison of current to desired state), acting to correct any errors (control), and optimizing models (learning from reaction to those actions and adjusting goals), creating not only a self-correcting feedback loop but also a self-improving system—the sort of "data refinery" powering leading tech companies today—and coming soon to most organizations. Already, portions of some of the tech companies are at least partially self-driving (operating semiautonomously, like the vehicles several of them are building), and Amazon, Google, and Microsoft have begun offering the necessary software-as-a-service (SaaS) for any organization to have its own data refinery. (For even scarier examples, see Palantir Technologies— "algorithmic policing" in the surveillance state—and the Chinese social credit system.)

Yet Cybersyn and Cyberfolk were not designed for private interests/the market; they were designed for a Solicialist government/purposes and therefore point to alternative futures—"algorithmic socialism"—the sort of information-driven management that may be required to ensure justice, avoid climate disaster, and keep healthcare from bankrupting us. The socialist dimension of information-driven management shouldn't be a surprise. After all, the internet

---

extract from Eden Medina's *Cybernetic Revolutionaries: Technology and Politics in Allende's Chile* (Cambridge, MA: The MIT Press, 2011).

[9]   Ibid.

has become critical infrastructure (as indispensable as other public utilities), and "information is a national resource," as Beer noted way back in 1975.[10]

The ulmers saw the information revolution early—information was one of five tracks at the HfG.[11] And Bonsiepe, in particular, got on board and helped lead the way.

"INTERFACE: AN APPROACH TO DESIGN"

In 1960, Bonsiepe worked with Tomás Maldonado (who was then rector of the HfG) to design an alphabet and "sign system for the display and control panels on an Olivetti ELEA 9003 mainframe computer." (The computer's industrial design was done by Ettore Sottsass, well before his Memphis period.) About the Olivetti project, Bonsiepe wrote, "Without having a name for it, we were working on the subject that is now called interface design."[12] (For perspective, consider that Xerox PARC, a source for many of the features of today's standard computer interfaces, was founded in 1970, and ACM's SIGCHI, the main professional organization concerned with Computer-Human Interaction, was formed in 1982.)

To date, Bonsiepe has worked on thirty-five interface design projects. He has also taught interface design at Köln International School of Design (KISD, 1993–2003), at the Jan van Eyck Academie in Maastricht (1997–9), and at Escola Superior de Desenho Industrial (ESDI, Rio de Janeiro, 2003–5).[13] Among those projects, Cybersyn stands out, but its story did not end with Allende's murder in the CIA-backed coup of 1973. The junta imprisoned Flores for three years. In 1976, with the aid of Amnesty International Flores obtained release to the United States, where he worked as a researcher in the Computer Science Department at Stanford. There, by way of Chilean biologist and cybernetician Francisco Varela, Flores met Terry Winograd, who was a professor of computer science.[14] Flores and Winograd collaborated on several projects, including a software application called the "Coordinator." Flores founded a company (in Berkeley) to release the application—Logonet (later Action Technologies). And in 1987, he invited Bonsiepe to Berkeley to work on interface design and documentation design for a subsequent application (the MHS Message Handling System and the Mail program). For this work, Bonsiepe used a Macinosh SE, MacPaint software, and the then new HyperCard application—the first generation of tools for interface designers before the web.[15]

---

[10]   Ibid.
[11]   Under the 1951 revised curriculum model, the HfG organized around five areas: architecture, city planning, information, product form, and visual design.
[12]   Gui Bonsiepe, "The Exotic and Interests," in *Interface: An Approach to Design* (Maastricht: Jan van Eyck Academy, 1999), 12.
[13]   Personal e-mail between the author and Gui Bonsiepe, November 2020.
[14]   Personal email between the author and Terry Winograd, November 2020.
[15]   Personal email between the author and Gui Bonsiepe, October 2020.

In 1992, Bonsiepe presented a paper, "Design: from the material to digital and back," for the Cultura y Nuevos Conocimientos symposium, at the Universidad Autónoma Metropolitana, Azcapotszalco, Mexico. This paper (published here as Chapter 3.2) became a cornerstone of Bonsiepe's book of selected essays *Interface: An Approach to Design*. The book was completed in 1994, published in Italian in 1995, German in 1996, and English in 1999. During this period, most writing about interface design (also CHI, HCI, UI, UX, web design, interaction design, or even experience design) was of three types:

— picture compilations, show-and-tell, with sample icons and screens
— guides on how-to-code, in Metafont/TEX, Postscript, HyperTalk, Lingo, ActionScript, DBN (later Processing), HTML/CSS/JS, Java, and so on
— rules-of-thumb or "heuristics," perhaps from small-sample "experiments" or case studies, sometimes dressed up as "principles"

These types of publications met the immediate needs of designers to see what was new in technology/new technologies and then how to dive in and make things on their own. A few authors were able to "pull up" and offer ideas of a broader, more lasting sort, an emerging set of principles for approaching designing for the digital realm. Rarer still were authors who offered ideas about what it all might mean—how the new technology might change the way we think about design. (Remember, in the early 1990s traditional designers, like Paul Rand, still claimed, "The computer is just another tool, like the pencil.")[16] The ideas understanding the computer as more than a tool (i.e., a new medium, with its own vocabulary, grammar, and rhetoric) and that its many roles were already reshaping design were not widely understood—and were difficult to see before the public internet burst forth in the late 1990s.

Bonsiepe saw the implications early, and he was one of the very few who offered something like a theory of design. Richard Buchanan was another. But where Buchanan simply identified "interaction" as a new frame, Bonsiepe dove in and explored.

Perhaps most profound, though, were Terry Winograd and Fernando Flores, in their seminal book, *Understanding Computers and Cognition: A New Foundation for Design* (1986). (Bonsiepe's review, "Through language to design" (1986), is included in this volume as Chapter 21.) Winograd and Flores connected design (in the context of computer science and artificial intelligence) to linguistics and philosophy. Heidegger and Maturana figure prominently; not exactly accessible-fare for the average designer or even for most design teachers. (There is some irony in Flores writing a book on design—and a difficult one—in that Flores told Bonsiepe that most designers are what he terms "confusionistas," makers of confusion.)[17]

Bonsiepe again provided a bridge. He noted, "The interface is the central domain on which the designer focuses attention. The design of the interface determines

---

[16] Personal communications between the author and Paul Rand, 1992.
[17] Bonsiepe, personal e-mail to the author.

the scope for action by the user of products. The interface reveals the character of objects as tools and the information contained in data. It makes objects into products, it makes data into comprehensible information. . . . The interface creates the tool. . . . Without interface there are no tools" (Chapter 22, p. 231).[18]

The boldness of these claims and Bonsiepe's purpose in making them may not be evident immediately; further explanation and reflection may be needed.

WHAT BONSIEPE MEANT BY "INTERFACE"

In the domain of computing, an "interface" is a communications link—a physical connection (a plug or cable) or a protocol (a set of rules) for requesting data (or both)—a bridge between two systems, sometimes between two devices but often between a human and a machine or more specifically between a human and a software application running on a computing device.

Beginning with the traditional two elements (the person and the tool), Bonsiepe added a third key element—an action or "a task which the user wishes to perform." He defined the "interface" as linking all three: person, tool, and action. He also emphasized "that the interface is not a material object, it is the dimension for interaction between the body, tool and purposeful action."[19] The designer's role is, in large part, "structuring the action space (topological structure)" for the user.[20] Later, Bonsiepe added, "An interface can illuminate connections or leave them murky and opaque. It can open up possibilities for effective action or obstruct them" (Chapter 26, p. 273).

Bonsiepe began with the narrow case of the human-computer interface (software as a tool). Then he applied his model (person-tool-action-interface, what he called "the ontological diagram of design") to the broader case of material objects (to physical tools), for example a pair of scissors, a thumbtack, "a bread knife, a lipstick, a Walkman, a beer glass, a high precision drill" (Chapter 22, p. 231). And finally, he also applied his model to "semiotic artifacts" or "sign-based" objects (communications tools)—to information. A traditional physical book or a digital multimedia piece (hypertext) is also a tool, which coupled with a user (reader/player), supports the action of learning, which itself supports other actions. He noted, "Typographic design is the interface to the text" (Chapter 26, p. 273).

Bonsiepe framed all of design as interface design, not just software design but also (and explicitly) product design, graphic design, and information design. Drawing on Heidegger, he concluded, "Design is the domain of transforming present-at-hand into ready-to-hand. The notion of ready-to-hand is constitutive of design—and in this central aspect it differs from both art and

---

[18] As with the other introductory essays in this volume that quote or make references to papers in the book, the pagination will be by the chapter number and page reference.

[19] Ibid., 29.

[20] Bonsiepe, "The Interface Design of Computer Programs," in *Interface: An Approach to Design* (Maastricht: Jan van Eyck Academy, 1999), 47.

science, constituting a domain of its own right. . . . I call this domain 'interface'" (Chapter 27, p. 279).

Drawing on Maturana, Bonsiepe noted that his ontological diagram of design suggests a "structural coupling" between the person and the tool. He also noted that the person needs the tool to complete a task, the action in his diagram. Of course, the action serves a purpose, the person's goal, an end for which the task is a means. Operation of the tool (its use) provides at each moment feedback to the user about the current state, which the user may compare to the desired state (to the goal), and so correct any errors. An aim of the designer (and the user) may then be that the tool "disappears" or at least becomes "transparent" "so that he no longer has to think about it and it recedes into the background" so that the user can focus on the task (Chapter 26, p. 273).

In this way, the ontological diagram of design echoes cybernetics and also connects Maturana and Heidegger. ·

Bonsiepe's original "onotological diagram of design"

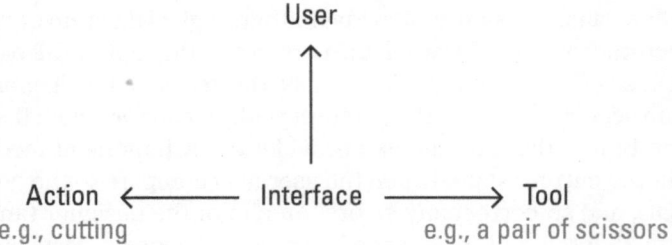

How Bonsiepe applied the diagram to software

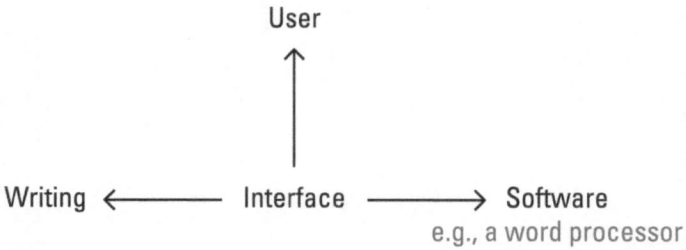

Concepts implied in Bonsiepe's "onotological design diagram"

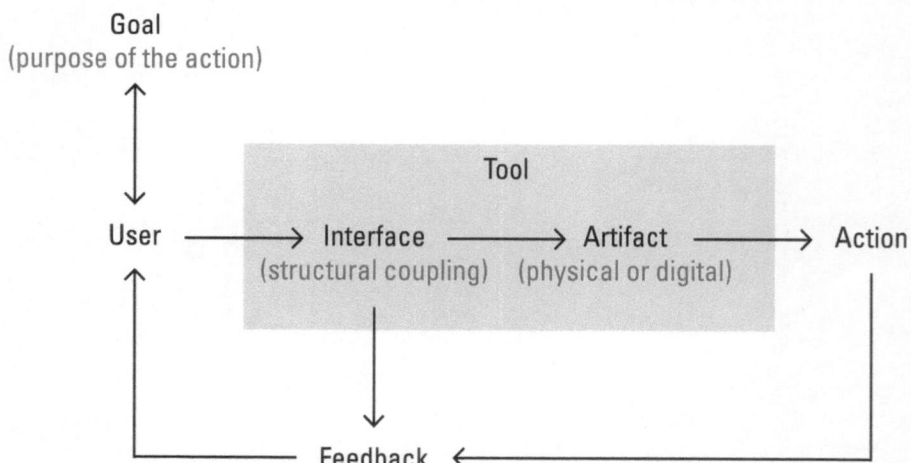

FIGURE 1 Bonsiepe's "ontological diagram of design" and its relation to software and interface

WHY BONSIEPE'S "INTERFACE" FRAME MATTERS

Bonsiepe saw "interface" as a new "approach to design"—the subtitle of his book. And in a 2003 interview (in Chapter 11 of this volume), he said, "I developed a reinterpretation of design as the domain of the interface where the interaction between users and tools is structured. I consider this not a minor contribution to design theory" (Chapter 11, p. 119).

In truth, it is a major contribution.

Here's why. In explicating his "approach"—his framing of design as "interface," Bonsiepe made several related "moves":

— Claimed design as a "fourth way," separate from art, science, and technology.
— Distinguished designing from engineering.
— Provided an alternative to frames which cast design as primarily about cosmetics, aesthetics, drawing, or form.
— Defined the need for theory in design practice.
— Connected design theory and design practice.
— Embraced the idea that language is a fundamental element of design practice.
— Linked the verbal and the visual in design.
— Provided a theoretical basis for a new design discourse.
— And perhaps even reformed the modernist ideal.

Recently, I asked Bonsiepe, "How is that you remain optimistic about the possibility of the modernist ideal (or perhaps 'the ideal of the project of modernity') that design might (or must) make the world better?"

And he replied, "I don't know whether I am an optimist; perhaps better to use the term 'constructive pessimist'. I am aware that we are living in a period of counter-enlightenment known under the term 'post-modernism'. But this approach is not convincing. Without the term 'Utopia' you don't get anywhere (as far as design is concerned)."[21]

In this other horrible year, 2020, one of the things Bonsiepe's essays bring us is a renewed sense that there might still be room in design practice for the modernist ideal that we can and must make the world better. For one thing, we might yet give algorithmic socialism a future.

Interface is out of print. However, this new volume brings forward many of the original book's essays together with a comprehensive selection of Bonsiepe's other writing. I'm delighted to see this new volume. And I commend Bonsiepe's essays (and his career) to design students of all types. Both deserve our study.

---

[21] Personal email.

Design and Language

The subtitle of this book is revealing and ambitious: "A New Foundation for Design." It is the result of the collaboration between a computer scientist and a management scientist and their approach to the domain of design that differs considerably from interpretations rooted in existing professional branches like architecture, engineering, industrial design, and communication design. With their new approach, Winograd and Flores interpret and characterize design as the interaction between understanding and creating, between knowledge and projecting: exactly, that is, what design is about.

21 THROUGH LANGUAGE TO DESIGN (1986)

The field of design theory is dry and empty as a desert. There are only a few orientation aids. The number of publications that can be regarded as relevant for design theory is shrinking to a handful.

Terry Winograd and Fernando Flores have changed this sorry state of affairs with their book *Understanding Computers and Cognition*. They formulate penetrating questions on computer technology, the essence of language and recognition, artificial intelligence, management, planning, and design.

They open new perspectives for design as an ontological human capability, subdivided into a number of disciplines like architecture, industrial design, graphic design, urban planning, and systems engineering. In other words, the authors start in depth, on a level from which a general design theory could develop in the future. Often the problems they raise are laid bare rather than solved.

Like every new approach involving a reappraisal and criticism of prevailing ideas, Winograd and Flores launch particular attacks on the tendency that was widespread in the 1960s to see design as a process of problem-solving underpinned by decision theory. Similarly, they question the popular iconography of the computer as a machine that can think. They also target the rationalist tradition which sees language as a system of signs arranged in structures that can depict facts and objects that exist in the world (what is known as the axiom of correspondences).

At first sight it may seem strange that a book in which philosophical questions are discussed related to the theory of recognition, theory of speech acts, semantics, the theory of perception, hermeneutics, ontology and artificial intelligence accords design so high a rank. But the passage in which the authors discuss new technology will explain their approach: "In order to understand the phenomena related to a new technology we must ask about its design—the interaction between understanding and creating. . . . We address the broader question of how society engenders inventions whose existence in turn alters that society. We need to establish a theoretical basis for looking at what devices do, not just how they operate."[1]

Book review of Terry Winograd and Fernando Flores *Understanding Computers and Cognition* in its Brazilian edition, titled *Computadores e teoria do conhecimento*. Originally published in: *Revista Brasileira de Tecnologia* 17, no. 1 (1986): 74. Published in English in Dawn Oxenaar Barrett (ed.), *Interface – An Approach to Design* (Maastricht, Jan van Eyck Akadamie, 1999) pp. 138–40. Translation by Gui Bonsiepe.

[1] Winograd and Flores, *Understanding Computers and Cognition*, 4.

In design work, contemplative thinking (understanding) is closely bound up with innovative action. Technical descriptions of new technologies (tools) are limited to listing the operational properties of those objects and decoupling the objects from the user, and his needs and interests. But a more in-depth approach asks: What are people doing with these objects? What can they do with these objects? The authors go to the heart of the matter concerning design: "We encounter the deep questions of design when we recognize that in designing tools we are designing ways of being."[2]

That statement reveals the reason why design is seen as an ontological category. For it is through the artifacts that we not only enter into a relation to the world, but that we also constitute our world. Tools in the wider sense must not be limited to their operational character, as means to a purpose. Like language, they must be seen as a constituent means.

In the view of Winograd and Flores, language is not only a medium for reflection, it is also a constitutive medium: "We create and give meaning to the world we live in and share with others. To put the point in a more radical form, we design ourselves (and the social and technological networks in which our lives have meaning) in language."[3]

This interlinking of language and design may seem unusual to designers, in that their competence results from handling nondiscursive codes, mainly drawings, and less discursive codes (language, mathematical calculations). But that is not the point; for language is not a coding system for an objectively guaranteed reality, it is a way to engage in commitments.

The theory of language as a theory of commissive readiness to act is based on the work of John Searle, who has developed a theory of speech acts based on commitments—signaling the readiness of the speaker to perform actions that will fulfill accepted obligations. Searle distinguishes between two kinds of world–word matching: matching the world to the word; and matching the word to the world. In designing, we are moving in the sphere of world-to-word match. Designing means entering into an obligation to ensure that the world meets our intentions.

Another central concept in the design theory of Winograd and Flores is "breakdown"— a term from Heidegger's ontology. Breakdowns are a necessary accompaniment to design, not because the designer lacks competence, but because of the nature of the design process. A breakdown is not a negative situation which one has to avoid. It is the situation of what is not self-evident. A breakdown reveals the tissue of the relations needed to fulfill the tasks. From this follows a clear target for design: anticipating the forms of breakdowns and opening up possibilities for action in case they happen.[4]

A breakdown, in the sense of interruption to the flow of what is self-evident, is not seen as a concept encumbered with negative meanings. The act of designing is interlinked with the inevitable possibility of breakdown. Design

---

[2]   Ibid., xi.
[3]   Ibid., 78.
[4]   Ibid., 165.

competence therefore proves to lie not only in the ability to imagine tools that will work (everyday objects, machines, visual information, organizational networks). Above all, design competence consists in the ability to imagine situations in which they will fail to function—and provide appropriate possibilities of effective alternatives for action.

Due to the diffusion of information technology starting in the 1980s, the range of issues that justify the intervention and competence of designers increased considerably. The new perspective that opened up reflected back to the established interpretations of design centered on form, function, and style. The notion of interface as a central issue of design and its implications were addressed in the article. The shift of the design discourse posed the question of how it would affect peripheral countries that were confronted with the advances of computer technology.

22 DESIGN

From Material to Digital and Back

(1992)

In the course of the last (six) decades, the term "design" has undergone a number of changes, which are reflected in changes in the central issues of design discourse. To put it simply, one can describe the change as follows: in the 1950s, the focus was on productivity, rationalization, and standardization. Industrial production, as exemplified by Henry Ford, was on the one hand the model for distinguishing design from the fine and the applied arts and on the other hand to give it credibility in industry as a new discipline. This debate grew more important in Europe during the reconstruction period after the Second World War. There was great demand for goods, and this could be met by mass production, which enabled consumer goods to be offered on the market at affordable prices. The times had not yet come when design mainly meant product differentiation.

In addition to this central issue of design, growing interest became apparent in design methodology, reaching a peak in 1964 with the publication of Christopher Alexander's book *Notes on the Synthesis of Form*, which has become a classic. The third issue in design discourse is the relation between design and the sciences, both the natural and the social sciences, and the humanities. At a very late stage, design entered the management and marketing discourse.

Any attempt to see design from the standpoint of engineering encounters difficulties, and it generally ends in the—hardly surprising—verdict that design is only a cosmetic exercise, in which a few decorations are merely added to the blueprints produced by product development departments. This narrow view is still found in software companies, where the contribution made by the designer is generally reduced to the idea of screen design and adding visual effects or "souping up."

If industrial production is seen within the categories of engineering, the designer is bound to appear as a makeup specialist, albeit one with the generally enviable ability to sketch and visualize. But design is not drawing. Design is also thinking, and thus a cognitive process. It is important to stress

Originally presented as a paper for the *Cultura y Nuevos Conocimientos* symposium, Universidad Autónoma Metropolitana, Azcapotzalco, Mexico, February 17–20, 1992.

Published in English in Dawn Oxenaar Barrett (ed.), *Interface—An Approach to Design* (Maastricht: Jan van Eyck Akadamie, 1999), 26–36. Translator Dawn Oxenaar Barrett.

In engineering-driven companies and in Latin American companies generally, design remained a fringe phenomenon, because it went beyond the traditional criteria of company management, planning, and engineering.

this, since the general public tends to closely associate design with the ability to draw.

The topic of cosmetic intervention has a long tradition in design discourse. In the 1950s, Max Bill was objecting to what he called the view of the designer as hairdresser. There can be no doubt about the negative connotations of such phraseology. The implication is that design is superficial, of minor importance, and that it need not be taken seriously. With differing nuances this attitude has survived in a tendency to see the aesthetic aspects—appearance and form— as the primary elements of design. The whole subject is then elevated to the level of an artistic and creative process shrouded in an atmosphere of mystery. When one does not know what to do, one can always hide behind the smoke screen of individual creativity.

Designers should not be astonished to see their activities interpreted in such a reductionist fashion. However, they could argue that their maligned design contributions are actually of central importance to very many people. The survival of entire firms is dependent on these supposedly "cosmetic" exercises.

Instead of the view that the designer creates wrappings for the technical structures evolved by engineers, a more differentiated approach may be helpful—it is the ontological design diagram.

This diagram consists of three domains which, as will be shown, are linked by a central category.

Firstly, we have a user or social agent who wants to realize an action effectively.

Secondly, we have a task which the user wishes to perform, for example cutting bread, putting on lipstick, listening to rock music, drinking a beer, and performing a root canal operation.

Thirdly, we have a tool or artifact which the active agent needs in order to perform this task effectively—a bread knife, a lipstick, a Walkman, a beer glass, a high-precision drill rotating at 20,000 rpm.

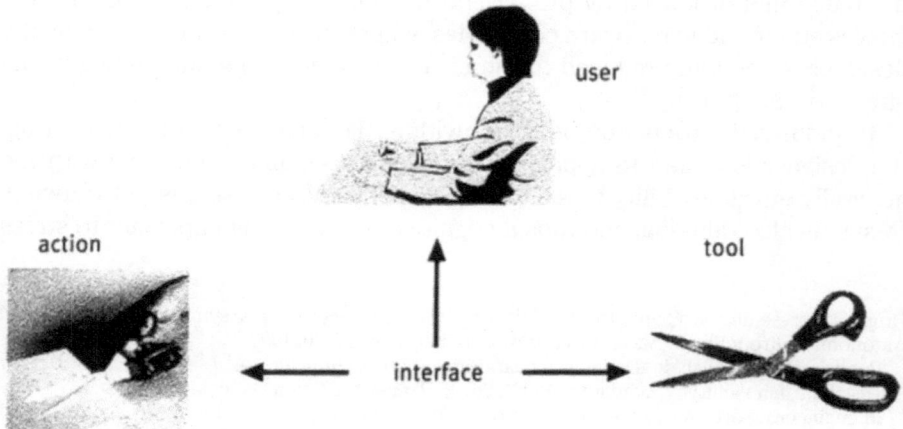

FIGURE 2 Ontological diagram of design.

It must now be asked how these three heterogeneous areas—a body; a purposeful action; an artifact, or information in an act of communication—are connected. They are linked by the interface. It should be emphasized that the interface is not a material object; it is the dimension for interaction between the body, tool and purposeful action. This is true not only of material artifacts but also for semiotic artifacts, for instance information in communicative action. This is the essential domain of design. This position is not meant to dismiss design as immaterial, and certainly not to dispel its materiality. On the contrary, the interface goes beyond the duality of material/immaterial; it covers what they have in common. It covers the design of a spanner just as it covers medical software for the purposes of diagnosing skin disease.

The interface is the central domain on which the designer focuses attention. The design of the interface determines the scope for action by the user of products. The interface reveals the character of objects as tools and the information contained in data. It makes objects into products; it makes data into comprehensible information and—to use Heidegger's terminology—it makes ready-to-hand (*Zuhandenheit*) as opposed to present-at-hand (*Vorhandenheit*).

Three examples will show what the interface achieves: a thumbtack, a pair of scissors, and a travel information kiosk.

The human body consists of a soft mass which is enclosed in a sensitive membrane that can easily be pinched. To use a thumbtack, we need a smooth surface provided by the head of the thumbtack. Without that interface, using thumbtacks would not only be painful, it would be simply impossible.

An object only meets the criteria for being called scissors if it has two cutting edges. They are called the effective parts of the tool. But before the two cutting edges can become the artifact "scissors," they need a handle in order to link the two active parts to the human body. Only when the handle is attached is the object a pair of scissors. The interface creates the tool.

The third example is from digital technology, which is where the term "interface" originates, and it makes the essential function of the interface and its design very clear. The digital data stored in the cloud are coded in the form of 0 and 1 sequences and have to be translated into the visual domain and communicated to the user. This includes the way commands like "Search" and "Find" are fed in, as well as the design of the menu, positioning on the screen, highlighting with color, choice of font. All these components constitute the interface, without which the data and actions would be inaccessible. As we know, the first generation of computer programs that worked with cryptic commands were so difficult to use that the term "user-friendly" was invented to describe the obvious fact that a digital product also has to be usable. Otherwise, it is a mere thing or nonthing—it is merely present-at-hand (*Vorhandenheit*) without being ready-to-hand (*Zuhandenheit*).

Without interface there are no tools. This fact makes interface a core concept providing a stronger argument for design than the culturally oriented interpretation that sees design as primarily concerned with aesthetics.

Let us return to our simplified account of the predominant themes in design discourse: the radical criticism of consumer society and alienation in the 1960s gave rise to hopes of an alternative design, a new product culture, and the possibilities for design in the planned economies that, for lack of a more appropriate term, are now characterized by the derogatory term "real existing socialism."[1] It seemed plausible that a society organized by different criteria could also create a different material culture, a world of consumption but without addiction to consumption.

The political processes since the end of the 1980s appear to have put an end to that idea. The product culture of the planned economies has been wiped out by the wave of commodities produced in market economies. Although design was promoted by government institutions, difficulties were encountered when integrating design into industry. Possibly this was due to the predominance of quantitative criteria in production. However, it may also be the result of a planning discourse where design and innovation remained foreign activities that would disturb the normal flow of production.[2]

In the 1970s, the subject of appropriate technology entered the debate. Moreover, for the first time the Euro-American concept of "good form" came under fire. Arguing on the basis of "dependency theory," Third World countries insisted on the development of their own design. Once the socioeconomic difference between central and peripheral countries was recognized and accepted, this in turn provoked doubts about the validity of a universal definition of design originated in the West.

It was not only the difference in GNP which grouped countries into two classes. To a greater extent, it was the debilitating effect of industrialization, which is evident in the gap between a minority oriented to consumption patterns of the central countries and the marginalized majority, vegetating at bare subsistence level. These wide gulfs in the peripheral societies inevitably give any debate on design in the periphery a political bent.

It is hard in central countries to understand this. In the periphery, the problems of design are primarily political, and only secondly are they technical and/or professional. This predominance of political factors can give the impression that the discussion on design in the periphery is politicized or— even worse—bound to an ideology. By contrast, the seemingly unpolitical and sublimely impartial attitude found in the central countries is bound to appear

---

[1] Elmar Altvater has commented on the inexact nature of the terminology used by the former socialist countries: "The term 'real socialism' came into use in the Brezhnev era and it is used to avoid problems with terminology. These problems would be even greater by using phrases like 'societies in transition' (transition from what to what?), Eastern European countries (there have been socialist experiments in other areas as well as Eastern Europe), 'post-revolutionary' societies (can one really speak of a revolution in many of the countries that are practising real socialism?), centrally steered economies (a term from the neo-liberal theoretical debate), planned economies (this blurs the specific quality of the social model) and so on." Elmar Altvater, *Die Zukunft des Marktes* (Münster: Westfälisches Dampfboot, 1992, 2nd revised ed.), 22.
[2] The product culture including the design culture of the GDR, for example, was simply not known in the West. This deficit is being remedied step-by-step. See the design history of the GDR between 1945 and 1990, https://www.industrieform-ddr.de (accessed August 21, 2020).

naïve or cynical. It is a contradiction on the one hand to proclaim the end of ideology and on the other to indulge in the mass pilgrimage to the cathedrals of consumer dream.

Peripheral countries' attitudes concerning design in the central countries have sometimes been ambivalent. The technical quality of design in the central countries was undeniable, and it often served as reference or model—acknowledged or unacknowledged—for the design that was aspired to. But the lack of technical know-how about processes and finish could easily mean that the design actually produced was second-rate. Attempts were made to compensate for this weakness, which was vaguely perceived, with a passionate search for a design identity—and this was occasionally combined with a nostalgic orientation to the formal codes of natives who had survived the massacres during colonialization. One may ask: Why not simply do design, instead of wasting time searching for an identity? The undertones of nationalism in the Third World can easily provoke the laconic and ironic verdict that nationalism is the last option left to the poor. But this negative assessment overlooks the link between identity and dignity. The search for identity is motivated by the wish for autonomy, and this means being able to have a say in determining one's own future.

In the 1980s, criticism of rationalism and functionalism, or to be more precise, criticism of a caricature of functionalism, revived in various guises. The time of personal gestures had arrived again. The question of the social relevance of design was doomed, and discussions on style and form again predominated in the design scene. Design objects acquired the status of cult objects. A neo-craft of small series production evolved, especially in furniture and lamps, with prices comparable to those on the art market. The slogan was that design should first and foremost be "fun." The customer was not paying for design but for a lifestyle signaled by the exhibition of design trophies.

Now, in the 1990s, environmental compatibility and design management are the main focus of design discourse. The talk is no longer of development generally, but of sustainable development, which readdresses the 1970s theme of appropriate technology, including its plea for development ecologically viable and suitable to the needs of different countries, and taking into account technical and financial resources available locally. Today we talk about self-sustaining growth, and this can be interpreted as a recommendation: the periphery should attempt to cope on its own, while the central countries focus on their own concerns, as long as the debtor countries pay the interest punctually on the loans given during the failed accumulation process. It failed, among others, because industrialization was conceived and implemented without the dynamic factor of innovation.

To judge from the design publications and media coverage, the subject is now being shoved into the limelight. Never before has it been possible to incorporate design as a decisive factor in discussions on the efficiency of firms and national economies. But this also reveals the contradiction between the widespread use of the term "design" and the lack of theoretical grounding.

Today design is a phenomenon that has hardly been researched theoretically, despite its omnipresence in our everyday lives and in our economies.

What is the explanation for this lack of theoretical research? Without attempting to give a definitive answer, one can assume that there is a mutual relation between the shallowness of design discourse and the lack of a stringent theory. So far, design has been an area without a proper foundation, where talk is "small talk."[3]

A reinterpretation of design which looks beyond the "good form" frame of reference and its inherent socio-pedagogical aims, may help to open a new perspective. This interpretation also looks beyond the concept of "lifestyle," where design functions as the supplier of interchangeable items in a scenario for disoriented acquisition potential. The reinterpretation is presented here in the form of seven theses on design:

> Thesis 1: Design is a domain that can be manifest in any field of human knowledge and practice.
> Thesis 2: Design is oriented to the future.
> Thesis 3: Design is related to innovation. The act of design gives birth to something new.
> Thesis 4: Design involves body and space, in particular the retinal domain.
> Thesis 5: Design aims to facilitate effective action.
> Thesis 6: Design is located linguistically in the field of assessments.
> Thesis 7: Design is concerned with the interaction between the user and the artifact—be it an object of daily use or software. The domain of design is the domain of the interface.

The first characteristic of design as a domain of human action takes it out of the narrow frame of disciplines with which the term "design" is generally associated, that is, industrial design, graphic design, fashion, and interior design. There is a risk of falling into the trap of vague generalizations like "everything is design." Not everything is design, and not everyone is a designer. The term "design" does refer to a potential to which everyone has access and which is manifest in everyday life in the invention of new social practices.[4] Everyone can become a designer in his special field, but the field that is the object of design activity always has to be identified. An entrepreneur or a manager organizing a company in a new way is designing, though he probably does not realize this. A systems engineer who works out a process to reduce the misdirection of luggage at an airport is designing. A genetic engineer who develops a new variant of corn that is resistant to external influences is designing. The inherent components of design are not solely concerned with

---

3   In the meantime, the situation has fortunately changed. Sharon Helmer Poggenpohl offers an overview of design-theoretical literature in her book *Design Theory to Go, Connecting 24 brief theories to practice*.

4   Regarding the extension of the concept of design, see Sasha Costanza-Chock: #TravelingWhileTrans, Design Justice, and Escape from the Matrix of Domination. (2020). Design Justice. Retrieved from https://design-justice.pubpub.org/pub/ap8rgw5e

material products, they also cover services. Design is a basic activity whose capillary ramifications penetrate every human activity. No occupation or profession can claim a monopoly on it.

The future is where design belongs. Design is only possible where confidence and hope are united. Where there is resignation, that is, no belief in future prospects, there is no design.

The terms "innovation" and "design" partly overlap. But they cannot be treated as synonymous. Design, as it is understood in this context, means a particular form of innovative action that focuses on the concerns of a community of users. Design without innovative components is an evident contradiction. But innovative action, which creates something new, something that did not exist before, is not sufficient to capture all the aspects of design. For that reason, the idea of concerns needs to be introduced, and this establishes a link with ethics.

It may be maintained that all design ultimately ends in the body. Perceptual space occupies a prime position, because people are first and foremost creatures with eyes. In the case of tools—both material and immaterial (software) tools—the task of design is to link the artifacts to the human body. That process is described by the term "structural coupling."[5]

The traditional interpretations of design use the terms "form," "function," and "style." Instead of linking design to these categories, it is to be more fruitful to see design as located in the domain of effective action. The answer to the question why products are invented, designed, produced, distributed, sold, bought, and used is simple: products are invented, designed, produced, distributed, bought, and used in order to enable effective action.

To assess an action as effective, the implicit standards always need to be identified. To an anthropologist a lipstick is an object for the production of a temporary tattoo, which is applied as part of a pattern of social behavior that we call seduction and self-representation. The criteria by which its effectiveness is judged are very different from those that would be applied to a text editor, a concert poster, or a bulldozer used in road construction. There is no point in talking about effectiveness without also stating the scale of values by which a product is judged as effective for a certain action.

---

[5] Humberto Maturana and Francisco G. Varela, *El Árbol Del Conocimiento* (Santiago de Chile: Editorial Universitaria, 1984), 50–1. Restarting teaching activities during the 1990s at the University of Applied Sciences in Cologne offered the possibility to develop a course in a new area labeled interface design, elsewhere also called Multimedia or information design. This course—the first at that time in Germany—was a reply to the popularization of information technology, particularly when in 1991 was opened the public access to the www. Obviously, advertising, marketing, and corporate design soon became clients for digital services. Another area making use of digital audiovisual resources is information design: making knowledge accessible and understandable. The corresponding theoretical discipline is known under the still-not-customary term "audiovisualistics." The course counted either with digitally literate students or students keen on experimenting with this new technology that permitted to confront intellectually demanding problems. The pedagogical approach chosen is known under the term "problem-oriented learning." The role of design does not consist in creating new knowledge but to make knowledge learnable and increase its comprehensibility.

The concept of interface will help to explain the difference between engineering and design, insofar as both are design disciplines. A designer looks at the phenomena of use with interest that focuses on sociocultural efficiency. Categories in engineering do not include user functionality; they are based on the idea of physical efficiency, which is accessed through the means of the exact sciences. Design, however, builds the bridge between the black box of technology and everyday practice.

Restarting teaching activities during the 1990s at the University of Applied Sciences in Cologne offered the possibility to develop a course in a new area labeled interface design, elsewhere also called multimedia or information design. This course—the first at that time in Germany—was a reply to the popularization of information technology, particularly when in 1991 public access to the www/World Wide Web was opened. Obviously, advertising, marketing, and corporate design soon became clients for digital services. Another area making use of digital audiovisual resources is information design: making knowledge accessible and understandable. The corresponding theoretical discipline is known under the still-not-customary term "audiovisualistics." The course resonated with both digitally literate students and students keen on experimenting with this new technology that permitted them to confront intellectually demanding problems. The pedagogical approach chosen is known under the term "problem-oriented learning." The role of design in this context does not consist in creating new knowledge but to make knowledge learnable and increase its comprehensibility.

23 DESIGN AS TOOL FOR COGNITIVE METABOLISM

From Knowledge Production to Knowledge Presentation

(2000)

ABSTRACT

The use of digital media for knowledge presentation in education and business (knowledge management) raises two fundamental questions: the relation between design and cognition, and the role of the still-to-be-invented rhetoric of audiovisualistics—the combined use of resources from different domains: sound, music, voice, type movement (animation), and images. The question is addressed of how design can help to reduce cognitive complexity and make transparent complex "*Sachverhalte*" (facts). The claim is made that a research policy should not exclusively aim at knowledge production, but take into account also the process of knowledge distribution and knowledge assimilation. In these two phases of knowledge socialization design can assume a decisive role by structuring and presenting knowledge in such a way that it can be effectively absorbed making use of audiovisual resources—including aesthetics as a constitutive domain and not simply as an add-on to usability.

DATA, INFORMATION, KNOWLEDGE

In the discourse on information technology and information design, there exists a "progressivist linguistic chain—from 'mere data' to 'processed data' (information) to 'verified information' (knowledge) to, perhaps, 'existentially validated information' (wisdom?)."[1] Though alerts have been voiced against a hidden ethnocentric bias behind this chain, I use the constellation of these four notions as a starting point in order to address the role that design can play in this process. In other words, I want to give tentative answers to the

Paper prepared for the international symposium on the dimensions of industrial design research *Ricerca+Design*, Politecnico di Milano May 18–20, 2000. First published as "Design as a Cognitive Tool: The Role of Design in the Socialisation of Knowledge," in *Design plus Research* (Milano: Politecnico di Milano, 2000 (May)) and as "Design as Tool for Cognitive Metabolism," *Image and Text*, no. 9 (2000/2001): 34–9. Department of Visual Arts at the University of Pretoria.

[1] David Hakken, *Cyborgs@Cyberspace: An Ethnographer Looks to the Future* (London: Routledge, 1999), 21.

question of how design is involved in this chain when data are transformed into information and when information is transformed into knowledge. The title of this text contains a claim for the crucial cognitive role of design for everyday life, learning, and management—a role that has become more evident with the expansion of information technology. Through arguments and evidence, I shall ground this claim. On this occasion, I don't touch the final philosophical question of the fourth level of how knowledge is transformed into wisdom.

The background for my exposition is provided by writings from various sources ranging from those who address the role of knowledge, and knowledge management in organizations, to those who deal with the role of visualization in enlightenment and the transition from verbal to visual culture. I am pursuing an eclectic strategy for outlining the contour of the issue of design in relation to cognition—an issue that draws on the contributions of disciplines like history, anthropology, computer sciences, and cognitive psychology, to name but a few.

A simple example serves to illustrate the process of transforming data into information and information into useful knowledge. Timetables are characterized as lists of data. These raw—and that means disordered—data about train numbers, departure times, arrival times, routes, and so on become information when they are structured, that is, when they pass from a state of high entropy to a state of low entropy. Already here design intervenes by presenting data so that they can be perceived and received. Once information is organized, it needs to be assimilated by an interpreter who knows what train connections are and—moreover—who is in a situation in which this information addresses a certain concern. The next step of transforming these bits of information into knowledge occurs when a user internalizes, interprets, and uses the information, that is, translates information into action. It should be evident that the way data and information are presented is of crucial importance for enhancing, understanding, and facilitating effective action.

In everyday understanding, knowledge is considered a phenomenon rooted in persons (knowledge in the "brain of persons"), which can also be externalized and deposited as text documents in data banks such as libraries. But two authors of management sciences go one step further and offer the following characterization: "Knowledge is a fluid mix of framed experiences, values, contextual information, and expert insight that provides a framework for evaluating and incorporating new experiences and information. It originates and is applied in the minds of knowers. In organizations, it often becomes embedded not only in documents or repositories but also in organizational routines, processes, practices, and norms."[2]

Though I have certain reservations against this definition of knowledge as mainly instrumental or operational knowledge—leaving aside the hermeneutical dimension, it brings into focus another feature that touches on design: knowledge as accumulated experience needs to be communicated and shared between individuals. The process of communicating and sharing

---

[2]   Thomas H. Davenport and Laurence Prusak, *Working Knowledge* (Cambridge, MA: Harvard University Press, 1998), 5.

knowledge is linked to the presentation of knowledge—and the presentation of knowledge is—or could become—a central issue of design. At first sight, it may not be obvious—or simply taken for granted—that the presentation of knowledge requires the intervention of design actions (*Entwurfshandlungen*); but without design interventions knowledge presentation and communication would simply not work, because knowledge needs to be mediated by an interface so that it can be perceived and assimilated. Otherwise, knowledge would remain abstract and could neither be accessed nor be experienced. Here is offered a leverage point for information design as indispensable domain and tool in the process of communicating and, at the same time, disclosing knowledge. Furthermore, the domain of information design is linked to the domain of education and learning—and, as R.S. Wurman put it, learning and the design of learning—may become a major business in the next century.[3] That is good news. But the bad news is that so far we don't have a coherent theory about information: "Today, in the Information Age, we are struggling to understand information. We are in the same position as Iron Age Man trying to understand iron. There is this stuff called information, and we have become extremely skilled at acquiring and processing it. But we are unable to say what it is because we don't have an underlying scientific theory upon which to base an acceptable definition."[4]

INTERACTION

Though we don't have yet an unequivocal and differentiated definition of "information," we have however a professional practice of information design in which contributions from cognitive psychology, linguistics, theory of perception, learning theory, theory of signs (semiotics), and last but not least, visual designs are integrated. In a recent publication about visualization, we find the following definition of information design as "design of external representations to amplify cognition."[5] Visualization is understood as a domain of computer-based interactive representations. One can go one step further and say that visualization means the transformation of generally invisible processes with the objective to facilitate and enhance understanding.

The various scientific fields that I have listed are grouped around the fundamental concept of communication that has been enriched with new possibilities by technological development: I refer to interactive digital media. Interactive presentation of information is the challenge that traditional graphic design and other research-based disciplines are facing today. Obviously, even a book is an interactive intellectual tool whose convenience has been proven during centuries, but interaction in the more restricted sense today refers

[3]   Richard S. Wurman (ed.), *Understanding USA* (Newport: Ted Conferences Inc., 1999).
[4]   Keith Devlin, *Infosense: Turning Information into Knowledge* (New York: W. H. Freeman & Company, 1999), 24.
[5]   Stuart Card, Jock D. Mackinlay, and Ben Shneidermann (eds.), *Readings in Information Visualization: Using Vision to Think* (San Francisco: Morgan Kaufmann Inc., 1999), 7.

to the presentation of information through digital documents in the form of interactive media.

I am aware of the dangerous appeal of buzzwords—and "interaction" is one of them. But I am using the term in a prosaic way. Interaction refers to a manner of presenting information to a community of users in a nonlinear way, that is, as hypertext or information in a form of branched structures composed of semantic nodes with choices for the user to move through this net of nodes. Here presentation taps the resources of different perceptual channels and enables new ways for presenting information, which allows selective access and a simulated dialogue format particularly scientific information that so far has been predominantly text- or print-based using static resources (typography and illustrations).

Dealing successfully with these multichannel aspects—sound, music, voice, type, images, film, motion—requires different competencies or "literacies" that are brought together in teams composed of the so-called content providers (i.e., persons with factual knowledge about the domain in question), representatives from cognitive psychology, specialists from music and sound design, illustration, programming, writing, and interaction design.

USABILITY FROM A DESIGN PERSPECTIVE

Taking the team approach as starting point for the development of digital documents and tools, we can ask how to characterize the professional responsibility of the designer in digital media. Looking at the numerous, sometimes conflicting, interpretations of design and its difference from engineering and sciences, we can perceive a set of basic features or constants. I shall focus on only two. On the one side, we have the concern for the user, and on the other side we have aesthetic quality. It is the focus on the user and her/his concerns from an integrative perspective that characterizes the design approach. In that aspect it differs from other disciplines (including ergonomics and cognitive sciences); furthermore, a comprehensive design approach does not put aesthetics into quarantine, but explicitly addresses the concern for aesthetic quality, including the dimension of play. At this point, we enter a contentious area, because the domain of usability is strongly claimed by well-known representatives of cognitive sciences that deal with web design and carry the banner of usability engineering methods. In order to formulate this exclusive claim on the domain of usability, a rather narrow vision of the world of web design emerges. "There are essentially two basic approaches to design: the artistic ideal of expressing yourself and the engineering ideal of solving a problem for a customer."[6] In this dichotomy between art and engineering, and between a self-centered focus and a client-centered focus, design does not even enter into consideration; it is simply swallowed up by usability engineering.

---

[6]   Jacob Nielsen, *Designing Web Usability* (Indiana: New Riders Publishing, 1999), 11.

Design vaporizes into the status of a nonentity and designers' expertise is usually dismissed as irrelevant to the process.

We may speculate about the reasons why this has happened. Perhaps it is caused by an understandable and justified reaction against "cool" pages that are user-hostile, though aesthetically captivating—the so-called sexy pages or killer sites. But that is hardly an issue, whereas an uncritical interpretation of usability is at stake that takes this complex notion for granted. Usability appears to be limited to what usability engineers can measure. No designer would deny the necessity of experimental testing of designs, but an understanding of usability that excludes the aesthetic, qualitative dimension becomes a blind victim of aesthetic choices that occur anyway. By the process of self-censorship, a constitutive aspect of use and daily experience in handling digital artifacts is excluded. This approach undermines its own relevance and usefulness for assessing web design projects. Concerns for formal quality cannot be disqualified as glitzy stuff and pushed under the carpet only because they are difficult to assess—they probably fall through the rough grid of usability engineering criteria. The claim that "the way you get appropriate design ideas (and not just good ideas for cool designs that nobody can use) is to watch users and see what they like, what they find easy, and where they stumble"[7] is anything but new—it is what designers do anyway in their profession. Furthermore, it does not explain how appropriate innovations in design occur—it is constitutively conservative and anti-dynamic. Having split up the world into two opposite domains—explaining away design— innovative solutions are explained by referring to the deus ex machina in form of "inspiration" and "creativity."

My final criticism is directed toward the unilateral interest in, for instance, the speed of finding information on a website, because it overshadows the central issue that the design of interactive media serves to communicate and to enhance understanding. Of course, fast access to information is a desideratum, and slow sites with excess of graphical components and distracting animations are a nuisance, but speed is not an absolute goal. Effective communication however is. And this would include notions of hierarchy, structure, and what may be called "sensory management," the expert choice of stimuli which guides readers and holds audiences captive and attentive.

AUDIOVISUALISTICS

Effective communication depends on the use of resources that are intrinsically connected with aesthetics. They can be grouped under the heading of rhetoric—of course, a revised and modernized rhetoric that reflects technological innovations. In classical understanding, grammar was concerned with formulating texts (speeches) in conformity with rules or formalized conventions, whereas rhetoric was concerned with embellishment (ornatus)

[7]   Ibid., 12.

and the reduction of tedium, that is, rhetoric as tool box for avoiding boredom, keeping the attention and maintaining the curiosity of the audience.

A characterization of the role of the designers who design information could state that their contribution consists in reducing cognitive complexity, in producing clarity, in contributing to transparency and understanding. This is achieved among others by judicious application of resources of visual rhetoric, or, as I prefer to call it: audiovisualistics.

FROM KNOWLEDGE PRODUCTION TO KNOWLEDGE DISTRIBUTION

Before presenting and commenting a case history that shows the role of design as cognitive tool or as intellectual technology, I want to present a quote from a specialist of literary studies who makes—according to my view—a bold proposal. We all have heard ad nauseam the lamentations about information overflow, information anxiety, information explosion, information saturation in our so-called information age and knowledge-based economies. But I will not indulge either in euphoric mantras of computer revolution (CR) or in the opposite of information dystopists. The author writes: "I am proposing that the great intellectual challenge of this Age of Information is not coming up with a grand unified theory as in physics or discovering the origins of human life. The great challenge is to be better served by what we already *know*" (emphasis in the original).[8]

Let me explain why I consider this proposal to be a bold proposal and why I consider it particularly relevant for design. It readdresses the priorities of scientific research. Scientists know—and perhaps suffer from—the career-enhancing rite of publishing. Though nobody would object to the production of new knowledge—and that is the main task of the sciences and scientific research—it should be kept in mind that this rite has also its negative side effects.[9] The different domains of knowledge escape any attempt to keep more or less up-to-date. Historians now have some 5,000 journals to carry and inform their work. Therefore, rather than investing huge resources unilaterally to produce new knowledge at an ever-increasing rate, we might redirect some resources to make existing knowledge available. Richard Rorty is quite explicit about that and recommends: "that sociologists and psychologists might stop asking themselves whether they are following rigorous scientific procedures and start asking themselves whether they have any suggestions to make to their fellow citizens about how our lives, or our institutions, should be changed."[10]

Exactly at this point designers ought to step in, because they have—or are supposed to have—expertise in reducing cognitive complexity and help to present information by designing the interface between the information

[8] John Willinsky, *Technologies of Knowing* (Boston: Beacon Press, 1999), 4.

[9] Pacchioni, *The Overproduction of Truth – Passion, Competition and Integrity of Modern Sciences.* New York, Oxford University Press, 2018.

[10] Richard Rorty, "Does Academic Freedom have Philosophical Presuppositions?" in *The Future of Academic Freedom*, ed. Louis Menand (Chicago: University of Chicago Press, 1996), 20. Quoted in Willinsky, *Technologies of Knowing*, 94.

source, the data, and the reader. This new creed of designers runs under different labels like "information architects" (a term that I consider misleading, because it is static) or "knowledge engineers" (a term that I consider even more misleading with its macho-style connotations). I prefer the term "information design," which is the preferred term in continental discourse. Its objective is to facilitate cognitive metabolism, that is, the assimilation of knowledge.

Designers are not known for producing new knowledge—though there are exceptions. In general, however, knowledge production is not the designer's expertise. But designers can play a significant role in the presentation of knowledge. Information technology offers perspectives that Otto Neurath— considered one of the founding fathers of information design in the 1920s— could not have dreamed of. Visual or information design can become a decisive discipline to counter the so-called information explosion and to contribute to information management. It could become a discipline of considerable social relevance replacing graphic design, which has become undermined by technological developments.

MAPPINGS

What are the epistemological and interpretative challenges that designers are facing when they get involved in information design? In order to answer this question, I want to make use of another term of central importance for information design: the notions of maps and the activity of mapping. Again, I want to quote a concise formulation that clarifies what mapping is about— and what it is not about: "the map is perhaps the most sophisticated form yet devised for recording, generating and transmitting knowledge."[11] Maps don't depict a reality—they are not mimetic devices, but they reveal or disclose a reality. The acts of mapping comprise "visualising, conceptualising, recording, representing and creating spaces graphically."[12] But not only physical spaces, but also and above all information spaces. Design faces here a cognitive task of mapping, for instance a loosely structured data bank of information in form of texts, sketches, videos, voice recordings, photos, illustrations, diagrams, and animations about a topic in education, onto an interface that can be perceived, understood, and acted upon by the final user who wants to learn something. Therefore, the design of information requires first giving structure to a mass of data and then translate these into visual and auditive formats with netlike pattern for navigation.

We can differentiate between searching for information and understanding of information. In both cases, maps can serve as devices for orientation and penetrating deeper in a knowledge area. Maps serve two different, though mutually dependent purposes: to facilitate access to knowledge and to assimilate knowledge—what I call cognitive metabolism. On the one side, maps

---

[11]   Denis Cosgrove (ed.), *Mappings* (Reaktion Books: London, 1999), 12.
[12]   Ibid., 1.

provide overviews of data structures and tools for finding, because a surfer is less interested in searching than in finding—we need "find" engines and not "search" engines. On the other hand, maps are devices for translating knowledge into an audiovisual space, which is a perceptual—material—space that makes knowledge tangible. It should be obvious that the multichannel resources offered by information technology increase enormously our possibilities for accessing and assimilating knowledge—and of course transmitting knowledge. The traditional procedure for storing and transmitting knowledge are writing and printing. The importance of graphic representation of speech for social development and education and the advantages of literate cultures compared to oral cultures has been brought to attention particularly by Jack Goody.[13] But nowadays we have audiovisual—and not only graphic—resources at our disposal. Thus, the complexity of knowledge presentation has grown. Handling this complexity is a design issue. Our study programs in primary, secondary and tertiary education are still text-based but will probably in the future be audiovisually based. I am not advocating the loss of the importance of texts; definitely I don't believe in the end of the book. What I am advocating is design practice and design research in audiovisuality.

A CASE: SOFTWARE FOR MEDICAL EDUCATION

In order to reinforce my claim of design as a cognitive tool I shall give a synthesis of one project of information design. I am using this example to indicate the many competencies the information designer needs to acquire. This project shows the approach, method, and contribution of designers for developing material in the form of a CD-ROM to be used in courses for students of medicine.[14] The topic is the function of nerve cells, more precisely of cell membranes which are subject to complex chemical and electrical processes. These complex invisible processes, occurring on an atomic level, are difficult to comprehend when relying on texts only with static illustrations. The understanding of these processes is crucial because they explain, for instance, the reasons why aspirin works.

At the beginning, a medical expert prepared a loosely structured data bank with sketches and texts and a general idea of a scenario (sequence in which the material might be organized in linear fashion without interaction). The designers analyzed and compared medical textbooks, collected material from sites, and analyzed CD-ROMs for education of medical students. Once having acquainted themselves with the subject matter, the designers structured the material in detail, planned animation sequences, sketched illustrations, and edited and rewrote all texts for better on-screen readability. A visual system

---

[13] Jack Goody, *The Power of the Written Tradition* (Washington; London: Smithsonian Institution Press, 2000).

[14] Doris te Wilde and Bina Witte, *Grundlagen der Nervenfunktionen*. Thesis (CD-ROM and documentation), Design Department, University of Applied Sciences (FH), Cologne, 1999.

was developed (color schemes, appropriate type for screen presentations, lines, textures, digital treatment of photographs, types of illustrations, components for animation processes, short movies), which I call audiovisual algorithms.

Thus, organized, the data were mapped on to an interface focusing on clear navigation, orientation, and hyperlinks. Different animations had to be designed in detail. Video sequences were filmed. Texts were recorded for commentaries, and sometimes needed to be rewritten by the content provider—a neurophysiologist—when he discovered that certain issues required more detailed treatment. A glossary of technical terms and a set of exercises for checking the understanding of the material by the medicine students have been added.

At the heart of the project virtual tests were designed, which allow the student user to measure electrical potentials inside and outside a model cell and to read the values on a display. A set of exercises has been added to check the medical students' comprehension of this complex subject matter. The digitized materials were imported into an animation program, with the corresponding programming. The prototype was tested with medical students to assess degree of acceptance, understanding, and quality level of usability. These observations provided essential feedback for an improved second version.

The whole project presented a challenge not only of visualization of complex processes, but of proposing solutions that would enhance understanding and achieve a satisfactory level of visual literacy. It went considerably beyond what is understood as "screen design." It started as an exercise from content, it continued with the transformation of knowledge into some form that could be communicated as shared knowledge. Knowledge has not only to be produced, but it has to be given a form and communicated. The example I presented shows the role of design for the process of knowledge assimilation, and moreover it shows that a successful research policy should not aim exclusively at knowledge production, but should include knowledge communication and knowledge assimilation.

TOPICS FOR A DESIGN RESEARCH AGENDA

I want to finish the tour on design and cognition with a research agenda. Compared with other domains, design is a scandalously underresearched field. I limit myself to mentioning three areas that can provide fertile ground for design research.

1) Design history, in this context of my presentation particularly the history of information design, not understood as a history of heroines and heroes, but as a history of innovations (in the literature about innovations, industrial design and graphic design are hardly considered as fields in which innovations occur apart from marginal aspects related with the form of products).

2) Audiovisualistics
   Classical rhetorical and semiotic studies are limited to text and
   language, that is single media, whereas modern technology offers
   multimedia, which tap multichannel resources and offer selective access
   to information. We see a new culture in the making, but our tools for
   analyzing and understanding the design aspects of this new culture are
   rudimentary and need to be updated. In this way we would build up a
   wall against the avalanche of "porridge speculations" that go under the
   heading of mainly rhapsodic and somnambulist speculations about new
   media.

3) Knowledge presentation, learning, and management
   Educational and business software analyzed from a design perspective,
   which would explicitly include the power of audiovisual rhetoric, could
   lead to a better grounding of design work in this fast-expanding field.

These are but a few topics that could be proposed to the institutions responsible
for financing research. But in addition to financing, we need institution building
that facilitates design research—I am afraid that our institutions of higher
education are not well prepared to deal with these challenges, because their
management structures suffer from hundreds of years of tradition. It is evident
that design research can only be done in interdisciplinary cooperation with
participants from different backgrounds. We will probably observe phenomena
of "intellectual migration": scientists moving into the field of design and
designers moving into the field of sciences. That is an encouraging perspective.

BIBLIOGRAPHY

Card, Stuart, Jock Mackinley, and Ben Shneidermann (eds.). *Readings in Information
   Visualization: Using Vision to Think.* San Francisco: Morgan Kaufmann Inc. 1999.
Cosgrove, Denis (ed.). *Mappings.* London: Reaktion Books, 1999.
Davenport, Thomas H. and Laurence Prusak. *Working Knowledge.* Cambridge, MA:
   Harvard University Press, 1998.
Development of an Interface for On-line Learning Environment Multileu.
   Unpublished research report in html format. Design Department: University of
   Applied Sciences, Cologne, 1999.
Devlin, Keith. *Infosense: Turning Information into Knowledge.* New York: W. H.
   Freeman & Company, 1999.
Goody, Jack. *The Interface between the Written and the Oral.* Cambridge: Cambridge
   University Press, 1993 (1st edition 1987).
Goody, Jack. *The Power of the Written Tradition.* Washington and London:
   Smithsonian Institution Press, 2000.
Hakken, David. *Cyborgs@Cyberspace—An Ethnographer Looks to the Future.*
   London: Routledge, 1999.
Hall, John R. *Cultures of Inquiry—From Epistemology to Discourse in
   Sociohistorical Research.* Cambridge: Cambridge University Press, 1999.

Kastely, James L. *Rethinking the Rhetorical Tradition—From Plato to Postmodernism*. New Haven, London: Yale University Press, 1997.

Klein, Gary. *Sources of Power—How People Make Decisions*. Cambridge, MA: MIT Press, 1999.

Nielsen, Jacob. *Designing Web Usability*. Indiana: New Riders Publishing, 1999.

Oswald, David. *Sound im Interface*. Thesis (CD-ROM), Design Department, University of Applied Sciences, Cologne, 1996.

Poli, Anna. "Impatto della modellazione e dell'analisi computerizzata nello sviluppo delle diverse fasi della progettazione." Unpublished manuscript.

Stafford, Barbara. *Artful Science—Enlightenment Entertainment and the Eclipse of Visual Education*. Cambridge, MA: MIT Press, 1994.

Stafford, Barbara. *Visual Analogy—Consciousness as the Art of Connecting*. Cambridge, MA: MIT Press, 1999.

te Wilde, Doris and Bina Witte. *Grundlagen der Nervenfunktionen*. Thesis (CD-ROM and documentation), Design Department, University of Applied Sciences, Cologne, 1999.

Willinsky, John. *Technologies of Knowing—A Proposal for the Human Sciences*. Boston: Beacon Press, 1999.

Winograd, Terry (ed.). *Bringing Design to Software*. New York: Addison-Wesley Publishing Company, 1996.

Wurman, Richard S. (ed.). *Understanding USA*. Newport: Ted Conferences Inc., 1999.

ONLINE REFERENCES USED AT TIME OF WRITING:

<http://www.DynamicDiagrams.com>
<http://www.cybergeography.org/atlas/info_spaces.html>
<http://www.useit.com/alertbox>
<InfoDesign@wins.uva.nl> InfoDesign Newsgroup
<infodesign-cafe@list.design-inst.nl> InfoDesign-Cafe mailing list.
<http://www.webdev.khm.de> Newsgroup on web development and design

(b) Design/Visuality/Theory

Since its origin, traditional rhetoric is dealing with language—for technological reasons or advertising. To explain just how it might be used in this way is the aim of this article.

The ancient Greeks divided rhetoric as the art of eloquence into three parts: the political, the legal, and the religious. It dealt with the speeches given before public assemblies, in the actualization of rhetoric by including the powerful visual domain. I started to address the importance of the visual domain in a seminar at the HfG Ulm in 1963, though rhetoric was identified as limited to the domain of advertising. The result of this approach consists in the set of definitions of visual/verbal figures or better visual/verbal patterns, published in 1964, in the journal of the HfG Ulm. Because the power of these patterns is linked intrinsically to a particular language, only a hint can be given in a translation. Resuming the topic of rhetoric in the 1990s, the particularities and rhetorical power of rhetorical patterns in digital media were explained. Once rhetoric was an obligatory course in university education. If this forgotten tradition were to be resumed, the quality of academic texts would possibly increase.

24 VISUAL/VERBAL RHETORIC

(1965)

A NOTE ON TERMINOLOGY[1]

Rhetoric has fallen not so much into disrepute as into virtual oblivion. It is but a shadow of its former self. It is taught little, if at all in the schools. And in the few advanced philological curricula which still include it, it is its literary aspects that receive the emphasis with the focus on poetry and drama rather than prose.

Rhetoric has come down to us from ancient times with an aura of antiquity about it. At first sight it seems unfitted for handling the message of the advertiser, which is the rhetoric of the modern age. Yet it can be shown that a modern system of rhetoric might be a useful descriptive and analytical instrument for dealing with the phenomena of legal pleadings and on solemn occasions and set out to show how they should be constructed, how they should be formulated stylistically, how they should be delivered, and what gestures should accompany them. It was primarily the politicians, lawyers, and priests who were adepts in rhetoric since it was their business to use speech to work on their public so as to obtain a definite decision, implant an opinion, or evoke a mood: a decision on a campaign of war, an opinion concerning the prisoner at the bar, a mood in a religious ceremony. To this end there were many means. "Rhetoric is par excellence the region of the Scramble, of insult and injury, bickering, squabbling, malice and the lie, cloaked malice and subsidized lie."[2] The domain of rhetoric is the domain of logomachy, the war of words.

Rhetoric divides into two kinds: one is concerned with the use of persuasive means (rhetorica utens) and the other with description and analysis (rhetorica docens). Practice and theory are closely linked in rhetoric. It is generally defined as the art of persuasion or the study of the means of persuasion available for a given situation.[3] The aim of rhetoric as a corpus of applied methods of persuasion is primarily to shape opinions and more specifically political opinions.

This article is based on a conference given before the AGW (Working Group for Graphic Design and Industry, Stuttgart) on March 25, 1965. First published in German and English as "Visuell-verbale Rhetorik / Visual-verbal rhetoric." *ulm—Zeitschrift der Hochschule für Gestaltung / Journal of the Ulm School for Design* 14/15/16 (1965): 23–40. Republished in Dawn Oxenaar Barrett (ed.), *Interface—An Approach to Design* (Maastricht: Jan van Eyck Akadamie, 1999), 69–82. Edited for length, excluding images and captions in the original.

[1] The term "relatum" and its plural "relata" are synonyms of the term "referent" and"referents," respectively.
[2] K. Burke, *A Rhetoric of Motives* (New York: University of California Press, 1955), 19.
[3] Ibid., 46.

The purpose of practical rhetoric is to exploit words so as to determine the attitude of other people or to influence their actions.[4] Where force rules, there is no need of rhetoric. For persuasion, the possibility of influencing and being influenced, presupposes the possibility of choice. "It is directed to a man only in so far as he is *free*. . . . Insofar as they (or, to borrow a term from Information theory, the recipient of information) *must* do something, rhetoric is superfluous."[5]

These conditions of choice are fulfilled by the situation in a competitive market where goods vie one against the other. The consumer is given a wide range of choice between goods and services, and it becomes desirable to influence him in the selection he makes. This is the function of advertising. And so a new partner joins the triad of politics, justice, and religion as the classic domains of rhetoric.

Of the cataloguing of rhetorical processes, there is no end. Shades of meaning have been set down with philological precision. Textbooks of rhetoric (and they are still textbooks of classical rhetoric) are as notable for their abundance of fine-spun distinctions as for their uncritical acceptance of traditional classifications. A hermetic terminology suited to Latin and Greek makes it difficult to manipulate and use the concepts. Rhetoric is weighed down by more than 2,000 years of ballast. The time has come to bring rhetoric up-to-date with the aid of semiotics or a general theory of signs and symbols. For apart from inconsistencies in the concepts it uses, classical rhetoric, which deals purely with language, is no longer adequate for describing and analysing rhetorical phenomena in which visual and verbal signs, that is, word and picture, are allied. Here the practice of rhetoric has far outrun its theory.

If one thinks of the unending spate of posters, advertisements, films, and television spots turned out by an industrial society with all the facilities of the communications industry at its command and compares it with the very sporadic efforts made to throw light on the rhetorical aspects of this information, the discrepancy stares one in the face.

Classical rhetoric is divided into the following five main sections[6]:

1) Rules for the collection of material, particularly the discovery of arguments
2) Rules for the arrangement of the material when collected
3) Rules for the linguistic and stylistic formulation of the material after arrangement
4) Advice on learning the speech by heart
5) Rules on pronunciation and gesture

For an analysis of advertising information, it is the third main section covering the stylistic features of texts which is useful. These stylistic features appear

[4]   Ibid., 41.
[5]   Ibid., 50.
[6]   According to K. Lausberg, *Elemente der literarischen Rhetorik* (Munich, 1949), 9.

primarily as rhetorical figures, which can be defined as "the art of saying something in a new form" (Quintilian) or as "changing the meaning or application of words in order to give the speech greater suavity, vitality and impact."[7] According to classical theory, the essence of a rhetorical figure consists in a departure from normal speech usage. The departure is made for the purpose of making the message more effective.

The figures can be divided into two classes:

1) Figures of speech which work with the meaning of words or the position of words in the sentence;
2) Figures of thought which work with the shaping and organizing of information.

The way the classical authors effect this division varies from case and case, which does not exactly make for a system in this field of knowledge. Furthermore, under the heading "errors of style" definitions of concepts and assignments of value are so mixed up that it is difficult to separate what there is from what is not acceptable.

The terminology of semiotics makes it easier to sort out these figures with greater precision. Starting from the fact that there are two aspects to every sign, namely its shape and its meaning, we arrive at two basic types of rhetorical figure, for such a figure can operate through the shape of the sign or through the meanings of signs. If we give our minds to the shape, we are moving in the syntactic dimension of signs. If we give our minds to the meaning—or the "relata," to use the term in semiotics—we are moving in the semantic dimension of signs. ("Relatum" is a term embracing everything a sign stands for. The class of the relata is divisible into three subclasses: designata, denotata, and significata. A sign can designate a designatum or denote a denotatum or signify a significatum) [After Tomás Maldonado "Beitrag zur Terminologie der Semiotik" Ulm 1961]. It follows therefore that there are syntactic and semantic rhetorical figures. A figure is syntactic when it operates through the shape of the sign; it is semantic when it operates through the relatum. On comparing the two road signs "Vorfahrt beachten!" (Give way to traffic from right) and "Reserviert fur Fußgänger" (Pedestrians only), we find that the contours, colors, and sign arrangements belong to the syntactic dimensions, whereas the meanings belong to the semantic.

If we sift and simplify the ultrafine distinctions of classical rhetoric (some philologists have drawn up catalogues with hundreds of different figures), we end up with the following classification:

The first main class is composed of the syntactic figures, divided into three subclasses:

1a) Transpositive figures. These are based on a departure from the normal order of words. Here is an example from copy written for a Gauloise

---

[7]   Ibid., 12.

cigarette: (the following specimens are taken from advertising copy
appearing in Western Germany during the past two years) "Herrlich
unkompliziert sind die Jungen—manchmal." (Delightfully uncomplicated
are young people—sometimes).

The normal order would be:

"Die Jungen sind manchmal herrlich unkompliziert" (Young people are
sometimes delightfully uncomplicated).

To give this admittedly incontestable statement more punch, the restrictive
word "manchmal" (sometimes) is singled out for emphasis by the simple device
of changing round the order. This figure is called anastrophe.

1b) Privative figures. These are based on the omission of words. Here is
an example from the same copy: "Sind Sie der Gauloise-Typ? (jung,
unkompliziert, lebensfroh)." (Are you the Gauloise type? [young,
uncomplicated, lighthearted])." The words "Are you . . . ,", which can be
supplied from the context, have been left out here. This figure is called
ellipsis.

1c) Repetitive figures. These are based on the repetition of words. An
example from the same copy: "Das Wesentliche erkennen und lieben.
Lieben überhaupt—lieben, lieben, lieben." (To recognize and love what is
essential. Just to love—love, love, love.)

In a sequence of ten words, the word "lieben" is repeated five times, partly in
direct contact. This figure is called anadiplosis (same word at the end of one
sentence and the beginning of the next).

The second main class is composed of semantic figures, also divided into
three subclasses:

2a) Contrary figures. These are based on the union of opposite relata.
An example from copy for Gorbatschow vodka: "Sie aber müssen
Gorbatschow trinken, um zu entscheiden, ob er nach mehr schmeckt.
(Weil Sie's ihm ja nicht ansehen)." (But you've got to drink Gorbatschow
to decide whether it has more flavour. [Because you can't tell just by
looking]). "More flavor" and "can't tell by looking" are the parallel but
opposite terms in this antithesis.

2b) Comparative figures. These are based on similitudes or comparisons
between the relata.

An example from copy for Brinkmann cigarettes: "Our new baby weighs 1,187
grams."

The word "baby" is being used here metaphorically or in a figurative sense.
It is taken for granted or suggested that between the two spheres of reference,
namely the family as the producer of babies and the factory as the producer of
cigarettes, there is a similarity of some kind or other.

2c) Substitutive figures. These are based on the replacement of one relatum by another. An example from the copy for Gorbatschow vodka: "Wir fragen uns: nach dem wievielten Glas werden Sie sich entschieden haben?" (We wonder: How many glasses will it take you to decide?). Here the contents, that is, vodka, is replaced by the vessel from which it is drunk. This figure is called metonymy.

Each of the six subclasses contains a number of special cases. The most important will be explained by a few examples and catalogued. The subtle distinctions of classical authors, that is, whether a word is repeated at the end or beginning of a sentence or in a direct sequence, have been disregarded.

CATALOGUE OF VERBAL RHETORICAL FIGURES

A. Syntactic figures
   1a   Transpositive figures
      1) Apposition (Latin), epergesis (Greek)—The sequence of the sentence is broken by the insertion of explanatory matter.
      2) Atomization—Making dependent parts of sentences into independent sentences.
      3) Interposition (Latin), parenthesis (Greek)—Insertion of a dependent sentence into another sentence.
      4) Reversion (Latin), anastrophe (Greek)—Departure from normal word or for purposes of emphasis.
   1b   Privative figures
      1) Omission (Latin), ellipsis (Greek)—Leaving out words normally required for a complete sentence, but which can be supplied from the context.
   1c   Repetitive figures
      1) Alliteration (Latin)—Repetition of the same initial letter in words forming the same sentence.
      2) Isophony (harmony)—Repetition of words having similar sounds or of part of a word in a series (e.g., words with the same ending).
      3) Parallelism—The same rhythm in parts of a sentence or in a sequence of sentences.
      4) Repetition (an omnibus term for anapher, epipher, anadiplosis).

B. Semantic figures
   2a   Contrary figures
      1) Antithesis—Confrontation in a sentence of parts having opposite meanings.
      2) Exadversion (Latin), litotes (Greek)—Assertion of a fact by double negation.

      3) Conciliation (Latin), oxymoron (Greek)—Coupling together of contradictory, mutually exclusive relata.

  2b Comparative figures
      1) Gradation (Latin), climax (Greek)—Words in an ascending order of forcefulness.
      2) Superlation (Latin), hyperbole (Greek)—Exaggeration.
      3) Metaphor—Transfer of a word to another field of application in such a way that a similarity (of no matter what kind) between the two fields is assumed and given expression.
      4) Understatement.

  2c Substitutive figures
      1) Denomination (Latin) metonymy (Greek)—Replacement of a sign by another, the relata of both being in a real relationship.
      2) Synecdoche (Greek)—A special case of metonymy. Replacement of one sign by another, the relata of both being in a quantitative relationship.

C. Pragmatic figures
      1) Fictitious dialogue (Lat. percontatio)—The maker of the signs asks and answers himself.
      2) Direct speech.
      3) Conversion of an objection of negative purport into an argument in one's own favour.
      4) Asteism—Irrelevant replies to a question or argument.

In the following, an example of a rhetorical analysis is given. It has not been translated because English and German differ in their rhetorical resources. In recent years, copywriters have indulged in a so-called "chop-style," characterized by short paratactic sentence structures (similar to infants' language).

Figuren
Metapher (neues Baby)
Atomisierung (Genau)
Wiederholung (Baby) Direkte Anrede

Isophonie (Band/Hand)
Atomisierung

Atomisierung, Wiederholung (sie) Parallelismus (—)
Metapher (Kontrollen machen Zigaretten) Atomisierung
(Normalsatz)
Omission (. . . gehört dazu . . .)

Direkte Anrede
Partielle Wiederholung (Sorta)
Atomisierung/Isophonie

(. . . iert)

Omission (. . . wurde aufgewendet . . .)

(Normalsatz)

Atomisierung

Reversion (ginge die ganze Packung zurück)

Fingierter Dialog

With the aid of these definitions from the art of rhetoric, advertising copy can be analyzed and described in terms of its rhetorical characteristics. In this way its persuasive structure can be laid bare. It is the usage among philosophers of language to contrast persuasion with information, opinion shaping with documentation and instruction, and everyday speech with scientific language.

In the eyes of orthodox representatives of a purified and unambiguous scientific language, rhetoric is merely a handbook of verbal tricks which is unworthy of the true scientist. Flexibility of language as seen, for example, in the metaphor (transition from one universe of discourse to another) is construed by them as a defect, a negative quality deserving of censure, which no scientist worth his salt should ever have recourse to. They pillory rhetoric as an avoidable evil, a source of obscurities, ambiguities, misunderstandings, sloppiness, and misrepresentation. Such a purist in language would entitle a book not "Ornament and Crime" but "Rhetoric and Crime."

According to this school of thought, rhetoric is simply ornamentation, the frills with which pure information is decked out. In reply to this the champions of rhetoric argue that the systematic ambiguity of language signs flows inevitably from the genius of language and forms an indispensable part of man's means of communication.[8] It was in the eighteenth century that the view was first advanced that figures of speech are mere decoration or adventitious beauty and that what counts is simple, dehydrated information which the patient and tolerant recipient can absorb without the use of rhetorical figures.[9] In thrashing out the theoretical question whether there can or cannot be any communication without rhetoric, the arguments seem to favor the second alternative. Informative assertions are interlarded with rhetoric to a greater or lesser degree. Information without rhetoric is a pipe dream which ends up in the breakdown of communication and total silence.

"Pure" information exists for the designer only in arid abstraction. As soon as he begins to give it concrete shape, to bring it within the range of experience, the process of rhetorical infiltration begins.

It would seem that many designers—blindly engrossed in straining to impart objective information (whatever that may be)—simply will not face this fact. They cannot reconcile themselves to the idea that advertising is

[8]   I. A. Richards, *The Philosophy of Rhetoric* (New York: Oxford University Press, [1936] 1950), 40.
[9]   Ibid., 100.

information aimed at a recipient and that its informative content is often of subsidiary or no importance at all.

It is hard not to feel a little sympathy for this view, mistaken though it may be, for it is the expression of a certain unease and dissatisfaction with the role of the visual designer in our competitive society where his abilities are often wasted on the mere representation of the imaginary qualities of goods and services. And this representation often strikes a grandiloquent note of glorification which is often in flagrant contrast with the crashing triviality and banality of the product offered. The exultant superlative produced to order is humbug. It is just as much humbug as "objective" information in advertising, which is ashamed of its promotional purpose and tries to dissemble itself.

As examples of information innocent of all taint of rhetoric, we might take the train timetable or a table of logarithms. Granted this is an extreme case, but because it is an extreme case, it is very far from representing an ideal model. Fortunately, communication is not tied exclusively to the perusal of address books. It would die of sheer inanition if these were to be its exemplar.

Once the point is yielded that there are various grades of rhetorical infiltration, then the question arises how these different grades can be assessed in terms of quantity. Mensuration and numerical data are the order of the day. They are flaunted as the proud achievements of science. Despite a certain suspicion of that fetishism of figures which will accept new knowledge on the sole condition that it is in numerical terms, we can broadly trace out a means of determining the rhetorical content of a text. In measurements one must keep to the ascertainable. And what is ascertainable in a text is the number of rhetorical figures of various kinds which it contains. The ratio of rhetorical figures to normal sentences in advertising copy is an index of its persuasiveness. If ten rhetorical figures and five normal sentences appear in a text, it may be said to have the degree of persuasion 2. But no one actually says what persuasiveness is. It is not even defined. All that is given is the data needed to measure what is called persuasiveness.

Dividing the rhetorical figures into their various classes can also be of use in characterizing a text. This gives the rhetorical profile of a text, that is, the ratio of syntactic to semantic and to pragmatic figures. The quoted example shows the following values:

Persuasiveness: 10 (20 rhetorical figures to 2 normal sentences)
15 syntactic figures
2 semantic figures
3 pragmatic figures
Rhetorical profile of the text: 100/13/19
(15 syn. I 2 sem. / 3 pra.)

Verbal rhetoric paves the way to visual rhetoric. As we said before, classical rhetoric was confined to language. But most posters, advertisements, films, and television spots contain linguistic and nonlinguistic signs side by side. And these signs are not independent but interact closely. So it makes good sense to ask about typical picture/word combinations, typical sign relations, and visual/verbal rhetorical figures.

Visual rhetoric is still virgin territory. In what follows, we shall make some tentative efforts to explore this new country. Our discussion is based mainly on interpretations of the analysis of a series of advertisements made during a course on visual rhetoric in the Visual Communication Department of the Hochschule für Gestaltung at Ulm in the first quarter of the study year 1964–5.

With the results attained with verbal rhetoric to guide us, we dissected out figures having exclusive reference to the interplay of word and picture. The terms of verbal rhetoric were used to designate the concepts of this new rhetoric. New concepts were introduced where necessary. In this first approach, the visual/verbal figures were simply noted.

The work of classifying and systematizing them still remains to be done.

To define a visual/verbal figure, it is no longer enough to apply the criterion of the "departure from normal usage" as in verbal figures, for no one can see what relations between verbal and visual signs establish the standard from which one can depart. For this reason, we should probably do better for purposes of definition to fall back upon the possible interactions already inherent in the signs. Thus, a visual/verbal rhetorical figure is a combination of two types of sign whose effectiveness in communication depends on the tension between their semantic characteristics. It is no longer a question simply of adding up the signs; they interact and their final effect is a summation.

BIBLIOGRAPHY

Barthes, R. "Elements de semiology," in *Communications 4*, Paris, 1964.

Barthes, R. "Rhetorique de l'image," in *Communications 4*. Paris, 1964.

Black M. "Metaphor," in *Models and Metaphors*. Ithaca, 1962.

Brooke-Rose, Ch. *A Grammar of Metaphor*. London 1958.

Burke, K. *A Grammar of Motives*. New York, 1955 (2.ed.).

Burke, K. *A Rhetoric of Motives*. New York, 1955 (2. ed.).

Carpenter, E. "The New Languages," in *Explorations in Communication*, eds. E. Carpenter und Marshall McCuhan. Boston, 1960.

Cassirer, E. *Wesen und Wirkung des Symbolbegriffs*. Darmstadt, 1956.

Dorfles, G. *Simbolo Communicazione Consumo*. Turin, 1962.

Durand, G. *L'imagination symbolique*. Paris, 1964.

Empson, W. *The Structure of Complex Words*. London, 1952 (2. ed.).

Graver, J. N. "On the Rationality of Persuading," *Mind* LXIX, no. 274 (April 1960).

Joseph, M. *Shakespeare's Use of the Arts of Language*. New York, 1947.

Keim, J. A. "La Photographie et sa Legende," in *Communications 2*. Paris, 1963.

Klotz, V. "Leo Spitzers Stilanalysen," in *Sprache im technischen Zeitalter*, Nr. 12. Stuttgart, 1954.

Lausberg, K. *Elemente der literarischen Rhetorik*. Münchеn, 1949.

Lausberg, K. *Handbuch der literarischen Rhetorik*. München, 1960.

Maldonado, T. *Beitrag zur Terminologie der Semiotik*. Ulm, 1961.

Perelman, Ch. and L. Olbrechts-Tyteca. *Traite de l'Argumentation*. Paris, 1958.

Richards, I. A. *The Philosophy of Rhetoric*. New York, 1950 (2. ed.).

Richards, I. A. *Speculative Instruments*. London, 1955.

Spitzer, L. "Amerikanische Werbung als Volkskunst verstanden," in *Sprache im technischen Zeilalter*, Nr. 12. Stuttgart 1964.

Volksmann, R. "Rhetorik der Griechen und Römer", in *Handbuch der klassischen Altertumswissenschaften*, Vol. II, herausg. von I. von Muller. München, 1890 (2. ed.).

Volpe della, G. *Poetica def Cinquecento*. Bari, 1954.

Vonessen, F. "Die ontologische Struktur der Metapher," *Zeitschrift für philosophische Forschung* XIII, Nr. 3 (1959).

In the beginning of the 1960s, the company Olivetti developed a main frame computer ELEA 9003 (begun in the 1950s) and approached Tomás Maldonado to design a sign system for the control panels of this complex new tool.[1] *At that time, the term "interface" was unknown and not used to name this new topic. The users of the main frame computer, to deal with voluminous quantity of data (for instance, in banks), were computer specialists interacting through a command-line interface with the system. Obviously, the space limitations on keyboards and control panels put a limit for the use of linguistic resources. Furthermore, it was considered inadequate to depend on the use of a particular language in a context of international use. Therefore, a sign system was designed for this special situation. Twenty-five years later in Berkeley, computer technology had advanced enormously so a less cumbersome mode of interaction with computers through graphical interfaces could be realized. Step-by-step the field of interface design developed.*

---

[1] Elisabetta Mori, "Sottsass, Maldonado and the Sign System for Olivetti," in *UMR Unité Mixte de Recherche. Savoirs, Textes, Langage, Université de Lille* (Lille: Université de Lille, 2020).

25 THE INTERFACE DESIGN OF COMPUTER PROGRAMS

(1990)

INTERPRETATIONS OF THE TERM "INTERFACE"

Interface design for computer programs is a new field of operation for designers. In computer sciences, the role of interface design is occasionally dismissed as a cosmetic exercise, or "souping up."[1] Such an idea is no different from other short-sighted attitudes to design, which see the dimensions of use and aesthetics as a gray area, with aspects that are often difficult to grasp in rational debate.

In a current interpretation, the human user interface is a "means by which people and computers communicate with each other."[2] That paradigm also appears in the following definition: "the interface is the totality of all communication between the computer and the user. It offers the user information and in turn receives information from the user."[3]

Although there is an understandable tendency to see the relation between the user and the computer as a communication process in which information is exchanged, this runs the risk of overemphasizing the communicative aspect, placing not enough emphasis on the program as a tool.

As well as the communication paradigm, one finds in specialist literature the statement that the central function of interface design is to help the user construct a model in his head, which is a replica of the in-depth knowledge of the programmer. In this view, the user is understood to have learned a program if he can create his own replica of the programmer's model. Any difficulties the user experiences in learning and using the program are thus attributed to either the lack of a model, or to clinging to a faulty model. Following that notion, the quality of interface design would be indicated by how quickly and correctly the user can construct a replica of the program and the way it works. This view of the interface is based on assumptions regarding the learning process that

Originally published in *Visible Language*, xxiv, no. 3/4 (October 1990): 262–85. Republished in Dawn Oxenaar Barrett (ed.), *Interface—An Approach to Design* (Maastricht: Jan van Eyck Akadamie, 1999), 42–9. Translated by Gui Bonsiepe.

[1] That the interface can no longer be regarded as a face-lift is evident from the cost of its programming, although this gives no indication of the expenditure on design. In recent years, the interface has accounted for on average 48 percent of programming cost. And this excludes design cost altogether. (Myers and Rosson 1992, quoted in Jacob Nielsen, *Usability Engineering* (Boston: Harcourt Brace & Company, 1993), ix.). But the dimension of use is particularly difficult to grasp within the conceptual framework of engineering sciences.

[2] *SAA—Common User Access: Panel Design and User Interaction* (Boca Raton: IBM, 1987), 7.

[3] *Human Interface Guidelines: The Apple Desktop Interface*, xii.

should not be accepted unquestioningly. The user has learned a program if it is transparent to him, so that he no longer has to think about it and it recedes into the background.

The immaterial nature of software has possibly encouraged an inclination for such interpretations, since the user does not have direct access to the program. Using a computer is not the same as getting into a car and grasping the steering wheel. It may be assumed that the user is less interested in communicating with an object (the computer) or constructing an isomorphic model. His interest lies in effectively solving a number of tasks. He is interested in action, and not in how the program works.

Interface has also been defined as "a specification of the 'look and feel' of a computer system. This includes what types of objects the user sees and the basic conventions for how the user interacts with those objects."[4] That definition conveys the essence of an interface and interface design, because it deals with specifications and the design of the visual components as well as the rules for interaction with them through keyboard, mouse, track pad, and touch screen or via voice commands. Direct manipulation interfaces are composed of a number of graphic building blocks in the form of windows, icons, menus, and keys that serve as metaphorical constructs for users who are familiar with the office environment (files, documents in the form of data sheets). However, these graphic objects do not so much depict a world as they constitute that world. Proceeding from the ontological diagram of design, the interface can be characterized as the domain where the computer program and the user are structurally linked ("coupled" in the terminology of Maturana and Varela). The connection is formed mainly in retinal perception, and supported by the auditive channel. The "look and feel" concept refers to that dimension of coupling. Retinal perception is structured by graphic distinctions familiar to the designer: form, color, color position orientation, texture, and transitions or time transformations, as found in film and television.[5]

Action triggers, like the command options in a pull-down menu, are important in the interface. The selection of the commands; their names, grouping, and distribution on various levels; and their visual treatment are a major part of interface design. The quality of the interface can be measured by the new possibilities for action it offers to a group of users.

Occasionally, it has been argued that from the standpoint of the user the interface *is* the program. This is a strong claim. Admittedly, a program is assessed by its stability, speed, and power. These distinctions go beyond the core of the interface, although speed and performance of a program are perceived through the interface. That can be supported by the fact that today the development of software begins with the simulation of the interface. It explicitly takes the user into account, including how he will learn the program and be able to work with it. That is a radical change since the first generation

[4]   *The Open Look UI (User Interface), Style Guide* (Mountain View: SUN Microsystems, 1989), 1.
[5]   Jacques Bertin, *Semiology of Graphics* (Madison: University of Wisconsin Press, 1983) (First French edition 1967).

of computer programs, where questions of the interface were regarded as an irritating restriction, and only tackled when questions of computational functionality and the architecture of the program had already been solved.

DESIGN RULES, USE, AND USABILITY

Today a number of guidelines are available for interface design. Probably the best-known work in this field is by *APPLE*.[6] The arguments for creating and maintaining coherence both within a program and between programmes are convincing. Well-designed interfaces enable programs to be mastered more quickly and make them easier to use. In order to ensure coherence in the interface, the so-called style books are used to specify: general principles; definitions of building blocks; visual design of elements; auditive signals; rules for action and the arrangement of visual building blocks. The pioneer work on the *STAR* interface, developed by *Xerox PARC*, was highly influential on the development of style books for the operating systems of other firms.[7]

BUILDING BLOCKS OF DIRECT MANIPULATION INTERFACES

Unlike command-line interfaces, the most important components of direct manipulation interfaces are the windows with borders, corners, headings, and control elements to move the window and to enlarge, reduce, or close it. Common control elements are push buttons, radio knobs, check boxes, scroll bars, rulers for adjustments, palettes ,and text input fields. Visual cues are used in the form of changing cursor shapes, icons, process indicators, active/inactive windows, and blinkers. These building blocks serve as boundary conditions for the design of specific interfaces.

CASE STUDY OF AN INTERFACE DESIGN[8]

The work on the interface started with the general specifications of the program. The following functions were needed:

— Writing and reading messages
— Individual and group addresses

[6]   *Human Interface Guidelines, The Apple Desktop Interface* (New York: Addison Wesley, 1987).
[7]   *The Open Look: Graphical User Interface Specification* (Mountain View: SUN Microsystems, 1988);
      SAA—Common User Access: Panel Design and User Interaction, 1987; *OSF/MOTIF Style Guide* (Cambridge: Open Software Foundation, 1989).
[8]   This program was developed in 1988–9 in the software firm Action Technologies in Emeryville (California). The first sketches were made in mid-1987. The computer scientist heading the development team was Pablo Flores.

— Storing messages in a private data bank
— Recalling messages according to various search criteria (date, sender, subject, draft version)
— Various methods of dispatch (normal, urgent, registered)
— Simple deletion of messages no longer needed
— Adding an attachment to a message
— Forwarding a message with a comment
— Visual and acoustic indication that a new message has arrived
— Importing address lists
— Forms (in this case, only for a telephone message)

Proceeding from this list, the first concepts were sketched out, with horizontal and vertical palettes for the commands for writing, reading, addressing, and sending. One consideration in the design was that the user should have direct access to all the tools in the immediate vicinity of the central work area on the monitor screen, instead of having them concealed in a menu. But that approach was rejected, when creeping featurism once again caused more and more functions to be added, with the result that the palette became overloaded. Because two- or three-level dialogue boxes proved impracticable, the HyperCard[9] principle was used, which means that any point on the monitor can be an event trigger. For example, one button was built into the head of the message as a hot spot, indicated by a change in the cursor shape. Since reading and answering are complementary actions, the writing/reading field was divided into two panes. A message received is opened with the appended answer field.

User manuals are generally regarded as a necessary evil. Their necessity becomes clear if one looks at the considerable number of textbooks available for popular programs. In a maximalist view, the programs should be so easy to use that manuals are not needed. It is easy to say, but almost impossible to achieve. Complex programs in particular need detailed documentation. In an integrated design approach, this material should be included right from the start, as part of the design work, and not written after the program is finished.

When the development of software changes from a program-oriented to a design-oriented approach, it will then be possible to break through the limitations of the mono-disciplinary approach. Then, immaterial tools can be designed that are not contrary to the user's requirements (thus, user-unfriendly).

Design contributions to interface development:

— Observing, analyzing, and interpreting work processes
— Formulating the functionality of use *(Gebrauchsfunktionalität)*
— Structuring the command options
— Sketching the story board
— Structuring the action space (topological structure)

---

[9]   NA. This program launched in 1987 by Apple was one of the first multimedia tools. https://al ternativeto.net/software/hypercard/ (accessed August 21, 2020).

— Designing the graphic building blocks (including the color palettes)
— Designing the documentation
— Designing the tutorials
— Designing the marketing material

Knowledge required for interface design development:

— Knowledge of interface standards
— Basic computer science and programming concepts
— Familiarity with prototyping programs
— Animation techniques and story board writing/sketching
— Scripting languages
— Basics of graphic design in digital medium
— Procedures for user tests
— learning theories.

The digitalization of the writing process and correspondingly the reading process attracted considerable interest in the 1990s, both of writers, language theoreticians, information scientists, and archivists. The role of language and reading were reassessed and new literary forms were propagated in the form of hypertext, though these existed already in predigital form, but the new technology offered increased possibilities to write and handle texts. It should come as no surprise that the role of design and its relation to language had therefore to be reassessed.

26 DESIGNING INFORMATION
(1993)

WRITING AND DESIGNING

"Writing is a special case of design, one of the most arbitrary and indeterminate."[1]

"Unfortunately, the more comprehensive subject of the design of multiple diagrams, in which pictures, figures and words are combined, has not, to my knowledge, been thematized anywhere as yet. . . . All these forms of communication, taken together, will form the main medium of the future."[2]

In the dynamically developing field of media informatics, the contours are still blurred. The field can be described with the terms "computer graphics," "multimedia", "interactive hypermedia," "cyberspace," "virtual reality," and "telepresence." Interface is one concept that is of fundamental importance for all these areas.

In computer science, the word "interface" has two meanings. Firstly, it means a piece of hardware, for example, between a central processing unit (CPU) of a computer and a printer. Secondly, it means the almost ethereal dimension containing everything the user sees on the screen, and what is heard from the loudspeaker when the programme is used. It is this definition of "interface" that leads us to revise the traditional understanding of design.

A few years ago, if one looked for a "Design" section in a book, shop one would not find it. One had to look under "Architecture", "Arts and Crafts," or "Hobbies." Design is now an independent and central field: there is a specific section for design. That must be seen as positive. However, it is questionable whether clarity has kept pace with popularization. Unlike the growth in the literature on design, the theoretical basis hasn't grown at all.[3] This situation is starting to change thanks to the creation of master's and doctoral degree programs. We still lack great discourse. There is no equivalent in this area of human activity to what Newton did for physics, Lavoisier for chemistry, Reuleaux for mechanical engineering, Gauss for mathematics, Freud for

Extended version of a paper originally given during the Cologne Design Days, October 25, 1993. Published in English in Dawn Oxenaar Barrett (ed.), *Interface—An Approach to Design* (Maastricht: Jan van Eyck Akadamie, 1999), 57–63. Translated by Dawn Oxenaar Barrett.

[1]   Ted Nelson, *Computer Lib—Dream Machines* (Redmond: Microsoft Press, [1974] 1976), 36.
[2]   Ibid., 37.
[3]   The situation changed during the first decade of the twentieth century. See the publication *Design Theory to Go*, by Sharon Helmer Poggenpohl.

psycho-analysis, and Panofsky for art history. They were formative influences in these fields of human knowledge and practice; indeed, to some extent they created them. That is, they opened great discourses.

The traditional interpretations of design move repeatedly around a group of concepts—form, function, symbolic function, aesthetic quality, economy, and differentiation on the market. Recently, a new term has been added to this arsenal—the ecological category of "environmental compatibility." The scope for interpretation is largely exhausted, with the exception of ecological design.

For almost thirty years, the International Council of Societies of Industrial Design (ICSID) has maintained the formulation of industrial design given by Tomás Maldonado in 1964. While this is evidence of great foresight by Maldonado over thirty years ago, technological change now enables us to move beyond it.

According to the theoretical approach presented here, the instrumental potential of material and sign-based artifacts is accessed solely through the interface. Only through the interface is the tool made a tool. The artifacts do differ by the degree of their connection to the human body, but it is contradictory to speak of artifacts without an interface, if a user handles them. A researcher in the fashionable field of cyberspace and virtual reality has put forward the thesis that in cyberspace the interface disappears: "In cyberspace appearance *is* reality."[4]

From the ontological diagram of design, one reaches the very opposite conclusion: in virtual reality everything is interface, everything is design—the rest has vaporized. Cyberspace and virtual reality are the climax of design, because the interface has become all-embracing. The techniques to produce virtual reality are—as Tomás Maldonado has shown—simulation techniques in real time with very high degree of iconicity.[5]

The ontological diagram, which was first developed for material artifacts and software, can be applied to what is traditionally understood as graphic design. Instead of the artifact, we have the immaterial artifact, which we call "information." A typographer designing a book layout not only makes the text visible and legible, the interface work also makes it interpretable. Competency in handling visual distinctions like size and type of font, negative space, positive space, contrasts, orientation, color, and separation into semantic units makes the text penetrable to the reader. Typographic design is the interface to the text (see Figure 3).

The growth of the information society and the information glut calls for a revision of the traditional view of the graphic designer as primarily a visualizer. In addition to visualizing concepts, the designer organizes information with the aim of reducing cognitive entropy. The infodesigner structures and arranges information elements and provides orientation aids to enable the user to find a way through the maze of information. In this situation, the graphic designer

[4]   Meredith Bricken, "Virtual Worlds: No Interface Design," in *Cyberspace: First Steps*, ed. Michael Benedikt (Boston: MIT Press, 1992), 363–83.
[5]   Tomás Maldonado, *Reale e virtuale* (Milano: Feltrinelli, 1992).

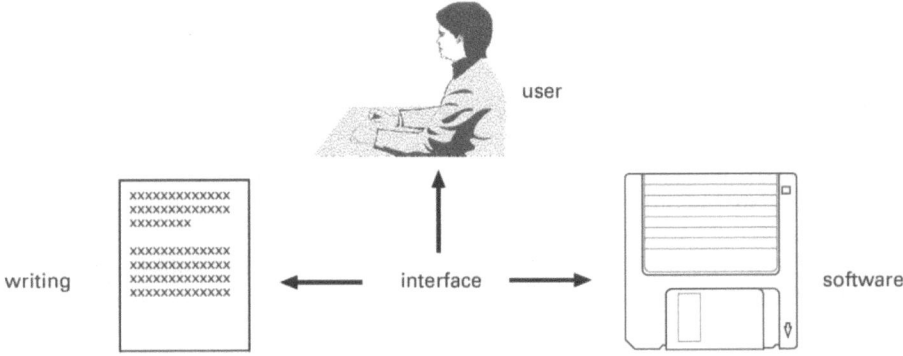

FIGURE 3 Ontological diagram of design in relation to software and information.

becomes an information manager. This shift presupposes cognitive and organizational competence that is generally neglected in design education today.

Info management is not as concerned with producing information as with isolating components from the mass of data, and coordinating them hermeneutically to provide aids for action for others to orient themselves. These new tasks call for a revision of graphic design education programs still determined [1990] mainly by the technology and demands of print media.

The specific new competences of infodesign include the following:

— Finding, selecting, and subdividing information in order to create coherent bodies of knowledge
— Interpreting the information and transferring it to the visual domain
— Understanding the interaction between language, sound, and picture in the dimension of time
— A command of the programs for digitizing pictures, text, animation, and sound
— A knowledge of learning theories
— Handling the constituent components of retinal perception (color, texture, size, orientation, contrast, transitions, rhythm)
— A knowledge of visual and verbal rhetoric
— A knowledge of the methods for checking communicative efficiency experimentally
— Research work (the cognitive dimension of design and teamwork ability)
— Coordinating projects

The entertainment industry, with its strong interest in the multimedia, is eroding the traditional difference between entertainment and education, and has created a special name for the result, infotainment. This blurring of the boundaries between entertainment and learning should not be dismissed as the outcome of ruthless economic interests, although their virulence cannot be underestimated. It may also be appropriate to remember that learning also requires making an effort. It *could* be that infotainment is a new form of regressive information consumption. But it *could* also be that it is a new cognitive technology with emancipatory contents that meets the challenge of making learning interesting and stimulating as well.

Infodesign has been used in learning software (for medical training); workplace productivity software (interfaces for computer programs); online presence in the net (e.g., "www" sites in the Internet); scientific representations and the visualization of scientific data (diagrams, charts); business graphics (digital presentations and business-to-business communication); administration (digitalization of administration processes, digital forms and their handling); and information kiosks and orientation systems. In everyday language, the term "multimedia" is used, although the new element in digital technology is not so much a combination of word, picture, and sound as it is the networking of information arranged associatively, and not on a linear basis. These are the basis of hypermedia and the concept of interactivity. Hypertexts are not radically new inventions. There are literary examples of associative combinations of words that are similar to hypertext. The main work of the Cuban writer José Lezama Lima contains the following paragraph: "History of fire that starts by describing its struggle against the element of Neptune, or water, how fire spreads, fire in the tree, colours of the flame, the stake and the wind, Moses and the burning bush, the sun and the white cock, the red cock for the Germanii, in short, the transformation of fire into energy, all these subjects, the first that have occurred to me and which people today need in order to penetrate new depths."[6]

This quotation shows associative thinking. Such associative processes are a guiding principle in the construction of hypertexts or hypermedia. In view of the exaggerated publicity given to multimedia, we should carefully avoid a naively optimistic approach to technology remembering the 1960s with its boom in programmed learning and later machine translation, complete with high-sounding promises that were never fulfilled.[7] Today there is no more talk of these technologies. The key question that multimedia and hypermedia face is: What about the cognitive content?

Today we have a flourishing doomsday cottage industry, all forecasting the end of the text, the end of the book, and the end of reading, indeed the end of Western culture altogether—although most of them choose to use the book as medium, despite the fact that it is supposed to be facing extinction. A simpler explanation for these complaints suggests itself: hypermedia call established canons of literary culture into question.

If it is proclaimed, by technophile supporters, that cybernauts will soon be able to converse in virtual spaces with synthetically produced hominoids, one may ask whether the priorities are being set correctly and whether this energy could not be better employed in conversing with real people.

---

[6]   José Lezama Lima, *Paradiso*, trans. Curt Meyer-Clason in collaboration with Anneliese Botond (Frankfurt: Suhrkamp, 1984), 428–9.

[7]   Clifford Stoll has put a damper on the often exaggerated and overblown promises of the golden age of the multimedia and the internet in his book *Silicon Snake Oil* (New York: Doubleday, 1995).

Julio Cortázar sketched a negative Utopia of the modern age thirty years ago:

> The realm will be plastic, that is clear. And the world certainly does not have to become the nightmare Orwell or Huxley foresaw; it will be much worse, it will be a wonderful world as its inhabitants see it—no flies, no illiterates, with giant chickens that will have at least eighteen legs, each one extremely tasty, with remote-control bathrooms and different water for each day of the week—a gentle and attentive gesture of the national water supply service—television in every room, huge tropical landscapes for the inhabitants of Reykjavik and views of igloos for Havana, subtle compensations to soften any tendency to revolt, and so on. In other words, it will be a world to satisfy reasonable people.[8]

This literary vision has now moved closer to feasibility, thanks to digitalization and genetic engineering. Whether the possibilities of digitalization are exploited to create the negative utopia sketched by Cortázar depends to a certain point on the quality of intervention of design. The interface can have either positive or negative effects. The interface is a medium that can frustrate and annoy; it can make learning easier or more difficult. It can be fun and make us relaxed at handling information, or it can be boring and stressful. An interface can illuminate connections or leave them murky and opaque. It can open up possibilities for effective action or obstruct them.

---

[8]   Julio Cortázar, *Rayuela* (Frankfurt: Suhrkamp, 1987), 439 (first Spanish edition 1963).

When the word "theory" is used, it inevitably has its counterpart in the form of practice. Design theory versus design practice: this can be understood as irreconcilable opposition, as mutual indifference or as dialectical mediation. The domain of design can hardly be described as theory-friendly. On the contrary: a gap opens up between the theory and practice of design. Where does that come from? The tradition of skill-oriented design training is probably having a significant impact. On the other hand, the primacy of practice over theory is—initially and ostensibly rightly—postulated. After all, design means: intervening in reality, that is, it is practice oriented and interwoven with practice. However, one should not leave it at that. For the fabric of practice is indispensably interwoven with theoretical threads.

27 VISUALITY|DISCURSIVITY

Design, The Blind Spot of Theory; Theory, The Blind Spot of Design

(1997)

> Since practice is an irreducible theoretical moment, no practice takes place without presupposing itself as an example of a more or less powerful theory.
>
> —*Gayatri Chakravorty Spivak (1990)*[1]

The title of this conference "Design—the blind spot of theory" can be reversed into "Theory—the blind spot of design." So both issues will be addressed. The first part will focus on design theory. The second part deals with New Media. Both questions apparently are disconnected. But as will be shown, the consolidation of design in the field of New Media depends heavily on theoretical contributions because the issue we face in New Media is complex and cannot be addressed successfully within the standard frame of reference of graphic design.

DESIGN THEORY AND DESIGN PRACTICE

The relationship between theory and practice in design is a thorny question that generally provokes visceral negative reactions when the topic of theory appears on the agenda of practicing designers. Theory and practice are considered as opposites. Therefore, one might be inclined to replace the word "and" with the word "or." Either theory or practice . . . as mutually exclusive activities. But theory and practice are not as separated in self-contained domains as common sense claims. Practice that considers itself unaffected by theory suffers from a strong error of perception. Theory permeates practice, though generally unnoticed.

Only at a late date did design become a subject which philosophers and scientists reflected upon. For reasons yet to be explained, they did not direct their attention to one of the central phenomena of modernity: the issue of design understood here emphatically in the sense of *progetto*, *Entwurf*, *ontwerp*.

Conference text for a semipublic event of the Jan van Eyck Academy, Maastricht, April 21, 1997. Translated by Dawn Barrett-Oxenaar.

[1]   Spivak, *The Post-Colonial Critic*, 2.

As early as the late 1960s, Herbert A. Simon published his fundamental work on design theory by positioning design within a general theory of artifacts.[2] He set the standards for deliberations on design theory from a scientific and therefore precise viewpoint. The approaches from other worlds of discourse have a harder time of it, particularly approaches that want to access the domain of design within the categories of art history.

CHARACTERIZATION OF DESIGN

Approaches to treating design as manifestation of art should today be considered questionable given that after seventy years the central philosophical concepts to distinguish between the two are available. I refer to Heidegger's notions of ready-to-hand (*Zuhandenheit*) and present-at-hand (*Vorhandenheit*). Design is the domain of transforming present-at-hand into ready-to-hand. The notion of ready-to-hand is constitutive of design—and in this central aspect, it differs from both art and science, constituting a domain of its own right.

Borrowing a notion from computer sciences, I call this domain "interface." I interpret design as interface design, that is, a domain where the interaction between users and artifacts is structured, both instrumental physical artifacts in form of products and semiotic artifacts in forms of signs. Admittedly, each instrumental artifact has also a semiotic facet, but nonetheless the instrumental value is the core for effective action. Interface is the central concern of design activities. I consider the venerable notion of designers as form givers outright obsolete. Particularly in the domain of New Media, we can observe a shift from the concern for form to the concern for structure. Designers thus structure action spaces for users through their intervention in the material and semiotic universe.

CHARACTERIZATION OF THEORY

Theory as contemplative behavior turns the object of contemplation into precisely that: an object. There is something of the voyeuristic trait about theory. What Walter Benjamin said of polemics, namely that they treat an object as lovingly as a cannibal treats an infant, is also true of objectifying theory. It voraciously consumes actual design. Theoretical discourse is also a discourse of power, a discourse of appropriation. Thus, theory constantly gets caught up in a compulsion to legitimate itself. It emerges in the duality of contemplation and action. Theory presumes the materiality of what it is theorizing about. Practice therefore initially has priority over theory. In other words, theory at first impression always arrives too late. But this impression is misleading, for theory affects all design practice. There is no design practice without theoretical components.

---

[2]   Simon, *The Sciences of the Artificial*.

Obviously, theory and practice are different. Theories are not directly applicable to practice, and practice is not an application of a theory. The relationship between these two fields is more complex and makes mutual instrumentalization prohibitive. Theory needs to avoid the danger of abstractness and head for the purported lower levels of practice. Practice, in turn, must not isolate itself in contingency and one-sided directness. Precisely action which obstinately insists on practice and practice only and sets itself as the imperial standard, succumbs to blind opinionating. This is all the more the case when practice blushes as it hears the word "theory." Anyone who barks against theory in fact unconsciously falls victim to it. Anyone who thinks that theory is some leisure-time occupation for the discerning bereft of any relevance for practice shunts himself onto the sidings of history with the signal on "No Future." Any demand that theory should be simple, in keeping with the motto of Apple's computers for the rest of us, is likely to take on board a populist prejudice. Theory is as differentiated as the practice on which it reflects. This is, as is well known, a decidedly complex matter. Were it not to be, then theory would be unnecessary.

LEGITIMIZATION OF THEORY

Why do we need theory, let alone design theory? What is theory good for? Why not spare practice of all theoretical considerations? From where does theory's legitimation come? Does design need a theory specific to it? What can one hope to get from it (and what should one not hope for)? What criteria are there for deciding the relevance of theory?

We cannot expect there to be unanimous answers to these questions. However much the meaning and purpose of theory may be doubted in design, there is at least one firm argument in favor of design theory. All practice is embedded in a world of discourse, a domain of linguistic distinctions that form an indispensable part of practice, even if many repress or deny the fact. Worlds of discourse vary in terms of degree of differentiation and stringency. Things are not good when it comes to design. Compared with other realms, the design discourse stands out neither through differentiation nor through stringency.

Theory can be characterized as the domain in which distinctions are made that contribute to practice having a reflected understanding of itself; in other words, it can help practice be regarded as a problematic issue. Put in a nutshell: Theory renders that explicit which is already implicit in practice as theory. This is why theory is irksome: it casts into question things taken for granted. An approach of this kind does not produce broad sympathy. Theory can be rather discomforting.

In his book *Che cos'è un intellectuale?* (What is an intellectual?), Tomás Maldonado introduced a subtle distinction between "*pensiero operante*" (operational thinking) and "*pensiero discorrente*" (discursive thinking). Based on this distinction, we can put forward the following interpretation: design practice *as pensiero operante* is rooted in the domain of social production and communication. Design theory as *pensiero discorrente*—as discursive thinking—is rooted in the domain of social discourse and thus, in the final instance, in that of politics, where the question is: In what sort of a society do the members of that society wish to live? Let me stress that this emphatic concept of politics in design theory has nothing to do with notions of professional politics or party politics.

VISUALITY

Given that theory constitutes itself in language and lives in discursivity, it has a tense relationship to visuality, a central category of design. This is the case, although epistemology has, since the beginning of classical philosophy, been permeated with visual metaphors—a fact that has been termed the "imperialism of an ocular-centric philosophy."[3]

Things might easily be given an anti-visual bias if theory privileges language and possibly declares it the only form of cognition. At latest since the recent visual turn in the natural sciences, resulting from the development of digital technology, the visual domain has been recognized as a domain that helps constitute cognition. This undermines language's claim to absolute predominance as a primordial basis of knowledge, thus attacking a powerful tradition of discursivity. The latter has a difficult time dealing with visuality.

One can only hope that a New Academy, a New University, will overcome the division between discursivity and visuality. Design theory could be brought to bear fruitfully in investigating the links between visuality and discursivity. Then words would be brought to images, and images to words; discursive intelligence and visual intelligence would be brought together.

A new approach to design education would then probably emerge. We are still in the prehistory of design, in a transition period. Design might—and with caution I would say—will become one of the foundations of higher learning in the New Academies of the twenty-first century, establishing itself as a fourth domain in addition to science, technology, and art. There is a reason for this ambitious—perhaps too ambitious—claim: project-oriented action is possible in all domains of human experience. It is an ontological cornerstone of our existence, in the same way as language.

NEW MEDIA

The breathless expansion of New Media poses some questions about design, language, visuality, and theory. The concern with these issues, particularly design for online and off-line digital documents—known under the popular

---

[3] Levin, *Modernity and the Hegemony of Vision*, 18.

marketing term "multimedia"—has been voiced repeatedly by design educators and professionals.

There is probably a consensus that design education, in particular graphic design education, is not in its best shape today and needs some drastic overhauling. Recently, a group of young media designers found hard words for our educational system, characterizing it as disqualification mills. Why this provocative statement? I suppose that it results from the recognition that in the field of New Media professional practice is advancing so fast that the courses in design departments simply cannot cope with the rate of innovation and are already obsolete from the moment that they are inaugurated.

Sometimes in a mood of resignation the declaration is made that anybody twenty years or older has already passed the phase for mastering the new realities—the Net as the arena for whiz teens and whiz subteens. I would prefer empirical studies to generalizing statements without proper evidence. Certainly, a generation that has grown up spending hours in front of staccato-like MTV with 100 visual changes per minute, gaining a mastery in vision/body reactions in video games, and hacking around days and nights in front of a computer monitor, has gained a particular experience that is literally engrained into their bodies. Nobody will deny that. However, a question not yet answered so far is, whether that base of experience is suited for understanding what is happening and to develop a critical stance against the technology so passionately employed. Only through reflection the danger is averted to credulously swallow everything that is propagated through the megaphones of New Media magazines and media conglomerates with their insatiable appetite to privatize the public domain, if not simply move it into oblivion with only one institution as all-over regulating institution: the Market.

We have been told that the Great Narratives are dead. That is the hallmark of the postmodern condition. But where formerly we had various competing narratives, we face now—on world scale—the propagation of One Meganarrative, called The Market. As any totalizing and universalizing claim, this is a cause for concern.

New Media pose an interesting question with regard to the relation between graphic design and cognition. A new category of graphic design is going to gain its proper profile step-by-step. This is known under various names: infodesign, information design, and information management. It is still in the making and not yet clearly defined.

Infodesign can build on small but exemplary tradition, among which I would quote Otto Neurath, who made fundamental contributions to what I call the visual rhetoric of cognition. He stated—in the beginning of the 1920s—that a visualizer alone is not sufficient and that—as he called her/him—a transformer would be required.

Traditional graphic design is characterized by a strict division between verbal and visual, between text and image. The visual domain is predominant and the capacity for visualization considered the core of graphic design. This paradigm went unchallenged until the New Media appeared. New games are played today. New players have entered the arena of what traditionally has been considered to be the exclusive domain of graphic design. We find gut reactions:

> Multimedia will never go anywhere until the amateurs take over, until the primitives rule and the designers are driven back into their holes.
>
> —David Thomas 1996[4]

Is this a manifestation of bad humor of a musician? Maybe, but it misses its point: what it predicts in conditional form is today already happening.

What are the new competences required from the graphic designer today? Though the term "interactivity" is exposed to overuse, I quote it as the central issue of New Media. Of course, a book, too, is an interactive device, and a brilliant one at that. But interactivity in hypermedia goes beyond the degree of interactivity as it is materialized in books or printed works. Interactivity in digital documents means that the user can choose his own path through a nonlinear structure made up of text in visual form, text in audio form, images, video sequences, animations, music, and sound. And not only choose her or his path, but also choose between different levels of complexity. To write a book for different publics is counterintuitive, but in digital documents this is possible and mandatory. That is new and exceeds the boundaries of traditional graphic design, and of filmmaking and writing. It touches issues of user scenarios (in that aspect similar to theater play and film) and the handling of perceptual and aesthetic variables other than letter form, composition, printed color—though nobody would deny their importance and sophistication.

THE NET

Compared to traditional forms of media, the Net is a fundamentally different medium. As early as 1970, H.M. Enzensberger characterized the difference between the New Media and Old Media (Print) in the following way: New Media are action- and short-term oriented, whereas Old Media are oriented toward contemplation and tradition.

> Their (the New Media) relation to time is opposed to that of bourgeois culture that wants possession, that is permanence, at best eternity.[5]

Established media are basically monologic and one-directional forms of communication that distinguish between producer and consumer, between sender and receiver, between author and public. The Net on the other hand—known also as the Matrix (William Gibson)—is a dialogic medium.

---

4   David Thomas, "It bytes," *emigre* 39 (1996): 45–6.
5   H. M. Enzensberger, "Baukasten zu einer Theorie der Medien," *Kursbuch* 20 (March 1970): 159–86.

The Net with its browsers for home pages and sites changes the predominant paradigm of the designer who controls the variables that make up a design. Design was once controlled from the center; now it moves to the periphery. As is well known, in the Net the user has the choice among variables that determine how type and color appear on his monitor screen. The role of the designer as the controlling instance of all design variables is thus changing. We do not yet have a special term for this situation that designers are facing. Perhaps we could call it open-ended design or fluid design.

"Print stays itself, electronic text replaces itself." (Michael Joyce, 1995)[6]

Documents distributed and made available in the Net undermine the traditional paradigm of the monumental closed text. The printed book is frozen, closed. Electronic documents however have a fluid character. Accordingly, the role of the designer becomes more "fluid," less imposing.

"The reader of a hypertext not only chooses the order of what she reads, but her choices, in fact, become what it is."[7]

"data is just noise if we can't decide what matters and what doesn't."[8]

What makes hypermedia interesting? Certainly not the feverish clicking from one screen to the next, but the play of visuality and discursivity. This we find also in other media, for example film. But what we don't have there—and even less so on TV—is a dialogic interaction.

Let me quote two characterizations of hypertext and hypermedia:

"Hypertext has been called the revenge of text on television since under its sway the screen image becomes subject to the laws of syntax, allusion, and association, which characterize language."[9]

"Print literally gives way on hypermedia screens to digitized sound, animation, video, virtual reality, and computer networks or databases that are linked to it. Thus images can be "read" as texts, and vice versa. Any hypertext holds the prospect of representing on the screen the sights, sounds, and experience of movement through virtual worlds that language previously only evoked in the imagination."[10]

This is a strong claim: the possibility of substituting language-based literature with hypermedia; and I would say that it runs the peril of overpromising. But it points to an important tendency: the importance of the visual domain for text,

---

[6]   Michael Joyce, *Of Two Minds—Hypertext, Pedagogy and Poetics* (Ann Arbor: The University of Michigan, 1995), 232.
[7]   Ibid.
[8]   Ibid.
[9]   Ibid., 23.
[10] Ibid.

not simply as a translation into the visual domain, but as constitutive for the meaning of text.

This claim puts into question the canon of literary standards and education that treats, for example, typography as an addition to the text, but not as constitutive of the text. To this paradigm shift the notion of postliteracy is referring. In this new environment, visual competence is mandatory. We should however be careful and not get hooked up on our skills, because professions based predominantly on skills are very vulnerable when inserted in an environment with fast pace of technological innovations. So, we need to go further if we want to consolidate our profession.

DESIGN-RELEVANT RESEARCH

Design is not known for being an area where new knowledge is produced. This deficit is dangerous, because professions that do not produce new knowledge are pushed to the margin in an innovation-intensive period such as ours. Research generally does not form part of our design education programs. We would need to set up an agenda of relevant design research. This would require a more intensive contact with other domains of human knowledge and experience to create a sensibility or *Problembewusstsein* researchers to focus on design issues. This proposal does not mean to transform design into a science. Such endeavors have not recognized the fundamental difference between design innovation and scientific innovation. But it would require that designers and design students get more literate and develop research and reading practices that would permit them to participate more actively in the design discourse and the broader cultural discourse.

An example can serve as illustration. If we take a look at the booming field of educational software, we discover that the immanent possibilities of digital networks are hardly used. The most comfortable—and secure—way is to continue on the tracks of the traditional paradigm: networks functioning for the electronic delivery of documents where the teacher fulfills the role of an information provider and the student the role of an information consumer. Networked education or network-based educational environments open up— and I would say require—a new understanding of teaching and learning. The teacher would less be a knowledge provider than a coach who orients the student to find and gather information and knowledge. These new learning environments will have to be invented and designed. Here the designer could come to terms with his mission: to be a provider of tools. But in order to cope with this issue, he needs to offer more than visual expertise if he does not want to run the risk to be pushed into an ancillary role of visualizing concepts provided by others.

Today we often hear the complaint about information glut, about too-extensive information exposure with its anaesthetizing effect on the public. Richard Wurman coined the term "information anxiety" for this phenomenon. Here a new area for professional action turns up for designers. They could use their competence in visual distinctions to reduce the complexity that produces perplexity in the public. The reduction of cognitive overload could become a major field of professional action. For lack of a better term, we could use the already-mentioned term "information design." The designer would work as an information vacuum cleaner (as Bilwet has named it).[11] Obviously working in this still-undefined area requires cognitive endeavors that opposes the all-to-often self-referential habitus of our profession. The cognitive rhetoric of graphics is still a speculative possibility, but inevitable if we want to understand the interaction of text (discursivity) and image (visuality), not excluding sound. But it certainly would contribute to "breaking down the barriers between form and content" as Lorraine Wild saw the promise of multimedia.[12]

Our Western culture is characterized by a deep schism between logocentrism and pictocentrism. This split is deeply engrained in our institutions of higher learning. Today for the first time in history it has become technologically feasible to overcome this divided culture. It might be worthwhile to pursue this possibility.

[11] Bilwet Agentur, *Der Datendandy. Über Medien, New Age und Technokultur* (Mannheim: Bollmann Verlag, 1994).
[12] Lorraine Wild, "That Was Then, and This Is Now: But What Is Next?" *emigre* 39 (summer 1996): 18–33.

Design and Crisis

The rather gloomy situation in which this chapter was written in 2011 has not improved—on the contrary. Today the issues of climate change, sustainability, and migration have increased the areas of concern and urgency to address them. In Latin America—and not only in Latin America—the revival of authoritarian forms of government eroding the already-fragile democratic structures of conviviality is alarming. These areas of concern require to address the question of social responsibility of the design profession. The public manifestations and protests are a clear symptom of the depth of the different crises and an indicator that populations, groups, and individuals are no longer willing to tolerate traditional forms of domination. Feminism is one of the currents that acts against patriarchal claims to power and privileges. These are encouraging symptoms of crisis that in the case of design require a rewriting of design history recognizing the role of women in the development of design.

28 DESIGN AND CRISIS

(2011)

Although the term "crisis" first surfaced in media headlines in the wake of the financial cataclysm of 2008, the symptoms of a far greater crisis were discernible well before that—a crisis in international politics, economic policy, the environment, climate, nutrition, genetics, energy, unemployment, and design, too. Seen from the perspective of the center, these crises have been termed the "four horsemen" of the apocalypse of the twenty-first century. Slavoj Žižek lists them: the eco-crisis, the consequences of the biogenetic revolution, the imbalances within the system, above all the exploitation of natural resources, and exploding social divisions and exclusion. Seen from the periphery, there is a fifth horseman, namely neocolonialism, who rides roughshod over the needs of the local populations on the periphery. Perhaps precisely because they dare to protest against how the basis for their lives and their survival is being undermined and thus oppose big business and, among other things, water and ground-polluting open-cast mining, water usage to extract metal ores, and agricultural monocultures entailing the massive application of pesticide substances. These symptoms are invariably reflected in design, too, in design education, in the design profession, and in design theory. The list of the phenomena attesting to the crisis here will not follow some taste for apocalyptic predictions or for offering forecasts that as a rule do not turn out to be true, but instead relies on the original meaning of the word "crisis": a crisis in the sense of an inexorable decision in order to square up to the undeniable turbulence (and often more than turbulence) of the current epoch. These crises are probably not likely to get noticed or accepted by the minds that are still wallowing in the period of the digital Belle Époque. Even if the turbulences may prompt feelings of uncertainty, concern, disorientation, and discouragement to the point of creating an impression of inescapability, a crisis enables us to understand its antagonisms and reasons more clearly and then outline possible alternatives to overcome it.

A crisis suggests that you should revisit the prevailing referential system in which you have hitherto moved. If we cast a glance back over the past three decades, then we will note the gradual erosion of public space, an increasing atrophy of public interests versus private interests and the undermining of the

A talk presented on occasion of receiving an honorary degree from the Universidad Autónoma Metropolitana, México, September 9, 2011. Published in Gui Bonsiepe, "In Theory (Design und Krise)," grid—Zeitschrift für Gestaltung Graphik + Industrial Design, no. 1 (2012): 96—100. This chapter is based on the revised version published in Gui Bonsiepe, Design como prática de projeto (São Paulo: Blucher, 2012). Translated by Gui Bonsiepe. Publication courtesy of Blucher, São Paulo.

notion of democracy. While individual possessivism is deeply rooted in current Western society, it seems to be coming up against its limits now. In the field of design, it is reflected among other things by the hectic innovation carnival in which innovation is celebrated as a value in itself without asking what the content is or above all what the relevance of the innovation is. And this is done in an autoreferencial manner on a scale previously unknown. The person of the designer seems to be accorded more weight than the design itself. To paraphrase Andy Warhol, we could say: Many people today believe the name of the designer *is* design. The way design feeds back into itself is mirrored in the inflation in the symbolic function of a form of design that bows down to the one-dimensional criterion of the market. While hardly no one will pretend that the market does not exert pressure, it is one thing to accept the market as a reality and another to celebrate it as the only reality. Justifiable doubts grow by the day whether the social invention called the market (coupled with an unprecedented compulsion to privatize) is the right instrument (if it ever was) to counter the challenges of today and the near future with any prospect of success. The economists' neologisms seek to sidestep this questioning of the market, raised to the status of a sacrosanct institution, by serving up results flavored with truly poetic inspiration. Instead of baldly talking of a recession, the helpless crisis managers now juggle with the vacuous shell of "negative growth rates." Without doubt, there is an ever greater number of voices in the design discourse (and not just there) calling for a revision to the notion of growth. A glance at the social upheaval and the eco-strain of the Western style of production and consumption that is spreading across the globe justifies this critique and the need for revision. A life practice that at heart amounts only to competition and individualism, and obeys the rampant cult of benchmarking and excellence, yields a great number of undeniable symptoms of increasing social and ecological unviability.

The German philosopher and author Jürgen Habermas asks: "whether a civilization as a whole should allow itself to be drawn into the vortex of the forces driving only one of its subsystems, even if the latter has become the pacemaker of evolution."[1] This sentence reflects the European context in which it was formulated, which is criticized by those authors who use the neologism of "coloniality" coined by Peruvian social scientist Aníbal Quijano,[2] to designate the concealed and dark side of the Modernist project and emancipation. The term goes beyond the customary notion of colonialism by highlighting the way in which labor, knowledge, and interpersonal relationships are articulated in the framework of the capitalist global market and colonial difference.

The same frame of reference is involved with the approach termed "epistemic disobedience,"[3] which has hitherto hardly been taken into account

[1] Jürgen Habermas, *Kritik der Vernunft—Philosophische Texte*, Vol. 5 (Frankfurt: Suhrkamp, 2009), 97.
[2] Quijano, "Notas sobre la teoría de la colonialidad del poder y la estructuración de la sociedad en América Latina."
[3] Mignolo Walter, *Desobendiencia epistémica—Retórica de la modernidad, lógica de la colonialidad y gramática de la descolonialida*d (Buenos Aires: Ediciones del Signo, 2010).

in design as opposed to the practice of developing creative projects. One of the results of the current crisis in design could be such a practice, reconfigured and driven by epistemic disobedience—to date it still seems an unclear possibility, but that is not to say we should exclude it happening. Of late, the issue of unemployment and of exclusion/inclusion has entered the domain of design, which therefore sees itself confronted direct by social problems, which for understandable reasons prompts a vehement response among the mainstream that continues to cling to the idea of a socially neutral and, as it were, aseptic notion of design. The proponents of the latter reject the hope innate in the former as erroneous and naïve, not to say absurd, that the process could impact on the fabric of social relationships. The mindset that allows itself to even wish to reshape social relationships is castigated as outmoded. What you think of this is probably a matter of personal discretion. But we would be blind if we wished to deny the social effect of the activity of designing, especially on everyday life practices. The programmatic dismissal of social context and the conditionality of design activity is tantamount to putting socially relevant projects in quarantine. This political dimension to designing can be illuminated by taking the contrast between center and periphery seriously, and I would like in this context to draw on the example of the history of South America.

In the last decade, the majority of Latin American countries celebrated 200 years of independence. In this context, there is talk of a "second independence." While the countries may have formally gained independence, they have not yet reached the stage of autonomy. And there are various explanations for this. I would like to limit myself to the international division of labor as a result of which the peripheral countries have a role of functioning primarily as exporters of commodities, that is, of materials that no longer need to be processed, such as minerals, petroleum, wood, energy, soybean, meat, and grains, which then serve as inputs for industrially differentiated economies. Latin America exports basically nature. These are products without any added design component. Local design trends seek to oppose this trend by asking to what extent design can serve to strengthen autonomy and reduce heteronomy. From the Eurocentric central viewpoint, it is hard to accept this approach.

There are different sides to the aforementioned question, including the sociopolitical aspect, which designates the constitutive difference between design in the periphery and in the centers. In the periphery, design fluctuates between two poles, on the one hand that which is driven from outside and, on the other, the self-determined design practice, that is, a politics to manufacture the second independence. It should be clear that there is no value judgment involved in defining this difference. I am interested in distinctions, not value judgments. We have to keep in mind that diversification in design has in recent decades led to a brace of new specializations, training programs, and course offerings. The enthusiasm for experience design has already abated and was attractive solely owing to the pipe dream of opening up new worlds for design. In the debate on the subject, two camps emerged: on the one hand, the champions of experience design and on the other the critics who considered the notion of wanting to design experiences to be utter humbug. The material substrate that is evidently indispensable to create "experiences" causes the experience design clear headaches. You cannot design experiences but can by means of the suitable design of material and semiotic artifacts trigger experiences.

To conclude, another element of the crisis in design bears mentioning, namely the efforts to make design art by eliding the distinction between design objects and art objects. Art historian Andrea Giunta writes: "the artistic avant-garde asked itself the programmatic question as to the relation between art and its social setting."[4] Examples of this avant-garde, this "great adventure of Modernism," were the Constructivists during the early days of the Soviet Revolution, who tried to translate art forms into forms of life. Today we can witness the reverse process, with design being interpreted as artistic activity, with the result that art galleries and art museums open their doors to exhibit "design." The industrial product that was once not deemed worthy of note is gentrified by being given the status of a cultural object and is ideal as an investment asset. There was a scandal in the past if industrial products were submitted to an art exhibition as art objects, for example in 1917 in New York when Marcel Duchamp succeeded in submitting an anonymous mass-produced object, to top it all a urinal. Of course, Duchamp was not interested in the design, and even less in blurring the lines dividing art and design, as transitional design today attempts; but he did want with his action to undermine the traditional concept of art and spotlight its arbitrary nature. Compared to this, new attempts today to renew design through art seem pretty harmless. The new and not so new class of designs claim to produce a special class of products as "artistic design" that stands out not only for an explicit indifference (not to say animosity) toward the irritating criterion of utility, but largely limits itself to the variation of customary products for living rooms, such as seating, tables, luminaires, and accessories. Design art has nothing to do with the critical, revolutionary zest of a Duchamp, and should instead presumably be read as an expression of neoconformism that with a radical gesture leaves everything as it was. And no doubt the "design as art" presumably works into the hands of those affirmative curatorial interests that are on the lookout for new exhibition themes which go beyond the established, well-trodden typology of exhibition objects in the form of panel paintings, sculptures, installations, and digital arts.

Literary scholar Edward Said once asked whether, apart from the synchronicity, there was a link between the politics of Ronald Reagan in the 1980s and the flood of literature programs in US universities with the poststructuralist textualism imported from France.[5] Textualism treats texts as self-contained units, with no extra-textual references to society, politics, and economics. Said proposed three questions for what he called a politics of text interpretation: Who is writing? For whom was the text written? Under what

4   Andrea Giunta, *Escribir las imágenes—Ensayos sobre arte argentino y latinoameriano* (Buenos Aires: Siglo Veintiuno Editores, 2011), 58.
5   Edward Said, "Opponents, Audiences, Constituencies, and Community," *Critical Inquiry* 9, no. 1 (September 1982): 1–26.

conditions? If we adjust the content of these questions, then we can transpose them onto design for the purposes of design interpretation, training, and practice: Who is designing? For whom was a design developed? Under what economic, social, and technological conditions?

Starting from these questions we could then probably pinpoint a link between the media-inflated auteur design and the neoliberalism that has dominated things since the 1980s, and on to design cynicism and even design nihilism. Perhaps on the basis of these questions, we could derive the criteria as to the relevance of design and thus find at least one possible way out of the crisis of design—and not just design. The chances of this may be slender at present but of one thing we can be certain: the crisis cannot be outsourced.

As was hinted earlier in Chapter 1.4 ("The invisible aspects of the HfG Ulm"), slowly the importance of Hannes Meyer in the history of the Bauhaus and beyond gets recognized against the traditionally cultivated picture of this institution in which Hannes Meyer hardly appears. One of the features of Meyer's approach to design education has been the acceptance of science as part of design education. This trait was emphasized at the HfG Ulm. It should not cause surprise that this process generated conflicts that became manifest in the basic or foundation course, both at the Bauhaus and later in at the HfG Ulm. Recognizing the role of Hannes Meyer does not mean to diminish the importance of Walter Gropius. A differentiated approach to the history of the Bauhaus is desirable.

29 CONVERGENCES/DIVERGENCES: HANNES MEYER AND THE HFG ULM

(2018)

The discussion of the relationship between Hannes Meyer and the HfG Ulm was given impetus by Tomás Maldonado in an article published in the HfG Ulm journal in September 1963, whose title took the form of a question that sounded provocative at the time: "Ist das Bauhaus aktuell? / Is the Bauhaus Relevant Today?"[1] Maldonado, who played an influential role in the second phase of the HfG Ulm, beginning in mid-1957, was calling for the recognition of the achievement of Hannes Meyer—a step that was by no means popular at that time.

Hannes Meyer had been shunted to the gray zone of oblivion by the writers of the official history—under the influence of Walter Gropius and above all of Nikolaus Pevsner. Just as there is a heaven for the canonized figures, there is a hell for heretics and dissidents.[2]

Gropius responded to this article without delay in two letters in rapid sequence from October 22 and November 24, 1963.[3] Despite the harsh judgments that Gropius makes there about Hannes Meyer even ten years after his death, he at least recognizes Meyer for strengthening the scientific disciplines in the curriculum during his two and a half years as director of the Bauhaus, from April 1, 1928, to August 1, 1930.

On precisely that point there were convergences between Hannes Meyer and the HfG Ulm, although one cannot speak of a direct influence. It was rather a confirmation ex post facto for a curriculum and corresponding pedagogy conceived from mid-1957 onward after the departure of Max Bill.

In writing his article Maldonado was not seeking to pick a fight with stale polemics; rather, he was concerned with revising the history of the Bauhaus with respect to this question. He pointed to an article by Hannes Meyer that was

Paper given to a conference at the Bauhaus Symposium "Hannes Meyer as Pedagogue," March 15–17, 2018. Published under Bonsiepe, Gui. "Konvergenzen / Divergenzen – Hannes Meyer und die hfg ulm," in *Hannes Meyers neue Bauhauslehre - Von Dessau bis Mexiko*, ed. Philipp Oswalt. Volume 164 of *Bauwelt Fundamente* (Basel: Birkhäuser, 2019). Translated by Steven Lindberg. Special thanks go to Philipp Oswalt (University of Kassel), David Oswald (University of Applied Sciences, Schwäbisch-Gmünd), and Christiane Wachsmann (archive of the HfG ulm) for the support during the research of this issue, especially in providing source material.

1   Tomás Maldonado, "Ist das Bauhaus aktuell? / Is the Bauhaus Relevant Today?," *ulm – Zeitschrift der Hochschule für Gestaltung* 8/9 (September 1963): 5–13.
2   Antonio Toca Fernández, "Héroes y herejes: Juan O'Gorman y Hannes Meyer," *Casa del Tiempo* III, IV época (Junio 2010): 18–23.
3   "Stellungnahmen zu 'ist das Bauhaus aktuell,'" *ulm: Zeitschrift der Hochschule für Gestaltung* 10–11 (May 1964): 62–73.

first published in Spanish translation in Mexico in 1940[4] and not until twenty-five years later in German in the book on Hannes Meyer edited by Claude Schnaidt, who at the time was a lecturer in the department of industrialized building at the HfG Ulm.[5]

Tomás Maldonado had formulated the revised concept for the HfG Ulm in 1957–8 and provided conceptual clarity about the design disciplines in the framework of technological, scientific civilization. In a two-page typescript, he sketched the tasks and objective of the HfG:

> It is advantageous to divide the world in which the human being lives as a social creature and in which he produces and produces himself into two parts:
>
> 1) Into the sector of primary or direct artifacts (not the totality but that class of objects whose use occurs with the objects themselves, e.g., tools, apparatuses, instruments, fittings, furnishings, etc.)
> 2) Into the sector of secondary or indirect artifacts (objects whose use occurs by means of the objects themselves, that is, those that fulfill a function of conveyance.)[6]

The Department of Industrial Design and the Department of Building were assigned to the realm of material artifacts, while the design of immaterial artifacts was the responsibility of the Department of Visual Communication and the Department of Information (verbal communication).

From this typescript it becomes clear how the HfG characterized itself, beginning with a turn away from a diffuse improvement of life that had once been proclaimed and with a hypothetical, abstract culture as a whole and instead focusing on concrete, currently existing problems. Maldonado took up the concept of the artifact, which was not common in the discourse on design of the time but used in anthropology, and the category of use, which is central to design. Until that point, the Bauhaus, whether updated or not, functioned as the point of reference. From that moment, the break from the terminologies of the Bauhaus had been made.

Using the example of Hannes Meyer's texts, it is possible to recognize convergencies and divergencies, differences and affinities, between his position and that of the HfG Ulm concerning what constitutes contemporary education

4   Hannes Meyer, "Bauhaus Dessau: Experiencias sobre la enseñanza politécnica," *Edificación* 5 (July–September 1940): 13–28; Hannes Meyer, "Bauhaus Dessau, 1927–1930: My Experience of a Polytechnical Education," in *Hannes Meyer: Bauten, Projekten, und Schriften / Buildings, Projects, and Writings*, eds. Claude Schnaidt, trans. D. Q. Stephenson (New York: Architectural Book Publishing, 1965), 107–13.
5   Claude Schnaidt (ed.), *Hannes Meyer: Bauten, Projekte und Schriften / Buildings, Projects and Writings* (Switzerland: Teufen, 1965); Hannes Meyer, "Bauhaus Dessau: Erfahrungen einer polytechnischen Erziehung," in *Bauen und Gesellschaft: Schriften, Briefe und Projekte*, ed. Lena Meyer-Bergner (Dresden, 1980): 78–88.
6   Tomás Maldonado, "Hochschule für Gestaltung, Ulm (1) Aufgabe und Zielsetzung (1957/58)," typescript, Maldonado hfg (1957) ai az 3226.1, Museum Ulm/ HfG-Archiv Ulm.

of designers and what role design plays in the context of a technological civilization.

SCIENCE AND DESIGN

The subtitle of the article is revealing: "My Experience of a Polytechnical Education"; it is, explicitly not an artistic education. Although it now seems that the artistic aspect is particularly esteemed for cultural events of the Bauhaus, so that the central themes of designing artifacts of everyday use and the training of corresponding specialists—the specific task of the Bauhaus—is largely being ignored.

Hannes Meyer cites as his contribution to the Bauhaus the increased role of the exact sciences in the curriculum, and the suppression of the painter's influence and the development of a pedagogy based on work with actual commissions.[7] He also emphasized the orientation of design work around the needs of the people—a concept that seems alien today. Meyer used the expression "artículos de amplio consumo" (articles for broad or popular consumption) in a marginal note in Spanish in the German manuscript.[8]

With regard to the composition of the student body, he alludes to their class background and emphasizes their proletarianization. A concept as fashionable now as signature design and the effort to stand out as a brand and creative disciple would have been irreconcilable with his view of design.

The decision to update, expand, and objectivize the curriculum after the departure of Max Bill in mid-1957 by introducing scientific subjects quickly turned out to be a center of conflict between divergent interests. This conflict makes it clear that the domain of design lies outside of—if not at odds with—the traditional understanding of educational institutions: on the one side, a design relationship (project) to the world; on the other side—the side of the sciences—a cognitive relationship to the world. For these two perspectives to interact without reciprocal instrumentalization and without hierarchization, it is necessary to have elastic, unbureaucratic university structures—which does not come easily to the ministries responsible for university education. The HfG Ulm—which is neither an academy of art, nor a technical college, nor a university—fell outside the grid of the established universities. In the

---

[7] The following period, during which the architect Hannes Meyer was director, was notable for the emphasis placed on the social mission of the Bauhaus, for the increased role of the exact sciences in the curriculum, for the suppression of the painter's influence, for cooperative development of the workshop units, for making on-the-job instruction the basis of workshop theory, for developing types and standards to meet the people's needs, the democratization of the studies, and for closer collaboration between the students, the workers' movement and the trade unions. MEYER 1940, 8 f.
In: Museum Pedagogical Service Berlin: Materials for Bauhaus 6, enclosure documents. Document 4. Berlin 1989, 8 12 (p. 107).

[8] Hannes Meyer, "Bauhaus Dessau, 1927–1930: Experiencias sobre la enseñanza politécnica," manuscript, Deutsches Architekturmuseum (DAM), Frankfurt am Main.

university landscape, design—with the exception of architecture—represents a placeless category.

It is revealing that two of the institutions that strongly shaped design education in the twentieth century—the Bauhaus and the HfG Ulm—were outsiders. As such, they constantly found themselves in a precarious situation, institutionally and financially, so that they were only able to exist for a few years.

Whereas at the Bauhaus the scientific topics were covered primarily in guest lectures, in the program of the HfG they were anchored in courses and seminars, with a correspondingly higher number of course hours. The value placed on the sciences at the HfG Ulm is evident from the requirement that a thesis project had to have both a practical and a theoretical part involving research.

In order to understand the exceptional character of this conception of the second phase of the HfG, it is necessary to consider that in the mid-1950s designers were primarily educated at applied arts schools, which solely based on their name tended to have a curriculum oriented around the arts. The HfG Ulm set itself apart from that. In Hannes Meyer, it found a precursor and confirmation of the rightness of this step, which at the time deviated from the status quo. Whereas Hannes Meyer still had to carry the ballast of guild-oriented, craft training ending with a journeyman's exam, the HfG Ulm was freed of it, so it had considerably greater latitude of movement. That can be seen by comparing the organizational chart of the Bauhaus under Hannes Meyer to the structure of the HfG. The heterogenous, if not haphazard, structure of the Bauhaus became a problem in the project of trying to bring the workshops together into overarching units.

PRELIMINARY COURSE

As is well known, the invention and spread of the basic course was central to the history of the Bauhaus and to Bauhaus pedagogy. As an organizational unity, it took up the first year of study. The preliminary course (Vorkurs—also known as the basic course (Grundkurs or Grundlehre)) functioned foremost as a stopgap to address the deficits of the schoolyears prior to university study. These can be explained by the neglect of design themes in teaching in schools. The training of design competence, which is a matter of training nondiscursive intelligence, has never been truly been at home in university education either, which is primarily about training discursive competence and accordingly works with discursive codes. The exercises in the basic course at the Bauhaus were based on, among other things, the training of the aesthetic ability to differentiate and were called nonapplied design exercises—in contrast to the applied exercise of later years of study, which were about solving a practical problem.

In the formulation of the tasks of the basic course at the Bauhaus, words used in art theory come up, such as "balance"/"equilibrium"/"symmetry"/"cohe

sion"/"proportion"/"simplicity"/"density"/"homogeneity"/"rhythm"/"order"
—that is, concepts used primarily in the aesthetic discourse. At first, this
concept of a basic theory common to and required for all specializations
was adopted at the HfG. The basic course of Tomás Maldonado included ten
exercises: Sierpinski surface, Peano surface, Weierstrass curve, Black as color,
Symmetries, Accuracy-Incertainty, Imprecision-Exactness, Spatial effect,
Balance of three surfaces, and Disturbance (noise). The three-dimensional
exercises were supplemented by lectures on finite mathematics, symmetry
theory, group theory, and descriptive topology.[9]

The limits of an undifferentiated basic course soon became clear, however.
The leap from the general exercises of the basic course to the practical
design exercises in the departments was too abrupt. The basic course was
thus reoriented to be specific to the departments, though organizationally it
continued to be a unit that came first. In the next step, the concept of a basic
course was no longer used, and the formal and aesthetic exercises were spread
out over a course of study lasting four years. At the time, the exercises were
associated with considerable investment of time in the crafts, but thanks to
digitalization, they can be implemented more quickly today.

Today the subject matter of the nonapplied design exercises can be described
with the overarching concept "design patterns," which are oriented around
recurrent phenomena. According to Hannes Meyer, the basic course, too, had
a vague propaedeutic value as an instrument for finding oneself in that the
students in this experimentally structured course could explore their individual
strengths, weakness, and inclinations. Not all of the students understood this
sort of concept. The reproach that it amounted to pointless tinkering and a waste
of time was made by several students in the form of a call to action in 1929.[10]

Similar criticism—but in the opposite direction: against scientific
methodologism or methodolotry—was also expressed at the HfG Ulm. In
February 1962, the first-year students wrote a letter to the Geschwister-
Scholl-Stiftung (Scholl Sibling Foundation), the Rektoratskollegium (Rector's
Committee), and the teachers in which they emphasized that they had come
to the HfG Ulm to learn design and not to waste their time with unproductive
mathematical methods.[11] This reaction becomes understandable if one recalls
one of the exercises of the first year of study that consisted of drawing the
curve of a frequency distribution of the diameter of fava beans measured with
a micrometer in order to learn that a normal distribution takes the form of a
Gaussian bell curve. It is an open question whether that is a suitable method of
acquiring mathematical knowledge.[12]

[9]   Bill Huff, "Albers, Bill and Maldonado: The Basic Course at the Ulm School of Design (HfG),"
      in *Tomás Maldonado*, ed. Triennale Design Museum (Milan, 2009), 104–21.
[10] Bauhaus Archiv BHA_8529-1+2-Vorkurs-1929.
[11] "denkschrift der studenten des 1. studienjahres 1961/62, 3.2.1962 an den vorstand der
      geschwister-scholl-stiftung, an das rektoratskollegium, an alle festdozenten." HfG-Archiv, Dp
      5.7 (Depositum Berner).
[12] The methodolatry was criticized from the perspective of design. See Gui Bonsiepe, "Arabesken
      der Rationalität / Arabesques of Rationality," *ulm: Zeitschrift der Hochschule für Gestaltung*
      19–20 (1967): 9–23. [See Chapter 1.6 in this volume].

WORKSHOPS AND COOPERATION WITH INDUSTRY

Hannes Meyer wanted to extend the function of the workshops to include small production operations in order to approach concrete design tasks from industry, rather than leaving it at fictional design works. As he himself admitted, however, difficulties arose. It is difficult to reconcile an ordered curriculum with the random fluctuations of requests from companies. If one ignores the military connotation of the term "brigade" ("vertical brigade"/"work brigade"), Meyer was clearly striving for a pedagogical innovation: by having students in different years working together on one project, younger students can learn from the experienced students. Moreover, teamwork would be practiced from the outset.

By contrast, the development groups at the HfG Ulm created especially for that purpose worked on requests from industry. Coupling these tasks with the curriculum was avoided. The workshops were for building models and prototypes. Education was primary, though that did not mean cutting off the connection to industry. Whereas Hannes Meyer—and Walter Gropius— regarded architecture as the crowning, guiding discipline for all design disciplines, the HfG Ulm decidedly moved away from this hierarchy of values.

The development groups of the HfG performed four functions: first, it ensured the professional competence of the design instructors was up-to-date; second, it bridged industry and work in the departments; third, it improved the income of the teachers (since compared to the salaries at other universities in Baden-Württemberg, salaries at the HfG Ulm were as much as a third lower, so that the Geschwister-Scholl-Stiftung had problems attracting and keeping qualified teachers); and, fourth, it demonstrated to politicians that the concept of HfG was meeting with a response in industry.[13]

DESIGN AND POLITICS

Hannes Meyer's political stance was no barrier to working in public institutions in Latin America. In Mexico, Brazil, Argentina, Chile, and Uruguay, the reception of modernist architecture took place without any great delay relative to events in Europe and the United States. In a climate of relative permissiveness, it could also be discussed in terms of its political dimension and its utopian framework, with a strong attraction felt for the Russian Constructivists in particular.

Outside of the Federal Republic of Germany, the debate over modern architecture after 1945 revolved around the controversial call for mediating between the cultural avant-garde and the political avant-garde. Philip Johnson—who was no friend of leftist political commitment—confessed in an interview: "They were all Communists in those days. It's not popular now to call yourself that. . . . But Hannes Meyer was a Communist, and was a damned

---

[13]   I am grateful to Herbert Lindinger for the reference to the fourth function.

good architect and the more I see of Hannes Meyer, the greater man I think he was. But I don't like what he *said*."[14]

It is reasonable to draw parallels between political resistance to and attacks on the Bauhaus during its three phases and the developments at the HfG Ulm—resistance and hostilities to which Hannes Meyer in particular was subjected in his day, as he made no secret of his political stance. Just as the Bauhaus must be understood in the context of the crisis-ridden years of the Weimar Republic and the rise of fascism, the HfG Ulm must be seen in the context of the Cold War. The parallels are revealing. This led Tomás Maldonado to conclude that it was time for Germans to finally come to terms with *their* Bauhaus and ask what caused political forces to make the life of this institution more difficult and ultimately made its survival impossible.[15]

Hannes Meyer's decidedly antifascist stance was appreciated at the HfG Ulm, which was by no means common in West Germany at the time. In the cultural and educational policies of the Federal Republic of Germany, forces seeking to restore the old order continued to have an influence after 1945, gaining ground and ultimately being able to close the HfG Ulm.

SUMMARY

If one considers the Bauhaus under Hannes Meyer and the HfG Ulm in the context of their respective historical phases of development,[16] the Bauhaus belongs to the first phase of the liberal capitalism of competition that began in the nineteenth century—which in Dessau was interspersed with social democratic, reform approaches—whereas the HfG Ulm falls in the second phase of state-controlled capitalism in which the enormous growth of productivity during the Second World War was redirected at civilian goods. Faced now with the third evolutionary phase in the form of globalized financial capitalism since the beginning of the twenty-first century, it remains an open question as to what critical training in design should look like so that it is not abandoned to subordination to the dominant category of the market.

In conclusion, we are left with the question, what made Hannes Meyer attractive to the HfG Ulm? He offered a point of contact for a continuation of the unfinished project of modernity with the central category of design (project) in its social connection—a theme that was sometimes dismissed as outdated because it was linked to the concept of industrial development. Criticism of the concept of development grew in the face of the consequences

---

14 Philip Johnson, in John W. Cook and Heinrich Klotz (eds.), *Conversations with Architects: Philip Johnson, Kevin Roche, Paul Rudolph, Bertrand Goldberg, Morris Lapidus, Louis Kahn, Charles Moore, Robert Venturi & Denise Scott Brown* (New York: Praeger Publishers, 1973), 11–51, esp. p. 38. I owe the reference to this quote to the quoted work of Antonio Toca Fernández.

15 Maldonado, "Ist das Bauhaus aktuell? / Is the Bauhaus Relevant Today?," 5–13.

16 Nancy Fraser, "Contradictions of Capital and Care," *New Left Review* 100 (July–August 2016): 99–117.

of industrialization for the environment and in the face of the question of whom the development ultimately benefitted. This question is posed above all from the perspective of peripheral countries in light of the ever-increasing disparity of asymmetrical value transfer in the form of the export of commodities to core countries and the import of finished products to peripheral countries.

Moreover, in Hannes Meyer the HfG Ulm found a teacher who advocated a more profound relationship between design and science and asked the question what training in design should look like today—obviously now under different technological, industrial, and economic conditions. It was not so much that Hannes Meyer had a direct influence on the HfG Ulm as there was a coincidence of objectives and a commitment to design within an overarching social framework: working politically against authoritarian interests and efforts to restore the old cultural policies and encouraging forms that countered the continual increase in inequalities and the associated potential for conflict. Seen from that perspective, Hannes Meyer is more relevant than ever.

This article addresses, among others, the question why it is that designers in their different manifestations (architects, industrial designers, communication designers etc.) are expected to act as agents for social change. One possible explanation is given: these professions deal with the future and envisage the possibility of another future. This interpretation does not mean to attribute to these professions an excessive importance. In recent years, the term "design thinking" produced a considerable resonance, particularly in management circles. It is submitted to a critical revision.

30   THE DISOBEDIENCE OF DESIGN

(2015)

In one of Hans Magnus Enzensberger's more recent works, the main character launches a diatribe against modern design—particularly furniture and lamp design. He sarcastically notes that one has to imagine hell as a place that is completely furnished by designers.[1] The inflationary use of the term "design," which somewhat, but not entirely, overlaps with the German term *Gestaltung*, may contribute to the fact that even the mere mention of the word causes irritation.

The loaded question of whether design can change society raises the concern if we are going beyond the scope of what design can actually achieve. The answer depends on what we understand as change and what demands arise with this understanding. A likely, if quick, answer: society changes design rather than the other way around. Design does indeed profoundly influence and change day-to-day life in society, but how deep the changes go and can reach remains a matter of debate.

Why is design expected to change society? There are three possible reasons for this. First, designs live from the myth of creativity, which is sometimes pathetically overestimated.

Second, design brings the dimension of the future into play. Third, it implies that change equals improvement—the utopian component. Yet, we should address the reflective, restorative changes too. If we place this question in the context of modernity, a project that has sometimes been tossed onto the wreckage of history, it would be advisable to caution against an affirmative answer, because another question looms in the background, the question of a structural change in social conditions—possibly of revolutionary nature. Artists and architects faced such a situation after the Russian October Revolution of 1917. They were able to be involved in the process and see their creative activity transforming society.[2] Otherwise it would have remained a purely verbal self-reassuring expression of solidarity that could easily escalate to an ultimately ineffective verbal radicalism. For it is difficult to expect a direct change in society from the mere distribution of color pigments on canvas, despite the pathos that characterized the artistic manifestos and

Presented in the Symposium „*Kann Gestaltung Gesellschaft verändern?*" (Can Design Change Society?), Berlin, September 18–19, 2015, organized by Philipp Oswalt. Originally published in German in Arch+ 222: Kann Gestaltung Gesellschaft verändern? 3/20/16. Reprinted in English in *Can Design Change Society?* (Basel: Birkhäuser, 2019), 60–3.

[1]   See Hans Magnus Enzensberger, *Herrn Zetts Betrachtungen, oder Brosamen, die er fallen liess, aufgelesen von seinen Zuhörern* (Berlin: Suhrkamp, 2013).
[2]   Ibid.

proclamations of the time. When I explore the potential of design to instigate social change, I explicitly want to include the periphery—and how this point of view on the subject at hand differs. By "periphery" I do not mean the suburbs, but the former colonial states, which strive for more political autonomy and less economic-financial asymmetry. When I refer to "peripheries," I am referring to the majority of the world's populations. Why did the Bauhaus and its programs have such an attraction not only in Europe and North America, but also and especially in other parts of the world? From time to time, the Bauhaus has been accused of being an expression of Eurocentric cultural imperialism. These accusations have appeared in the context of the debate on de-colonialism and the reaction against the philosophic narcissism of occidental culture over the past two decades. It should come as no surprise that there have been movements to criticize and reject a supposed or real Eurocentric universalism of the Bauhaus concept—one united with the arts and design disciplines institutionally and programmatically under the hegemony of architecture. Yet these objections have not diminished the paradigmatic value of the Bauhaus for four reasons.

FIRST

It is well known that the Bauhaus placed its hopes in industry and technology—especially in the era of Bauhaus director Hannes Meyer (1928–30), who has received little attention in the predominant history of the Bauhaus, and who represented the socially critical wing of the Bauhaus. In this sense, the Bauhaus could be said to have been Saint-Simonistic.[3] The term "industry" had an almost religious connotation, but at least a positive meaning, since it was understood as a manifestation of "development." Especially when local governments in Latin American countries made industrial development a part of the agenda starting in the 1920s and have increasingly done so since the 1950s, primarily in order to satisfy domestic demand for industrial goods. This affected both consumer and capital goods.

SECOND

In addition to the economic political relevance of industrialization for improving the foreign trade balance, the political idea of emancipation also played an important role in industrialization: economic development was seen as a means of achieving political independence.[4] During the 200-year anniversary celebrations of political independence in many Latin American countries

---

3   "Saint-Simonianism: A socialistic system in which the state owns all property and the laborer is entitled to share according to the quality and amount of his or her work." *Merriam-Webster Dictionary*, www.merriamwebster.com/dictionary/ Saint-Simonianism (accessed May 17, 2019).
4   Timothy Brennan, *Borrowed Light: Vico, Hegel, and the Colonies* (Stanford: Stanford University Press, 2014), 26.

over the last few years, there have been calls for a "second independence" to which industrialization, including design activities, should contribute. While in Europe and North America the term "industrialization" has lost its appeal because of environmental damage and consumer oversaturation and is no longer regarded as a sacrosanct credo, it is universally accepted in most other countries. However, there is no lack of critical dissent in subaltern studies, for example. The political facet of design clearly emerges in the periphery, even though in recent decades the discourse on architecture and design has shrunk to a pale imitation of its former self. The political component is expressed by the fact that design is measured by the criterion of whether the design contributes to the strengthening of autonomy or promotes heteronomy. In the wake of the international division of labor, peripheral economies have been assigned the role of raw material supplier, for unprocessed goods such as ores, rare earths, crude oil, biofuel, gas, grain, soy, meat, wood, coffee, and, possibly in the future, even water. These export goods constitutively lack the design component; they are "designless" products.

THIRD

Since the Bauhaus, as a modernist project, was directed at the future and the associated horizon of design, it could be regarded as a historical reference point not only for the possibility, but also for the necessity, of changing the status quo. This is precisely the idea behind the notion of utopia, which now seems to have fallen into disrepute. Its absence or rejection leads to design nihilism or cynicism. The indispensable political component of design as a manifestation of utopian intentions and the desire to change the status quo or asymmetric power relations, in whatever way communicated lends design its explosive force. This interpretation stands in stark contrast to a turning away from utopia, as Rem Koolhaas, among others, seems to advocate: "There is no need for a new utopia, but rather for the creation of a better reality."[5] This formulation can be the expression of a resignative realism—or of a quietism that is comfortable with the status quo.

FOURTH

It is well known that design, and thus the design disciplines, hardly fit, if at all, to the conventional, innovation-resistant structures of higher education, be they universities, technical colleges, or art academies. Design falls out of the conventional tripartite category scheme of sciences, technology, and art. Industrial design and visual communication, which developed as

[5]   Reinhold Martin, *Utopias Ghost: Architecture and Postmodernism, Again* (Minneapolis: University of Minnesota Press, 2010), quoted in Lahiji, *Architecture Against the Post-Political*, 113.

professions and degree courses after the end of the Bauhaus, are usually treated as appendices to existing institutions or established disciplines. Design is therefore academically homeless. The sciences generally treat design with indifference, if not ignorance, for quite understandable reasons. Or they dismiss it as a frivolous sales-oriented and commercially "infected" field of activity and therefore culturally irrelevant. Or they upgrade design to an artistic activity, thereby misinterpreting the core of design. Design is its own "continent," next to the sciences and the arts, with its own systems of problematization and solutions. It is difficult to come to terms with the terra incognita of design from the perspective of the humanities and media theory for two reasons. Barriers exist due to their distance to the design projects as well as the material dimension of design. Until we become aware of these barriers and can uncover suitable accessibility to the "continent of design," the potential of an academic examination of design cannot be fulfilled and so lies fallow.

It is peculiar that the history of design is predominantly assigned to the discipline of art history with its two main categories, namely style and form. This explains the interest in style development and style comparisons, as well as the fixation on form, that is, on the aesthetic component. Generally, this overlooks the fact that it is difficult to come to terms with design without the history of industry and technology, which is foreign to art history. The intense, sometimes hagiographic interest shown in the Bauhaus and its history is probably primarily due to the fact that it is rightly regarded as one of the most important institutions of modern art of the twentieth century, linked to the emblematic names of great artists—despite the well-known fact that, official art lessons, like those at an art academy, were not given at the Bauhaus.

Yet the participation of influential artists suggested that an affinity between art and design was hinted, if not implied. It also reflects the hierarchical subordination of craft at the time, whose social standing was to be elevated to that of the arts. As far as the change in technological and industrial production methods since the Bauhaus is concerned, the following should be noted. The Bauhaus looked to serial production as the definition of a standard and its iteration of copies. This allowed critics of industrialization to decry the repetition of the depersonalized, anonymous, and selfsame. Mass production makes it possible to achieve the social goal of mass consumption by lowering the price of products; this is known under the name of Fordism. In the 1970s, this form of production was replaced by post-Fordism, which is characterized, among other things, by the flexibilization of production processes through the use of multipurpose machines. These make it possible to produce small series at a low cost while promoting product diversification through slight variations of essentially the same form; it is therefore more fitting to speak of a fictitious diversity. The most recent innovation in the production of goods is the highly-touted 3D printer, which enables the production of small to medium-sized objects, primarily from plastic, either at home or in a "maker shop" (the catchy name for these workshops). According[6] to an overly optimistic assessment, production would be ultracheap, ultra-flexible, and ultra-distributed.[7] As the size of the products is limited by the dimensions of the printer, the designs are likely to be designed and manufactured from assembled parts, that is, modular systems. At times, exuberant hopes are pinned on this method of production, which is intended to overcome capitalism with new postcapitalist working conditions. Designs can be downloaded from a website and produced at home.[8]

It remains to be seen whether this design and production method will popularize the activity of design, and, above all, what kinds of products will be manufactured. A mass production of knick-knacks cannot be ruled out. The objects that have been self-produced so far look rather conservative compared to applications in the medical field, such as the detailed reproduction of bones for prostheses. Apparently, this technology makes the dream that everyone is or can become a designer come true—and not just a designer, but also a producer. Certain computer programs enable one to move seamlessly from design to production. The categorical separation between design and execution is thus undermined. This separation goes back to Filippo Brunelleschi, who claimed and monopolized design for the architect and wanted to considerably limit the power of the guilds, as well as force a separation between mental and physical work. The profession of industrial design is also based on this separation between design and production.

Certainly, it seems wrong to want to see in the 3D printer a return to precapitalist, handcrafted production methods. It also seems appropriate to recall Antonio Gramsci's statement about intellectuals, which could be adapted to this context to explain that while everyone might be a designer, not everyone fulfills the social function of designer.[9] What characterizes a designer's methods? To borrow from the idea of the "system of attention" from neuroscience, it could be said that a designer pays attention to the use of materials and semiotic artifacts in order to structure the user's everyday scope of action, as it is determined by those artifacts. It follows that design should no longer be determined mainly by form, but rather by the scope of action enabled by the artifact. The ambiguous concept of utility, both as usefulness and as use, has been largely ousted from design discourse today. The symbolic value of artifacts has taken the place of utility, thus creating a link to marketing, which has largely taken over design. The ambiguity of the terms utility and function can also be used to identify the difference between design objects and art objects. Theodor W. Adorno gave a cutting response to the question of what the function of art is—it consists precisely in having no function.[10] He resisted

---

6   In a very short period, the size of components produced with additive manufacturing processes has grown to building components.
7   Rob Lucas, "Xanadu as Phalanstery," *New Left Review* 86 (2014): 142–50.
8   Ibid.
9   See Antonio Gramsci, "The Intellectuals," in *Selections from the Prison Notebooks*, eds. Q Hoare and G. N. Smith (New York: International Publishers, 1971), 21.
10   See Theodor W. Adorno, *Aesthetic Theory*, eds. Gretel Adorno and Rolf Tiedemann, trans. and ed. Robert Hullot Kentor (Minneapolis: University of Minnesota Press, 1997).

any kind of instrumentalization of art. As Charles Baudelaire put it, "To be a useful person has always seemed to me something hideous."[11] Although art can be functionalized as an object of capital investment, this does not mean that that is its function. Nobody climbs the slippery slope of the commodity with their character unscathed. An emphatic concept of art, as Adorno might represent it, refuses any form of co-optation. On the other hand, the notion of the instrumental is constitutively linked to the concept of design. This does not mean, under any circumstances, fixing oneself on the object in the design process and taking it as the starting point and end point of the design process, but rather looking beyond it to address a further area of concern: a chair as an answer to the problem of sitting, a lamp to the problem of lighting, a car to the problem of mobility. The French philosopher Etienne Balibar recently hinted at another direction: concentrate on use instead of consumption, as a civilization of use instead of a civilization of credit. This is precisely where design that aims to change society would have a starting point. This is precisely where Balibar meets the narrow but indispensable neuralgic zone of the intervention of design into the material and semiotic artifacts of everyday life.[12]

Finally, I should like to briefly mention the radical innovations that digitalization has had on information, on which the new media are based, including social networks. A closer look at the social response to this innovation reveals the oscillation between euphoria and disappointment. On the one hand, this technology is seen as a potential for liberation, while on the other it is criticized as an instrument for establishing domination by a few multinational companies and for dissolving privacy. Everyone knows that internet users offer up their data for supposedly free services, which is sold as goods to companies and agencies for advertising purposes, or to surveillance authorities, in order to compile offers tailored to the respective personal profile or to provide a starting point to surveil the user. Due to the ambivalent nature of information/communication technology—an instrument of both domination and liberation—the American computer scientist Jaron Lanier has suggested that each individual consider their personal data as their property, with the right to use this data commercially.[13]

According to this concept, everyone would be their own data capitalist and would have the power to protect their privacy—assuming this bourgeois category, created during the Enlightenment, is still considered valid today. As to the question of knowledge design, it nudges design into direct connection with scientific research, which is a novelty in this form. It is well known that scientific research generates knowledge and cognitive innovation—not only research that can be translated into technological innovation and exploited on the market, but research in general, including the humanities. While scientists

---

[11]   Charles Baudelaire, *The Parisian Prowler- Le Spleen de Paris, Petits Poemes en prose*, trans. and ed. Edward K. Kaplan (Athens: University of Georgia Press, 1989), 21–2, quoted in Brennan, *Borrowed Light* (see footnote 4), 203.

[12]   AA.VV. "Etienne Balibar: Pensador Latinoamericano". Review, Julio-Agosto (2015): *Lectura Mundi* 1-8, Universidad Nacional de San Martín.

[13]   See Jaron Lanier, *Who Owns the Future?* (New York: Simon and Schuster, 2014).

speak of "designing an empirical study," knowledge design means something else, alluding to the use of resources derived from audiovisual perception to enable the receptivity, assimilation, and understanding of the results of research,[14] not only outside research, but within the practice of research itself. Of course, aesthetic moments also come into play, which is not the rule in the largely nonaesthetic tradition of the sciences, typically characterized by the dominance of discursivity. It is obvious that this job is best tackled in teamwork between scientists and designers, just as a social network of players is necessary for a design and its implementation in particular. If one reverses the word order of the term "pair knowledge and design"—one arrives at design knowledge, that is, simply the expertise of a designer.

In this context, a term has made waves in recent years: design thinking. Not just in the close professional circle of designers but also in managing and marketing circles.[15] And if this concept of design thinking is not merely a modish synonym for creativity,[16] possibly it refers to the special way that designers tackle a problem using a holistic approach. The expectations therein are so high, it almost seems that a tool has been invented here to save capitalism. The concept of design as a panacea, for all of the wrongs and the manifestations of crises in capitalism (which can hardly be denied), must be motivated by a concern to retain the constitution of society as it is defined by the market and finances. Creativity and innovation are popular topics these days. So it should come as no surprise that not only in management circles is the idea on the rise of using the aura of creative design and design thinking to calm the global turbulence currently facing companies and nations alike.

After this perfunctory overview, the question may well arise as to what any of this has to do with the Bauhaus and whether it can be fitted in to the framework of the Bauhaus at all. I tend to want to answer the question affirmatively, because the problems the Bauhaus movement saw continue to have the urgency that they had back then, and in fact a new constellation of problems has emerged. In particular, this web encompasses questions around "development without growth," sustainability, and "public control of the use of resources and their distribution with a logic of which tends to kowtow to the oppression of any regulation."[17] On this point, the Bauhaus clearly remains topical—even if the way in which it addressed these problems may have lost relevance.[18]

14 See Gilles Rouffineau (ed.), *Transmettre l'histoire: contribution du design a la production des savoirs/Passing on History: Design Contribution to Knowledge Production* (Paris: Editions B42, 2013).
15 Tim Brown, "Design Thinking," *Harvard Business Review* (June 2008): 85–92.
16 See Don Norman, "Design Thinking: A Useful Myth," *Corell*, June 25, 2010, www.core77.com/posts/16790/design-thinking-a-usefulmyth-16790 (accessed May 17, 2019).
17 Etienne Balibar, "Etienne Balibar-Pensador latinoamericano," Review. *Revista de libros* 3 (July–August 2015).
18 I am grateful to David Oswald for his critical review of this manuscript.

PART FOUR
Design and Development/Projects

Gui Bonsiepe: The Possibilities of Designing

Constantin Boym

I first heard the name of Gui Bonsiepe, along with his mentor Tomás Maldonado and other ulm school protagonists, in 1985, when I was studying design at Domus Academy in Milan. At the time, Italian design attempted "to start again from scratch in search of utilitarian culture that is not worn out,"[1] in the words of Andrea Branzi, one of main characters of the new movement. It is not surprising that the work of preceding design movements was met with skepticism and critical attitude. Hochschule für Gestaltung at Ulm was no exception. Ulm school was thought of as a negative example of everything the new design wanted to get rid of: "hyper-rationalism, the international style, functionalism, the methodological leveling of the profession."[2] Specifically, the scientific foundations of design, which ulm founders worked hard to formulate and apply in practice, was an anathema to the intuitive and artistic approach of New Italian Design. In a paradoxical way, we studied ulm design as an example of how *not* to design. Noticing a clearly discernable visual impact of ulm-generated projects, Andrea Branzi —always a contrarian—attempted a revisionist reading. In an essay characteristically titled "These Monks on the Hill," he claimed that the ulm model essentially proposed "a formal code of great purity and correctness,"[3] a technological style of "pale instruments," which via the work of Dieter Rams influenced the appearance of most electronic apparatus from 1960s to the present day:

> In Ulm's case there has been the comic misunderstanding of treating designers as scientists, simply because they were talking about science. In reality that hill was home to a group of extraordinary artists who, in the guise of inflexible scientists, were looking for beauty.[4]

As a design student, I was confused. What about the social dimensions of the ulm project? What about their concern for the masses, the people who are "excluded from access to industrial products because these lie beyond the range of their economic possibilities" (Chapter 31, p. 325)[5], as Gui Bonsiepe would write in 1973? He believed that "industrial design could find one of

---

[1]   Branzi, *The Hot House* (Cambridge, MA: The MIT Press, 1984), 8.
[2]   Branzi, *Learning from Milan* (Cambridge, MA: The MIT Press, 1988), 40.
[3]   Ibid., 41.
[4]   Ibid., 42.
[5]   As with the other introductory essays in this volume that quote or make references to papers in the book, the pagination will be by the chapter number and page reference.

its noblest aims, and one of its very few really worthwhile justifications, in developing products for the needs of the poor majorities" (Chapter 31, p. 325). How could radical Italian designers who wholeheartedly embraced left-wing politics—many were members of the Italian Communist Party—ignore the social aspiration of the ulm program?

One country that *did not* overlook the progressive thinking of ulm was the Soviet Union. The oppressive country that I left in 1981, in the depth of Brezhnev's stagnation, was a different and more optimistic place in the 1960s. That period, known at the Thaw, saw a renewed interest in the material welfare of all citizens, and industrial design was seen as a means of improving the quality of life for the entire country. Just like Western designers from ulm, Soviet design intellectuals saw value in the social aspect of objects. Their overreaching goal was to produce "harmony in the material environment,"[6] which seemed within reach in the condition of planned socialist economy. The harmony was understood as unity of standard typologies for all consumer goods. An oft-repeated Soviet designers' joke about excesses of capitalist market cited 377 refrigerator models that were available to the Western customer in the 1960s. Soviet manufacture countered this with something like *three* models—basic, mid-range, and "luxury"—which were produced by the millions for distribution around the country. The joke, ultimately, was on the Soviet consumer. Still, the chosen designs were a result of thorough research of the needs of the country's population. Tomás Maldonado, the second rector of ulm school, was intrigued by the possibilities offered by statewide economic planning and the view of design devoid of commercial profit-making. I believe ulm designers viewed the Soviet approach as a case study, the lessons of which they were hoping to apply elsewhere. The relationship between the two groups, which started in 1963, reached an apogee during 1975 ICSID conference in Moscow, where Maldonado gave a series of lectures on socially oriented design.[7] Gui Bonsiepe never traveled to the USSR, but one can clearly feel reverberations of this cultural exchange in his design work in Latin America.

The leitmotif of all Bonsiepe's writings is his unshakable belief in the role of industrial design in an economic development, specifically in the course of industrialization of Latin American countries. This single perspective sheds light on Bonsiepe's own design projects. This book offers only a limited number of design examples that his various studios have developed in Brazil, Chile, and Argentina in 1970s–1980s. What is the overall impression of this creative output, seen from a perspective of half-a-century in 2020s?

First of all, the designs are very dissimilar. There is no such thing as Bonsiepe's "style," no attempt at establishing a consistent design language. He called it "a decided denial of personalist design" (Chapter 34, p. 363). The appearance of each product is uniquely determined by specific functions,

6   Cubbin, *Soviet Critical Design* (Bloomsbury, 2019), 37.
7   Ibid., 35.

requirements, budgets, available technologies. This is the only logical output for an approach that unreservedly values solving problems and needs. This design offers improvements rather than innovations. It is curious how Bonsiepe emphasizes those small improvements, describing his designs with great modesty. In design of an Air Compressor (Brazil, 1984), "by orienting the soldered seam about 45 degrees the new design requires one step less in the stamping process (4 instead of 5 steps)" (Chapter 37, p. 375). In a project for a Chopper (Chile, 1973), due to introduction of "components with modular coordination the weight of the chopper has been reduced by 35%" (Chapter 36, p. 371), which increased agricultural productivity. In design for a Spoon for Powder Milk (35, Chile, 1973), the designers' preferred solution had to be discarded "due to the cost of purchasing new machinery. Alternatives with two parts and internal spring were also discarded because they were too complicated. . . . This modest, far from spectacular project, reveals the crucial weight of economic and technological factors that often conditions the work of the designer" (Chapter 35, p. 367), concludes Bonsiepe, somewhat melancholically.

A second aspect of Bonsiepe's design relates to social and political dimensions that imbue his every project. This is especially evident in his work in Chile for the socialist government of Salvador Allende from 1971 to 1973. Every humble object was envisioned as a component of a larger economic goal of improving basic living conditions for Chilean society. Thus, the Spoon for Powder Milk, mentioned earlier, related to a program of rations of powdered milk given to children through the National Milk Plan (and which reached 800,000 families by 1973). It was essential that "these goods were simple in design, easy to construct, inexpensive, and of good quality, all important considerations for the majority of Chilean consumers" (Chapter 40, p. 387). A new generation of consumer products: inexpensive television sets and record players (such as the one that Bonsiepe designed and which is shown in Chapter 34) were produced to give working-class and low-income Chileans access to material goods that were previously out of reach.

A third aspect of Bonsiepe's designs relates to their visual quality. Visuality is a crucial part of all his design proposals, from a hyperfunctional spoon to a sinuous record player, a clear successor to "the pale instruments" of ulm. For Bonsiepe, industrial design had to be *seen*. This belief perhaps explains his enigmatic example of the transoceanic cable which he brings up as a case where "industrial design is of little importance" (Chapter 31, p. 325). I believe that visuality is at the core of the famous disagreement between Gui Bonsiepe and Victor Papanek, both proponents and advocates of design for the needs of the poor and underprivileged, both developers of strategies for design in the developing world.

Today, a design student will be hard-pressed to find reasons for a discord in their attitudes. And yet, Bonsiepe called Papanek's groundbreaking book *Design for the Real World* (1973) "a hodgepodge of conservative ideas" (Chapter 32, p. 337). The very first sentence of the first chapter of Papanek's book—"All men are designers"—rankled Bonsiepe. He was concerned with reducing the role of industrial designer to "the interpreter, the generalist . . . a jack-of-all-trades" (Chapter 32, p. 337) and with allowing people's direct participation in the design process. He was also disturbed with Papanek's embrace of local crafts and materials in designing for the periphery, mockingly comparing such contextual designs to sticking seashells onto a modern radio (Chapter 32, p. 337).

Forty years later, Bonsiepe commented on his earlier position:

> My criticism was directed against what seems to me an anti-industrial romanticism enchanted by the simple solutions with the aroma of do-it-yourself, very attractive particularly to young designers who maintain certain reservations or even aversion against the range of products that constitute the current professional practice. Papanek identified a problem, but the path he proposed to solve the problem seemed wrong to me. (Chapter 33, p. 349)

Why did it seem wrong? Was it because it did not agree with Bonsiepe's heightened sense of visual culture, the rational and pure approach, cultivated at ulm, which would be under threat of contamination with creative involvement of workers, consumers, and direct influences of the periphery itself?

Bonsiepe's own belief, in full accordance with his modern design upbringing, was in industrial designers' leading role, in their ability to chart the way for the masses. This creed found its full expression in his work on Cybersyn—an Operations Room for Cybernetic Management—the most ambitious of Bonsiepe's projects for the Allende socialist government in Chile. Conceived in 1972, the same year when Italian exhibition *New Domestic Landscapes* had a triumphant showing at Museum of Modern Art in New York, Cybersyn could rival most Italian entries on a visual level. Yet no Italian proposal was developed to such degree of detail towards realization, nor was it commissioned by a government agency for practical implementation.

The history of the project is well described in this book. Cybersyn had no precise brief simply because there were no precedents for such entities in the past. Bonsiepe recalls that the "design team was entrusted with the vague task of equipping a space with chairs and tables for about ten people, in which economic data were presented on a projection screen, so they could make political and economic decisions" (Chapter 40, p. 387). More challengingly, they had to design the interface and rules for visualization of all relevant information. It is perhaps not surprising that the first task was resolved at a much more inspiring and holistic level. From spatial symbolism, where a round room (later changed to hexagonal) was to emphasize equality of all participants, to ergonomic sophistication of futuristic fiberglass seats, to removal of all table surfaces (since writing presumably hindered direct communication)— the space projected a startling new image of Chilean modernity. The oft-evoked comparison with *2001: A Space Odyssey* movie classic (a connection that the designers firmly denied) underscores the impact the space could have had if it was completed.

The true function of the Operations Room—the visualization of data in order to facilitate analysis and decisions about the nation's economy—proved to be a more elusive goal. First of all, the technology to enable such real-time communications simply was not there. Thus, all information graphics for futuristic datafeed screens were to be hand-drawn, photographed, and projected as 35-mm slides. The interface itself, with its emphasis on large geometric "thump" buttons, was designed to look anti-elitist and comprehensible by a factory worker; instead, it ended up looking obscure and counterintuitive. There were other strange social and gender biases, such as the ban of keyboards to avoid association with female typists' work (and because government bureaucrats did not know how to type) (Chapter 40, p. 387). More importantly, the virtue of collecting and analyzing real-time information on a national scale—known today as Big Data—was never thought out in a comprehensive way. The politics of the project claimed a lofty goal of enabling workers' direct participation in newly nationalized economy. But already in the trial run of the system in late 1972, Cybersyn data was used to thwart a nationwide strike by truck drivers.[8] It is impossible to predict whom the project would benefit in the long run, since its full implementation never took place. Together with other agencies of centrally planned economy, Cybersyn fell prey to the 1973 coup, which put an end to socialist government in Chile.

Perhaps the experience of Latin America's political instability and controversial economic policies made Gui Bonsiepe take a more nuanced view on his lifetime strive for economic development. In a 2019 interview, he says:

> Today, I would adopt a more critical stance on the concept of "development" because it is not a neutral concept—it is always linked to interests that reflect constellations of power. I would ask: development for whom? (Chapter 4.3, p. 349)

This essential political question—*"who could do what to whom for whose benefit?"*[9]—will remain the cornerstone of social design in the decades to come. Today, the work and thinking of Gui Bonsiepe is timelier than ever before. Social design has entered university curricula and academic conferences, receiving serious scholarly and media attention. Young designers are eager to question, undermine, and rethink the global political, financial, and social system. It is symptomatic that "design as an attitude"—the definition conceived by the Bauhaus master Laszlo Moholy-Nagy (whom Bonsiepe quotes in his writings)—becomes a title of a new book by design author Alice Rawsthorn. In the prologue to the book, she writes about the priorities of the new design generation:

[8] Morozov, "The Planning Machine: Project Cybersyn and the Origins of the Big Data nation," *The New Yorker*, October 6, 2014.
[9] Geuss, *Philosophy and Real Politics* (Princeton: Princeton University Press, 2008), 26.

Designers are responding by planning and executing projects to tackle climate change; to reinvent dysfunctional areas of health care and social services; to provide emergency support for the victims of man-made and natural disasters; to help asylum seekers to settle into new communities; and to champion social justice.[10]

Inevitably, design students will be looking for role models, and for guidance regarding how to put the world together in a new way. The long and turbulent career of Gui Bonsiepe may provide them with a fitting example.

[10]   Rawsthorn, *Design as an Attitude* (JRP/Ringier & Les Presses du Reel, 2018), 9.

Design Policy/Design and Development

Editorial note. This report has been included as an instance of Bonsiepe's important, though far less visible, work as a consultant to governments and industry especially across the 1970s. It reflects the necessity at that time of instructing institutions as to the capabilities and limits of professional design activity in regard especially to economic development. Chapter 4.3, "Design and Development Forty Years on," gives some useful context of the origins and circumstances in which the report was commissioned.

31 DEVELOPMENT THROUGH DESIGN

(1973)[1]

A WORKING PAPER PREPARED FOR UNIDO AT THE REQUEST OF ICSID

UNITED NATIONS INDUSTRIAL DEVELOPMENT ORGANIZATION
April 18, 1973

FOREWORD

This paper was prepared in February 1973 in order to serve as a basis for discussion between representatives of UNIDO and ICSID.

The ideas and proposals expressed in this document do not pretend to reflect an official opinion, neither of ICSID nor of any other institution or group of industrial designers. The author considered the guidelines given by UNIDO, which refers to the items this paper should cover and to its purpose: to serve as an "eye-opener."

The author has tried to give an account of the role industrial design could play as a development factor in developing countries.

This paper is based on a five-year work experience as industrial designer in developing countries in Latin America, part of which was under contract of an international agency. Furthermore, there have been consulted publications dealing with the subject matter "Industrial Design / Developing Countries / Development."

There exist today different opinions (reflecting different interests) about the role industry could and should play in developing countries. As no opinion can pretend to possess universal validity, the author has made use of the right to present a point of view which he can stand for and which he considers to do justice to the interests of developing countries.

I. HISTORICAL NOTES

As a **profession** industrial design became recognized about two generations ago, when for the first time university courses in this specialty were offered in Europe and the United States.

---

[1]   The paper was presented in 1973 as described in the opening sentence. It has been edited for length for publication in this volume.

As an **activity**, however, it reaches back to the nineteenth century, when industrialization started to change the physiognomy of the material world.

As a **tool for development**, especially export promotion, its role was recognized as early as in the first decade of this century.

As a **tool for marketing**, it became used as the beginning of the 1930s.

As an **area of government promotion**, it became established in the mid-1940s.

II. DESCRIPTION OF INDUSTRIAL DESIGN

(a)   From art-inclined to technology-inclined interpretations

One interpretation of industrial design is as a marriage between art and technology or art and industry. This interpretation shows the difficulties which arise when a new phenomenon is explained in terms of already-existing, known phenomena. It does not give information about the character of industrial design to call it a mixture of art and technology.

In its beginnings, industrial design, which was not called by that name, was assigned the task of embellishing ugly industrial products. Art (with capital letters) was thought as a "civilizing" force for (brutal) industry.

According to another opinion, industrial design can be explained as a result of the failure of engineering professions, assuming that engineers possess neither aesthetical sensibility nor the capacity of synthesis, that is, to see the product as a whole. Whereas the industrial designer is considered as an especially gifted individual whose task consists of coordinating the efforts of other professionals. This opinion lacks empirical basis and results from wishful thinking. It has caused justified suspicion among industrialists and technologists and cannot be considered a tactically adequate way to introduce industrial design into industry.

The aforementioned opinions overemphasize the aesthetical and cultural aspects of industrial design, neglecting other equally or even more important aspects such as use value, productivity, and technological innovation.

In the history of industrial design, one can observe a shift from art-inclined interpretations to technology-inclined, and even science-inclined interpretations.

(b)   The core of industrial design

Industrial design is tightly interwoven with and dependent on the socioeconomic context in which it is exercised. Therefore, it is unsound to try to formulate a universally acceptable definition of industrial design. There might exist general agreement as to what the industrial designer is doing, but disagreement may arise when the question "what for?" has to be answered. Method and content of this activity may be quite similar in the different parts of the world where industrial design is predicated; objectives on the contrary will differ.

Nevertheless, during the course of the last two decades, there have been proposed various formulations of industrial design which coincide in certain traits of this activity and which might form a meeting ground for general consensus.

Industrial design centers around the following topics:

(i)   It is concerned with the improvement of usability of industrial products which forms part of the overall quality of a product. From the point of view of industrial design, a product is primarily an object which provides certain services, thus satisfying needs of the user.

(ii) It is concerned with "formal properties" of industrial products. Formal characteristics refer to the overall appearance of a product, including its three-dimensional configuration, its "physiognomy," its texture and color. (The term "formal" is preferred to the term "aesthetical," it being more descriptive than evaluative.)

(iii) It is an innovative activity. It is one special type of technological innovation.

(iv) It is concerned with the marketability of the product in that it relates the product to its sales market in terms of both raw material supply and product demand.

(c)   Technological conditions for industrial design

Industrial design is connected with the satisfaction of needs as far as these needs can be fulfilled by physical structures conventionally called "products." In order to be able to fulfil these individual, collective, or group needs, industrial design presupposes a technology of production (machinery, materials, labor force, methods of industrial organization, management techniques) and a technology of distribution (marketing including need analysis, product diversification, advertising, assortment policy, production evaluation, packaging). Without these two technologies, industrial design cannot exist.

(d)   Product areas of industrial design

Although the industrial designer deals with product innovation, it is not the total variety of industrial production he gets involved with.

The division—taken from economics—between consumer goods and capital goods does not serve for purposes of industrial design. The decisive factor determining when and where industrial design could and should become involved differs from existing product taxonomies.

The core of industrial design is formed by those products where there exists a direct perceptual and/or manipulative operative interaction between user and object.

For this reason, industrial design is of little importance when the design problem consists in the design of a transoceanic cable—though obviously that is an industrial product.

For the same reason, industrial design is called upon or should be called upon when the design problem is: a tractor seat, a plough, a tool, a kitchen appliance, a medical instrument, a light source, a milk container, a food package, a food-conserving device, a transportation equipment, a toy, a component for prefabricated housing, a device for educational purposes—to name but a few products.

The use qualities of this very broad spectrum of products are determined by a series of factors, such as comfort, simplicity of use, durability, functionality, safety, ease of cleaning, ease of maintenance, and repair.

These factors are related to costs which depend on technological factors, such as

complexity of assembly, degree of standardization, tolerances, utilization of right production method, utilization of right material, and finishing.

(e) What industrial design is not

From these observations, one conclusion can be drawn about what industrial design is not.

— Industrial design is not purely wrapping nice, attractive, fancy new
    shapes around (supposedly) ugly products.
— Industrial design is not face-lifting.
— Industrial design is not streamlining.
— Industrial design is not "sexing up" outdated unfashionable shapes.

Of course, industrial design can be all this, and it would be naïve to overlook or to deny that industrial design is not practiced in this way. But this orientation is hardly adequate for developing countries, and highly questionable in industrialized countries (waste of resources, ecological crisis).

III. INDUSTRIAL DESIGN AND NEIGHBORING PROFESSIONS

(a) Its relation to mechanical engineering and marketing

Industrial design is not only a multidisciplinary but also an interdisciplinary activity, that is, an activity realized in direct interaction, and not only parallel to other disciplines involved in the design process.

Except in rare cases (low-complexity products), it is not the industrial designer alone who designs the product in splendid isolation and who could be considered as the only authority responsible for design.

The industrial product, contrary to the craft product, results from a team effort, involving the participation of a number of other professions, including design professions such as mechanical engineering.

The direct professional neighbors of industrial design, especially in case of more complex products, are mechanical engineering and production engineering on the one side and marketing on the other side. All these professions influence the final design of a product. The industrial designer's share centers—as it has been mentioned earlier—around the use qualities, including the formal qualities of products.

The division between "guts designer" and "skin designer" is wrong and misleading because there does not exist a clear-cut borderline between inside and outside of an industrial product. Structure and form represent, or should represent, a coherent whole, and not a mix of separate, often incompatible components.

The difference between industrial design and mechanical engineering is not be to seen in method (art vs. science), but in the emphasis put on certain aspects of the design problem. The industrial designer is dealing predominantly with nonquantifiable aspects of design problem-solving.

(b)   Its relation to other design fields

Often the omnibus term "design" is used to cover a series of design specializations, such as the following:

Graphic design
Exhibition design
Packaging design
Interior design
Architectural design
Fashion design
Arts and crafts design (and crafts-based design).

Interior design is a discipline of its own. The same holds for architecture, graphic design, and fashion design.

Packaging design lies on the borderline between graphic design which is concerned with the bidimensional information on the package, and industrial design which is responsible for the three-dimensional shape, closure, stackability, and so on of the package. Indeed, industrial design offices often deal with package design and ICSID considers packaging as part of industrial design.

Thus, although industrial design primarily is concerned with the two areas, product design and package design, there are intermediate areas of concern such as the design of architectural components, which is a field where industrial design and architecture merge and exhibition design, which is often viewed as part of industrial design activity.

(c)   Arts and crafts design

Arts and crafts design is often, especially in developing countries without technological infrastructure, considered a forerunner or preparatory step of industrial design, and sometimes even identified as the complete entity or industrial design. The latter is misleading.

There is a distinguished difference between arts and crafts design and design for craft-based industries. The former refers to piece-by-piece manufacture in areas such as pottery, weaving, weaving with cane and reed, jewelry, leather goods with simple technological means, and work organization, while the latter refers especially to chinaware and textiles, produced on industrial scale, with corresponding machinery and work organization. These products may be included under the heading "Industrial design" but it would be a mistake to treat arts and crafts design as the only and primary strategy for developing countries. The development potential of arts and crafts design is quite limited. Industrialization is precisely a way of overcoming arts and crafts manufacturing methods and remaining on that level leads to a self-inflicted cut-off from development possibilities.

IV. INDUSTRIAL DESIGN AND ITS IMPORTANCE FOR DEVELOPING COUNTRIES

Industrial design should be used as a tool in the process of industrialization of developing countries. As a matter of fact, industrial design constitutes an indispensable instrument for endeavors toward developments.

Its importance—and the necessity to formulate a design policy in developing countries—is based on the fact that it can help to solve the following problems:

(i)   Dependent economies rely on the import of manufactured Goods (capital goods, consumer goods, and social service goods for hospitals and schools). These imports exercise a negative influence on the already-distorted balance of payments. By developing and producing their own designs, developing countries can use their hard currency reserves and incomes for productive purposes, that is, direct these financial resources to the creation of a diversified technological infrastructure.

(ii)   Products designed in developed countries do not necessarily fit the requirements and needs of the developing countries. For this reason, it is imperative that developing countries start designing their own products which correspond to their specific needs and which can be manufactured with the help of existing technology, or technology not requiring heavy capital investment and preferably local raw materials.

(iii) In developing countries, one of the most urgent problems to solve is the creation of jobs in order to integrate the population in productive activities. Industrial design in developing countries could be directed toward the development of labor-intensive products, instead of capital-intensive products which characterize the tendency of industrialized countries. This is important since the local labor market will have a shortage of qualified labor force in manufacturing countries.

(iv)   Developing countries suffer in general from the fact, among others, that their economies are not diversified, and often existing production capacities of manufacturing industries are not fully used for the lack of innovative designs. Industrial design can help to promote full use of these production facilities and diversify industrial output.

(v)   Industrial design as one type of technological innovation is a very effective means for export promotion. Internally developed designs with innovative character possess an export potential, especially in regions where trade arrangements between various countries have been established, particularly where the product design has been market oriented.

(vi) Industrial products or, in terms of anthropology, material artifacts constitute an ever-increasing portion of the man-made environment. They are an expression of a culture. Every nation has, to more or lesser degree, its own cultural identity. As far as industrial products form a part of a culture, industrial design can help to create cultural identity, overcoming the state of second-hand culture in developing countries.

(vii) There does not appear to exist a historical law or pattern that industrialization and development must follow. On the contrary, the ecological crisis caused by technology of "developed" countries raises the question whether it is justified to call these technologies "developed." Although industrial design is less concerned with the creation of new technologies than with the use of technologies to satisfy certain needs, it may nevertheless stimulate the development of alternative environmentally compatible technologies.

(viii) Income distribution is also one of the serious and explosive problems of developing countries. Often the majority of the population is excluded from access to industrial products because these lie beyond the range of their economic possibilities. Industrial design could find one of its noblest aims, and one of its very few really worthwhile justifications, in developing products for the needs of the poor majorities.

(ix) Developing countries need to utilize their limited resources in an optimal manner with the least waste possible. They are not well advised when they copy lifestyles and product assortments of industrialized countries. Confronted with the undeniable scarcity of means, the formulation of product policies and the definition of priorities becomes necessary, priorities of which needs are to be satisfied first, and which needs are to be satisfied at a later stage of higher development, with a higher level of productivity. Rationalization and formulation of product assortment policies could become one of the chief areas of industrial design in developing countries.

Industrial design is thus important for developing countries because it can help to solve nine basic problems.

1. Relieving the balance of payment
2. Fulfilling specific requirements and needs of the relevant market
3. Creating new jobs
4. Diversifying the industrial output
5. Creating export markets
6. Creating cultural identity
7. Stimulating the development of alternative technologies
8. Responding to the needs of the majorities
9. Rationalizing the output of industrial production

V. POSSIBLE FIELDS OF ACTIVITY OF INDUSTRIAL
DESIGN IN DEVELOPING COUNTRIES

Although in general public opinion industrial design is to a great extent associated with consumer goods, there are many product areas in which industrial designers can become, and already are, active.

These areas include, for instance, the following:

Passenger transport equipment (bicycles, buses, trains)
Cargo transport equipment (trucks)
Health equipment (mobile operation rooms for rural areas, surgical instruments)
Educational equipment (school furniture, kindergarten equipment, educational toys)
Agricultural machinery and tools
Building components for low-cost housing
Machine tools for medium and light industry
Food packages as well as methods for food distribution and conservation
Low-cost housing furniture
Consumer durable goods of all types.

VI. SOME GENERAL RULES FOR INDUSTRIAL DESIGN POLICY

When a developing country starts to work in the aforementioned fields, the list of which does not pretend to be exhaustive, it might be useful to have in mind the following observations concerning a design policy:

(i)   Industrial design activity in developing countries does not mean to develop cheap replicas or low-quality versions of existing designs developed in industrialized countries. Rather, it requires a definition and solution of the design problem in terms of existing (scarce) means and (abundant) needs.

(ii)   Design activity in developing countries should not derive its standards of evaluation from industrialized countries, but should take its points of reference from its own reality. Only that reality can yield standards of evaluation for design efforts made in developing countries.

(iii) Generally, it would not be viable to aim at complete design autarchy in developing countries. Therefore, it becomes necessary to establish priorities of design projects or design areas according to their global social benefits and development potential (multiplier effect).

(iv)   When exercising design transfer as one type of technology transfer, one can follow two strategies:

    a) The imported foreign design is adapted to the technological possibilities of the developing countries, without sacrificing qualities of usability. This adaptation requires redesign taking into account technological resources and parameters: machinery, materials, level of execution, possible tolerances, labor force, volume of production existing or attainable in the relevant country. The aim is to reproduce a foreign design with existing resources which requires modifications. Because of the redesign, the idea of fast and easy copying of foreign designs is an illusion.

b) The imported foreign design is adapted to the functional requirements and context-specific needs of the developing country. This adaptation implies a new formulation of performance specifications and may lead to major modifications of the existing design, and even to a development of a new product.

It is important to note that in both cases the foreign design serves as a starting point, and not, as in mere copying, as a terminal point. In adapting designs the adapting country needs to create a capacity for innovative work which helps to reduce its state of dependence. This practice is the opposite of reproductive technology transfer as in licenses or use of royalty schemes.

(v) Design transfer in the form of software or know-how, especially design methodology, should be made in a flexible way, that is, the software of industrialized countries should be adapted to the needs and contingencies of the developing countries, and not vice versa. Otherwise design know-how would tend to become superimposed on a reality which cannot assimilate the transferred knowledge. Design transfer both as hardware and software from industrialized countries to developing countries without modifications is hardly possible and would cause counterproductive effects.

(i) Developing countries which want to use industrial design as a strategy for development need to assign highest priority to the training of local manpower resources in the areas of the following:

Design management
Design research
Design projects.

The second priority refers to the logistical support, especially equipment of prototype workshops and laboratories with adequate machinery and equipment of design offices.

VII. THE SPECIFIC DIFFERENCE IN THE ROLE OF INDUSTRIAL DESIGN IN DEVELOPING COUNTRIES AND INDUSTRIALIZED COUNTRIES

(a) The situation of industrial design in industrialized countries

Generally, industrial design in industrialized countries has at its disposal a sophisticated technology with a great variety of materials, manufacturing processes, and skilled labor. Furthermore, there exists a highly diversified market structure, with a great variety of subtle consumer preferences.

The rate of obsolescence, both technological and psychological, is usually very high. In these "economies of abundance," industrial design has a decisive share in the creation of formal (aesthetical, visual) innovation, especially of consumer goods—a role which more and more gets criticized by a growing number of members of the profession. This critical attitude holds that formal innovation with its high turnover of merchandise has to be checked because of ecological considerations, not to mention social considerations.

Thus, enterprises and corporate organizations in industrialized countries use industrial design as an instrument in their comprehensive strategy of growth.

In corporate planning industrial design plays an important role, coordinating the many different manifestations of an enterprise in the marketplace and the general public. These "messages" which build up the corporate image can be emitted by the firm's products, interior and exterior architecture, packaging, graphics of stationary, vehicles, advertising, and so on. These components altogether create the so-called house style or corporate identity.

Industrial design in industrialized countries finds itself in a situation where the relation between means (technology) and needs (demand) is precisely the opposite of developing countries. In the former the volume of needs is smaller than the volume of productive forces or means, whereas in developing countries the volume of needs is bigger than the capacity of the productive forces.

(b)   The different approach to industrial design in developing countries

The fundamental difference between the two opposite contexts implies, of course, a different approach to industrial design in developing countries. The different approach can be described as follows:

(i)   Concerning the importance attributed to formal aspects: though undoubtedly important, and hardly to be eliminated from any industrial design effort, formal factors play a secondary role in developing countries compared with industrialized countries which can afford to indulge in aesthetical innovation and sophistication. Thus, "Good Design" has a secondary place in developing countries unless export marketing is envisaged.

(ii)   Concerning costs: If industrial design is aiming at the satisfaction of needs of the poor majorities, it is exposed to heavy economical constrains. Therefore, the problem is: How to get a good, and not shabby, use value at low cost and low price! Generally the flexibility of the price range in industrialized countries is bigger than in developing countries.

(iii) Concerning technological resources: Due to the lack of a technological infrastructure, the range of materials, manufacturing processes, and skilled labor force on the use of which the industrial designer can draw is quite limited. Industrial design in developing countries is forced to work under "imperfect" and restricted conditions.

(iv)   Concerning the production volume: Developing countries often have rather limited markets, whereas industrialized countries can count on enormous markets. The use of certain technologies is only economical when there is a great output and a market potential which can absorb the products. Industrial design in developing countries must therefore consider market limitations and possible economies of scale.

(v)   Concerning use value: The scarcity of means imposes the search for a maximum of use value for a relative minimum of costs. That does not mean design of cheapest products. "Cheap" designs are not necessarily products with least cost. They result often from false economies.

(vi)   Concerning utilization of resources: The restricted resources of developing countries require a rational approach which guarantees their optimal utilization. Concern for total social benefit of design activities is a prerequisite. Any error made in this field weighs much more heavily than in industrialized countries.

(vii) Concerning economic implications: Countries not yet industrialized still have the possibility, at least theoretically, to opt for a different pattern of industrialization which pays attention to ecological compatibility and which contains built-in preventive measures against environmental sell-out.

(viii) Concerning food problems: Populations of a great part of developing countries do not receive sufficient food, neither in quantity nor in quality, especially protein-rich food. Design imagination and design effort might focus on the solution of this basic problem: production and conservation of more and better food for hundreds of millions of people. Industrial design can contribute to the solution of this problem by development of adequate tools and machinery for agriculture, and in the future perhaps aquafarming or other such innovations.

The original version of this review of Papanek's book was published under the title "Cardboard Bombast" in form 61 (1973): 13–16. It was partially written in a polemical tone which can hardly be understood today and therefore does not add much to the comprehensiveness of the presentation of divergent positions. Nevertheless, it has been included at the insistence of the editors in order to illustrate the "historical climate" at the beginning of the 1970s, where two approaches to design in peripheral countries were facing off against each other. On the one hand, there was an emotionally appealing narrative about an alternative of design in the periphery, bound up with a critique of consumerism in central economies and the professional practice of designers. It hinted at the prospects of an ostensibly simpler lifestyle in a small community with artisanal and manual production methods. On the other hand, there was an approach that aimed explicitly at industrialization. The latter corresponded to a general development policy prevalent in Latin America, in the form of import substitution industrialization (ISI), which since the mid-1970s was gradually undermined and dismantled. As far as the term "Third World" is concerned, which was common at the time, I have retained it and not replaced it with the more contemporary term "periphery."

32 DESIGN AND DEVELOPMENT

The Debate with Victor Papanek

GUI BONSIEPE: REVIEW OF *DESIGN FOR THE REAL WORLD* BY VICTOR PAPANEK

(1973)

The author pays homage to a broad definition of design. Already the first sentence of the first chapter claims: "All men are designers." This thesis is advanced via the definition that design means any creative and goal-oriented effort. Because design as the "primary underlying matrix of life"[1] is elevated to an ontological constant bound up with the term "order,"[2] the author forecloses the chance to define a politically grounded, forward-looking design ethic.

This conservative attitude breaks through in several places in the book, for instance when a cologne bottle is glorified as an apotheosis of design experience. The author clearly accepts that the "telesic" content of form reflects certain contemporary myths according to which women are flowery expensive sex toys. Woman as a fetish: taking the negative givens as a measure of "telesis"[3] (the author defines telesis as the coherence of form and the historical conditions of its development which puts it close to the term *Zeitgeist*). This type of apologetic argumentation forfeits the right to take the high road over styling-designers from Detroit or to condemn their practice.

But the polemic goes into battle against more than just this group of designers. It takes aim at design practice in late capitalism in general. Papanek is in no way squeamish about doling out criticism of his own colleagues. Accordingly, the designer appears to be a first-order danger to society. He is one of the main culprits of environmental degradation and the waste of resources, a privileged dilettante who wastes his time and energy on a carnival of gimmicks to lure conspicuous consumers. There is no shortage of examples: mink-covered toilet seats, diapers for parakeets, bathroom floors covered in monkey fur, electronic

Review of the book *Design for the Real World: Human Ecology and Social Change*, by Victor Papanek (Toronto; New York; London: Bantam Books, 1973). The text was originally published in the German magazine *form* and after published in the Italian magazine *Casabella*, no. 385 (February 1974) with the tile "Design e Sottosviluppo."
Revised by the author in December 2019. Translated by Anke Grundel.

[1] Victor Papanek, *Design for the Real World: Human Ecology and Social Change* (Toronto; New York; London: Bantam Books, 1973), 23.
[2] Ibid., 24.
[3] Ibid., 34.

necktie selectors, baroque fly swatters, life-sized blow-up vinyl sex dolls—in short, paraphernalia of consumer society and excretions of a design practice beholden to laws of marketing.

Against this foil of false desires, the author offers a design for real-life needs: the needs of the underprivileged, the elderly, the injured, the disabled, children, ghetto residents, and the poor. The place of individual consumerism ought to be taken by medical diagnostic devices, hospital equipment, surgical instruments, teaching materials, and similar products for social consumption.

Certainly, one cannot ignore the fact that capitalism only satisfies as a desire if it is intelligible to the market. Acts of goodwill eventually shatter upon the rocks of production conditions. Even if the activity of design studios could be rerouted directly and without resistance toward socially relevant products, the deliverables would get caught up in the net of capital. Not a single word mentions the organization of the production conditions or the role of productive forces, especially the working class whose participation is an essential condition for any fundamental change of design practice. Hence, the simultaneously naïve and well-meaning appeal to introducing a kind of design token fades away in abstract and noncommittal moralization. In addition, the rule of allocating every tenth working hour pro-bono to real design would soon be abused as an alibi. The sins of design cannot be redeemed by the purchase of letters of indulgence—even if they are self-issued.[4]

These comments are not intended to defend design in the center. However, they take issue with the manichean juxtaposition of good design/bad design (good and bad depending on societal norms), as well as with a universal critique of the following kind: "I am questioning, then, the entire currently popular direction of design. To 'sex-up' objects . . . makes no sense in a world in which basic need for design is very real."[5] One could agree with that. But not even the most ardent critic of capitalism would dare to claim that the entire design practice in late capitalism is reducible to the formal revamping of products.

Among the solutions for the ills the author denounces is "design for developing countries." He outlines three different possibilities:

1. Working in the urban centers for the periphery, which typically produces exotic souvenirs. These provide the spaces of a high-technology environment with an air of pretechnological quaintness.
2. Temporarily working in underdeveloped countries as a visitor.
3. Relocating to the periphery and primarily training local workers.

Fittingly, Papanek believes the third point to be the most worthwhile. The solutions for the third world will hardly be worked out in the center, but in the respective context of the vulnerable country. Had the author drawn the

---

[4]   Note: This translation is closer to the original German. However, the reference to Catholicism may not be understood by everyone. An alternative would be this: The sins of design cannot be washed away so easily—even if designers do the cleansing themselves.
[5]   Papanek, *Design for the Real World*, 178.

necessary consequences from this insight, he would not busy the students in the center with "design projects for the Third World." These design efforts do not solve the problems of the third world; rather, they are a symptom for the problematic relationship between center and periphery, thereby becoming a problem for the center itself. This is not to say that problems tailored to the third world should not be part of the design curriculum. Nothing, however, can hide the fact that the third world in no way benefits from this. Faced with the third world, the first and second world are becoming problems for themselves. It would be better to address the resulting guilty conscience instead of shirking responsibility by flaunting an image of self-identification with the poor of this world.

Concerning its design-theoretical savviness, the book reveals an impoverished understanding of functionalism. Papanek joins the chorus of those who identify functionalist design, or whatever they determine as such, as a "bankruptcy of spirit." Singing the song of "human values"[6] has always been the favorite pastime of anti-enlightenment champions. The short formula "form and function are one" by Frank Lloyd Wright is not at all as obsolete or devoid of meaning as Papanek would have it. Instead, it still harbors—defying all seemingly progressive developmental ideas—a potential which only theory can unlock.

The conceptual pair function/form allows for three schematic interpretations, which are here represented in abbreviated form:

1. "First form, then function." Depending on context, this type of design practice is called academism, formalism, or styling. The worry about giving shape dominates, shape in turn being bound up with the business of increasing exchange value and the false promise of utility.
2. "First function, then form." Derived from a variant of the classic interpretation, this design maxim of neo-functionalism says that the highest determining factor of form must be its function or purpose. Aesthetics are reined in by use value.
3. "Function and Form are one." This formulation corresponds to the determinist doctrine of functionalism, according to which form and function fully map onto each other. In addition, it holds that appropriately fulfilling a form's purpose determines its aesthetic quality which is little more than an additional bonus.

Insofar as the author complains that the last point is untenable, he tends toward a nondeterministic interpretation of the relationship between form and aesthetics on the one hand and functions on the other. Certainly, in claiming that the Solaris desert dwelling is the inevitable outcome of tools, material, and working method, he falls back on a technological determinism no less restrictive as functional determinism. Subjectivism often wears the garb of

---

[6]   Ibid., 96.

trenchant technological objectivity, thus hiding idiosyncrasies with the aura of the universal.

After Papanek has once again raised the old straw man of orthodox functionalism just to effortlessly knock him down—how else could it be?—he falls back on the widely known thesis that aesthetic quality must be an integral part of functionality. In other words, the dimensions of a product's use value must incorporate aesthetics. Thus, he once again treads the worn-out path of the Bauhaus and the institutions that succeeded it toward which he does not have particularly warm feelings.

This also extends to the HfG Ulm, which he lashes out at through an anecdote. Certainly, there are many particularities to be criticized about the HfG. However, the alleged Ulm proposal to paint the casing of the 9-Cent-Tin-Can-Cow-Manure Radio gray for developing countries fits too neatly into cliché ideas about the institution used by its critics to avoid any differentiated analysis.

This example of design practice—the cheap radio for the third world—is doused in the cliché of the needy savage, who can be mollified with simplistic technology developed just for him by the designers of the center. Concerning the paint job, the author acts demure ("I feel that I have no right to make aesthetic or 'good taste' decisions that will affect millions of people in Indonesia, who are members of a different culture.")[7]. By contrast, he offers Indonesians the illusion of adapting imported design to their context by simply adding surface decorations. That is, he wants to overcome the foreignness of the tin can radio, of the technology smuggled into their culture, by sticking seashells on it.[8] The author celebrates this practice as "a new way of making design both more participatory and more responsive to people in the Third World."[9]

The professionalization and specialization of design practice are for Papanek a bone of contention. Similar to Moholy-Nagy, he defends the opinion that design is not a profession but a world view. He relegates design to a kind of stopgap for the weaknesses of other disciplines in a team of planners. In the role of the interpreter, the generalist, the designer acts mostly as a mediator if he is not busy with other things. As a generalist, he made his lack of specialization precisely his specialty. Thus, Papanek advocates for the education of a multifaceted, "horizontal" generalist instead of a strict "vertical" specialist.

Horizontality versus verticality is a wrong alternative only explainable by the apparent lack of knowledge the author has of contemporary curricular research on horizontal and vertical specialization. The weakness of argumentation is compensated by the threat of catastrophism: "the price which a species pays for specialization usually is extinction."[10] Thus, human history is reduced to the level of natural history—an ahistorical method cultivated by Buckminster

---

[7] Ibid., 192.
[8] This translation slightly loses the snark of the original German. An alternative translation would be "bedazzling it with seashells."
[9] Papanek, *Design for the Real World*, 192.
[10] Ibid., 291.

Fuller, whom Papanek worshipped as a mentor. The de-specialization of the design professions degenerates the designer to a jack-of-all-trades.

Papanek recommends "Integrated Design" as a cure for the splintering of design. With theatrical gestures he creates a tabula rasa, broadcasting his intention "to replan and redesign both function and structure of all the tools, products, shelters, and settlements of man into an integrated living environment, an environment capable of growth, change, mutation, adaptation, regeneration, in response to man's needs."[11] Integrated design as the instrument of this global enterprise spans from regional planning, via city planning and industrial design, to visual communication. The architect is chosen for the office of the all-talented generalist. This resurrects Gropius' doctrine from the first Bauhaus phase of architecture encompassing all artistic fields.

Integrated Design does not only refer to amalgamation of practices from different professions but beyond this it introduces a cabal of "designers able to deal with the design process comprehensively."[12] Concerning methodology, this means determining the level of complexity of the planning problem while acknowledging its historical dimensions along with human factors. Finally, it means not forgetting about the societal perspective. An example should serve to illustrate how the author imagines this process: "As more and more methods of social classification, stratification, and class identity break down, there is a ready market for products used to express social ambition and strivings for status."[13] Here, therefore, the designer labors over recycling status objects. Only a few lines on, Papanek voices his indignation over styling-designers being accessories to maintaining the separation between income classes in consumer society. Confusion and the absence of a clear thread, first and foremostly, mark this work.

The delusions of all-powerfulness of the architect-designer are lived out in a ludic vision of the future, free of the worries of subsistence. In it, agricultural and industrial labor are carried out by automated facilities and design is apostrophized as "the only meaningful and at the same time crucial activity left to man." The designer would take on the purpose of "help[ing] to set goals for all of society."[14] We don't need designers for that; but we do need a revolution. In light of this a technocratic self-understanding cannot imagine anything else but establishing an "International Council of Anticipatory Comprehensive Design"[15] (a phrase from Buckminster Fuller's intellectual treasure trove). Papanek parrots this socially and politically blind topos, according to which there are more than enough resources for all of humanity if only these would be appropriately planned, distributed, and consumed. At fault for war, hunger, deprivation, exploitation, imperialism are politicians whose irrationality now has to make way for the rationality of total design. The technologist is showing the politician the door. Here the world view of technocratic utopianism is

[11] Ibid., 284.
[12] Ibid., 286.
[13] Ibid., 289.
[14] Ibid., 322.
[15] Ibid., 336.

showing through. Incapable of detecting the real causes in the constitution of class society, it can only explain the deficiencies of the world as failures of a particular group of people.

The book does not shine with innovations for design training. A bucolic idyll, the dream of rustic communal bliss, is painted with the brush of the most modern communication technology (a new world of work/life, according to the author, should best take place in the countryside and best beginning in an old barn, but not too far away from the city).

This institution understands itself as mission control for a new lifestyle, and for none other than the "peoples of the world."[16] Structured curricula are a horror for him. He uses the word in scare quotes, to demonstrate his distance to it, propagating an open learning style without strict study plan, learning in the style of commune life, learning in the form of a permanent experiment. According to this planless approach, content and form of teaching "would evolve organically out of the needs of society,"[17] which dispenses with the effort of building a systematic pedagogical project.

Public misery is not alleviated with the escape into private subculture. The retreat into artisanal crafting methods and cultlike social forms, manufacturing amulets and health sandals, does not change a social order of technological making but at best serve as cute decorations. Uncritically, he welcomes the rebirth of arts and crafts without recognizing it as a form of busy work for the big children of affluent society and therefore a regression into the private sphere which only serves the ruling interests.

Even the boastful radical language of the book fails to cover up such a hodgepodge of conservative ideas. One would have wished for the author to not merely wave the flag of the ecological problematic, but to actually take it seriously—or to say it in his voice of arrogant humility, "not wasting paper printing books such as this."[18]

---

[16]   Ibid., 342.
[17]   Ibid.
[18]   Ibid., 343.

VICTOR PAPANEK: REPLY TO BONSIEPE'S BOOK REVIEW

(1974)

About a year ago, Bonsiepe reviewed my book *Design for the Real Word* for the German magazine *Form*. I marveled at his discovery of translation errors and found material for many meditations in his critique of specific points of the text. I therefore didn't reply to the review, but now the situation has changed. His reviews were reprinted on *Casabella* (no. 385) to coincide with the release of the Italian edition [of the book], and I am convinced that the review will travel around the world following the book's various editions.

Better to answer before it becomes an international "case." It is difficult to establish a starting point. I will try to examine the various points rationally. In his review Bonsiepe continually uses semantically "loaded" words and carefully isolates some sentences in quotes to increasingly confound the readers. The system is simple: first, write what I would have said (but what I have not said) and then amass a great deal of dialectical indignation about the things I have never said. Second, take quotes out of context and then misrepresent them (not always successfully). Third, resort to the old system of insulting "show me your friends and I will show who you are." Bonsiepe seems to believe that my friendship with Fuller is rather adoration. I am also accused of respecting Max Bill's work (and I respect it indeed) and of loving Frank Lloyd Wright. On the other hand, he is unsure whether to judge me because I do share some of the ideas of Walter

Gropius or because "I do not" share them.

Some of our differences seem to have a semantic origin: he does not like that I use the word "order" and declares that by using this word in relation to design, I have "forever destroyed my chances of dealing with the revolutionary anthropological concept of need." However, reading my book it is clear that the basic concept of the book itself is precisely the need—social, human, and environmental. (Incidentally, my background is closer to anthropology and its adjacent fields than anything else, except design.) Throughout the book I clearly use the word "order" as opposed to "randomness," rather than as "law and order," as Bonsiepe strives to reduce it.

To get to the point, Bonsiepe declares that the participation of the working class in the design process "is not mentioned even once." Unfortunately for Bonsiepe's credibility, they are mentioned on pages 55, 58, 59, 60, 76, 92, 93, 111, 112, 113, 118, 119, 125, 127 (and these are only the most important).

As far as I know, *Design for the Real World* is the first book that has ever suggested worker participation in the design process. I have a long list with the indispensable members of a design group: a client representative, and then "the people the group works for must be part of the group" (list on page 276).

Published in *Casabella* magazine no. 396 (December 1974) with the title "La risposta di Papanek a Bonsiepe. Prima l'amicizia, poi la polemica. Replica a una recensione" (Papanek's Response. First the friendship, then the polemic—Reply to a book review). The original text, probably written in English, was not located. What follows is a translation from the Italian text by Lara Penin.

In many places, too many to list, the book states that workers must directly participate in the decision-making process regarding the form and substance of the objects for them intended. Elsewhere in his review, Bonsiepe claims that "the architect was chosen to synthesize every form of design." But I have never suggested anyone for this super-synthesis role (a role he invented himself). I have repeatedly suggested that it is an interdisciplinary group (including workers) who must take on this role.

The truth that is clearly demonstrated in pages 152–7 is my disillusionment with today's architecture, and also that I have spent eleven years of my life so far working with residents of countries classified by the United Nations as "developing."

This issue takes me to an important point. Bonsiepe states that I have described "three different ways of doing design for developing countries" and goes on to list the three ways that he has managed to identify. In reality, there are four. My proposals are as follows (pages 71–2):

1. The worst, in my opinion, is to sit in your office in New York or Stockholm and design exotic Kitsch.
2. The second, slightly better, would be for designers to spend some time in an underdeveloped country, developing a type of design suitable for the needs of local people.
3. A slightly better system would be for the designer to go to the developing country and educate designers, and work together, staying in the country (not yet ideal, but it seems to summarize what Bonsiepe did in Chile and I have done in various countries).
4. (A point appropriately forgotten [by Bonsiepe]) The designer does everything mentioned in point 3 but moreover prepares local designers to teach other designers, and thus becoming a "human source of projects." The designer should also participate in an interdisciplinary group together with local workers, farmers, and designers (all this is said and discussed as a fourth independent point from the others on pages 71–2 and 63–71, 116, 124–6, 162–4, 171–8 etc.).

Unfortunately, the listing of three instead of four points becomes the starting point for further criticism. It is difficult to discuss his fantasy interpretations rather than the things actually written in the book. (Regrettably he does not even understand that many projects in my book that concern the third world were made either in important third world countries or on behalf of developing countries, or by design students from the third world.)

Should or should not design

1. make more sense in a country with a socialist economy than in a country with a capitalist economy?
2. in a developing market respond to real needs?
3. increase jobs, without destroying existing social relationships?

4. increase exports, without the consequent need to produce Kitsch-for-export?
5. produce new work and develop new skills?
6. contribute to the balance-of-payments of countries?
7. contribute to the development of intermediate technologies to prevent cultural setbacks, environmental imbalances?
8. [help] diversify the industry?
9. stop producing for an elite or a particular class?
10. create a national identity for the industrial countries?
11. be accountable to workers in developing countries?
12. give the workers the opportunity to influence and change the design and building/making processes?
13. contribute to the economy of nonrenewable materials and energy sources?
14. help create autonomous and decentralized ways of living and working?
15. demonstrate how to work with low energy availability?

While discussing an old project of a radio, Bonsiepe says: "There is no doubt about the client's intentions." There are however about 800 words (pages 162–5) about that project. I refuse to believe that he really thinks that the purposes of US military propaganda coincide with mine. His reaction to the project of that radio seems to me very similar to that of the US army who rejected the project out of fear and concern.

Although the book is written from a point of view completely opposite to that advocated by the US military, Bonsiepe ultimately achieves its senseless goal suggesting that "the author finds his friends in military circles" (!) referring to one of my 667 examples in the chapter on bionics.

Let's move from the third world to interdisciplinary design and worker participation. In 1968, Bonsiepe wrote in "Ulm" [magazine]: "To solve the problem, it would be necessary to create new versatile institutes, in which environmental design can be studied on largely interdisciplinary bases; there would be a field of research for collaboration between sociologists, psychologists, economists, technicians, doctors and designers." For my part, I wrote: "Interdisciplinary design also needs groups of specialists—specialists whose approach is not that of achieving profit, but a human and humanistic interest in man and his environment."

Such groups could be composed of a designer, an anthropologist, a sociologist, and specialized technicians. A biologist and experts in medicine and psychology should complete the group. And finally, most important of all, the people for whom the group works must have representatives in the group itself: no socially significant design can disregard the cooperation of its "clients." When students are faced with this problem for the first time, they try to avoid meeting "clients" by invoking alleged communication difficulties and the possibility that clients are too ignorant to fully understand their own needs. Such a lack of trust has no justification (page 266).

To further clarify this participation, I wrote: "I encourage students to travel a lot and to take on different jobs in offices, in industry, and in agriculture and commerce. This work is a mandatory part of their training; one year as a factory worker or as a farmer is useful" (page 265).

Moving on. Because I am against exploitation through design and particularly the exploitation of "status" or class, I wrote: "Much of current design needs to be re-examined as whether it contributes to keeping the class system and social status unchanged" (page 259).

To strengthen this point, two pages later I speak of the anonymous design of the Swedish wooden clogs and express my positive opinion, saying, "they (the clogs) exist beyond the concept of social class and income, rejecting any idea of status" (page 261).

Unfortunately, Bonsiepe correctly quotes this phrase of mine, but tries to convince the reader that my words relate to the fact that I am for the maintenance of the class system through design. After all, this jumping around the reader has the right to ask himself: What is this book *Design for the Real World*? It is a book that deals with *human ecology and social change* (subtitle of the American edition). It deals with the energy crisis, and proposes alternative solutions such as "a vehicle lighter than the airplane" for low-speed transportation, sailing boats guided by computers and with servo-mechanisms for navigation that requires a reduced crew, heating with solar energy, cooling through vaporization, as well as technologies based on wind and methane.

It deals with design for the elderly, the handicapped, the poor, the blind, deaf, fat, left-handed, and pregnant women; toys and school environments for children; medical instruments and hospital equipment; the needs of the third world; in short, the books deals with all that the industry and designers currently serving the industry have conveniently forgotten. It deals with how to make design within everyone's reach and how to directly involve workers, peasants, and ordinary people in interdisciplinary design groups, how to use the ideas and participation of workers in the decision-making. It deals with the political, social, ecological, and environmental effects of design itself, and the consequent responsibilities of design teams. It proposes a greater interest of design for agriculture. "Because it is important that the idea of design is taught in school and then connected to the farm . . . Student participation in design is an important part of their education" (President Julius Nyerere, Dar El Salam, Tanzania).

Finally, by presenting more than 100 photographs of objects (most of them students' work), the book demonstrates that an alternative design is possible.

Victor Papanek

Editorial note: This interview, made in 2015, relates directly to the 1973 report for UNIDO published in this volume as Chapter 4.1 and more generally to many of the papers in Part II. It offers Bonsiepe's reflections on forty years of "development" in Latin America and on the place and role of design.

33 DESIGN AND DEVELOPMENT FORTY YEARS LATER

Interview with Gabriel Patrocínio and José Mauro Nunes

(2019)

In 1971, a joint meeting between the United Nations Industrial Development Organization (UNIDO) and the International Council of Industrial Design Societies (ICSID) decided that UNIDO would produce a study, and subsequently recommendations, on the implementation of national design policies in peripheral countries. How could design be used to leverage development in these nations?

After a first attempt was discarded as unsatisfactory, a repeat request was issued in 1973, as a matter of urgency and with the recipient being the designer Gui Bonsiepe. His experience as a European designer with permanent residence in South America (Argentina, Chile, Brazil) since leaving the HfG Ulm in 1968 enabled him to speak as someone who practiced design—and, better yet, as someone who had practice in policy design in peripheral countries.

Four decades later, how does the author see his 1973 text today?

Interviewed in March 2015, Bonsiepe addressed issues such as design and development policies, design for basic needs, and globalization. The interview is in part a commentary on the original UNIDO report (published in this volume as Chapter 4.2).

PART 1: DESIGN POLICIES

Gabriel: In the essay "Design and Democracy" you mention that an up-to-date industrialization policy should include the contribution of information design. Would it be the main update to your national design policy document? ("Development Through Design," prepared for UNIDO in 1973. Published here as Chapter 4.1.)

An interview with Gui Bonsiepe by Gabriel Patrocínio and José Mauro Nunes. Originally published in *Design and Development. Leveraging Social and Economic Growth through Design Policies*, by Gabriel Patrocínio and José Mauro Nunes (Blucher, 2019). Republished by kind permission of Blucher. Translated by Lara Penin.

Bonsiepe: This document was written four decades ago in a very different historical context. It is unlikely that it kept the relevance or the same guiding value that it had when it was created. As we know, there have been radical changes in this period (besides the growth of biogenetics and neurosciences), such as the emergence of information and communication technologies, which constitute a broad new field for both development and design policies. When ICSID asked me then to draft a response to UNIDOS's request, the focus was on hardware, physical-material products. Even today I consider it indispensable that an industrialization policy be focused on the products and the processing industries, but at the same time give priority to technologies concerning media. We know the clinging resistances of information conglomerates against any attempt to democratize or reduce this enormous power they hold. Because they are so embedded in the political and social system, at times when voices dare to question this power and demand democratization, such voices are accused of being undemocratic and populist. These conglomerates do not hesitate to present themselves as the sole representatives of democracy—an unreasonable and therefore questionable claim.

Another new point compared to forty years ago is the relationship between design and science, and design and knowledge. I think it is essential that design be incorporated into the policies of scientific and technological innovation, collaborating in the research institutes with scientists, contributing to the sciences from the design perspective. It's a new area, but obviously there are historical antecedents. This new field is identified today by the still somewhat vague term—that I thus use with caution—of knowledge design. Knowledge design is of central importance in this change that has taken place over the years—leaving aside any political aspect—and I hope it does not become merely a fad used opportunistically.

In addition to these main points for updating, I allow myself a critical remark: the document was written with the idea that "development" in general and "industrial development" was "good." I have assumed—rather naively—that in one way or another this process would contribute to reducing inequality in peripheral societies. It was not clear by which mysterious ways that would happen. Today I would adopt a more critical stance on the concept of "development" because it is not a neutral concept—it is always linked to interests that reflect constellations of power. I would ask: Development for whom? I would also further inquire about the content of this process and the interpretation that social groups affected by industrialization give to this concept. I would certainly not limit it to a process to increase GDP. An economist interpretation seems too narrow. A community that lives in an extensive area transformed into a National Park where a government intends to build a road does not share the same conception of development as the politicians who decide for the construction that would negatively affect the life of that community [such a community] does not consider "development" to be the possession of a mobile phone or TV set and has good reasons to suspect that the goal of building such a road is less to bring in "progress" than to facilitate access to natural resources for financial interests.

Today, I would also take more into consideration the environmental and social effects of industrialization, as well as the danger of any form of top-down drive and alleged "anonymous restraints" so beloved by closed circles of experts outside of democratic control.[1]

Gabriel: Still considering what would be an updated view of the 1973 document ("Development Through Design"), do you think service design (in the public and private sectors) and public policy design (strategic design as a tool for developing, implementing, and evaluating public policies) would be contributions that could be added today?

Bonsiepe: Certainly. I am a strong advocate of public policies in the area of design—despite the perennial danger inherent in bureaucratization (an endemic risk in any administration). But if we look critically at the current state of public design policies, they exist nominally at best, in the form of declarative documents, but in practice, save for laudable exceptions, they are very small scale. This is because public domain, or a concern for common interests, has been practically extirpated since the 1980s, particularly in Latin America, which suffered a tsunami of privatization policies. Whatever existed in public design policies was also devastated. So I ask: Can we still talk about effective public policies, and not just declarative ones? For even the most outdated politician has understood that today, without design, one does not get very far.

Service design—a new area that emerged some twenty years ago—is primarily a coordinating activity. As a coordinating action, service design incorporates components of graphic design, visual communication, and industrial design, carrying out a supporting role to services. In 1973, this activity certainly existed—for example, in the corporate identity policies of the companies—although without this label.

It also refers to the processes citizens deal with to resolve procedures in public administration often created and implemented from the administrator's perspective rather than from the perspective of the citizen. Improving public services for the citizen is an undeniable need. And there design can contribute to reduce the irrationality of the often-humiliating processes that the citizen must face to solve problems in daily life. Design can help turn services, largely adverse to the citizen (citizen-unfriendly), into citizen-oriented (citizen-friendly) services.

---

[1] (NA) It is necessary to critically revise the term "development." Embedded in the fabric of energy crisis, climate crisis, environmental crisis is a consumption crisis that requires a fundamental reformulation of design. In view of the central societies bursting at the seams and saturated with consumption, the contributions by the social scientists from Latin America, for example, the contributions of the Grupo Alternativas al Desarrollo (*Más allá del desarrollo* [Quito: Fundación Rosa Luxembourg, Ediciones Abya Yala, 2011]) open perspectives to rethink and reassess the social function of design, and in doing so permit to examine among others the validity of the claims of green design and sustainable design. See also the works of Maristela Svampa, https://www.youtube.com/watch?v=6maaa6jS-ZY and https://www.youtube.com/watch?v=Zld7G_PcXg4 (accessed August 21, 2020).

The design of public policies (from a design perspective) is a task that still needs doing, and especially to be implementing. I think it is essential to reinforce public design policies in these countries that have left the "peripheral countries" category, such as Brazil, which is now incorporated into the BRICS and already has a more significant economic and even political role internationally. But as a professional body, designers should ask themselves if they were able to move to power centers where politicians and representatives of other professions—mainly economics, finances, technology, management, lobbyists (not designers)—take the decisions. I ask: Where are the designers in this complex system?

Gabriel: The insistence on "updating" your document comes from understanding that it remains cutting-edge in many aspects. There are, of course, a few dated subjects, such as when you suggest the importance of radio, film and television for communication and promotion of design. The media might have changed, but the idea remains the same.

Bonsiepe: I think the basic content of the document—to reduce dependency and increase autonomy—remains as relevant as it was four decades ago. This is the way in which development policies can or should, must or ought to be. Obviously a policy with such an orientation may conflict with antagonistic interests, both internal and external, who do not view these efforts with sympathy but with suspicion. In many aspects one deals with a reality that, for lack of a better term, could be described as "neo-colonialist." Therein lies the fundamental difference between design and design policies toward central countries.

Gabriel: In your 1973 paper on design policy, you stated that designers should reverse their priorities and focus their primary attention on the government, and only then reach out to entrepreneurs and ultimately to society. Professor John Heskett also wrote that designers need to learn how to talk to the government—understanding that it has another language, another way of operating, and other priorities, quite different from those designers usually master. Do you still think that this order of priorities should prevail today? How do you think designers should establish a dialogue with governments?

Bonsiepe: First they must know about politics and interpret the concept of citizenship, occupy the space of citizenship. I wonder: Do students of design courses today have a political background, say, the opposite of celebrating individual creativity? One should not passively wait for the government to guarantee work or to summon designers. Due to the trivialization the "design" concept suffered, including design education, I am afraid that today's politicians have a confused and even misguided idea of design or the possible contribution of designers to solving community and industry problems. It is possible that the understanding of design by most politicians is limited to a simplistic interpretation as a tool of political marketing, that is to say, the repetition of the known: instead of publicity for soap powder to wash clothes, publicity for politicians.

Gabriel: You talk about the demonization of state intervention in the economy—except when it comes to rescue broken financial institutions. Do you believe that this very denial, of the role of the state in regulating market processes and especially in industrial policies, is a barrier to a broader adoption of national management tools for design programmes? Here I specifically meant models such as the Design Council, adopted in several highly competitive European and Asian countries, to which funding and the acceptance as viable continue to be denied—in Latin America.

Bonsiepe: The state abdicated the role of guiding industrial policy, and it makes no sense to have a design policy without an industrial policy as a previous condition, or without them being connected to each other. Latin American countries in general had a deindustrialization process as corollary to unrestricted opening of imports and the *financialization* of a wide range of activities. Consequently, national design programs are rather circumscribed to tasks such as promoting events, registries, promotions, design seals, awards—namely, correlative activities, without intervention at the nerve endings of the industry. That is why [national design programs] sometimes choose to help artisanal design as alternative to the lack of design for the industry. Or to support entrepreneurial designers who at least launch a product to the market. We may ask whether amid the reindustrialization goals, the opportunity to target small and medium enterprises has been tapped. The reports mention hundreds of products developed with state support, of hundreds of seals, but it does not reflect on products with industrial design in the market. So, design promotion bodies with national design programs face a complicated problem: How to move beyond tangible activities and reach the difficult clientele of small and medium-sized enterprises? For design is a buzzword and thus some politicians consider appropriate to include it in the range of government activities. The formulation of a national design program is the necessary but not sufficient condition for an effective design policy.

PART 2: DESIGN TO MEET BASIC NEEDS OR TO GROW?

Gabriel: In 1973, ICSID had a Design for Need working group, which had the designer Victor Papanek as consultant and possibly as top exponent because of the great repercussion of his then newly released book *Design for the Real World*. Suddenly, however, in the same year ICSID received a consultation from UNIDO in this regard and assigned you to deliver a document with the subject of design policies for peripheral countries. How did this change happen? In that same year, you joined the ICSID board—was that somehow part of that episode?

Bonsiepe: You ask me why ICSID did not assign Victor Papanek, author of a best-selling book about design out of the market track, to write the report for UNIDO. I do not know. The fact was that a US colleague prepared a document that apparently did not meet ICSID's expectations. It is difficult to draw up a policy document without an extensive empirical basis. I had already worked for five years and gained concrete experience in the local context of a peripheral country, I didn't show up as "swallow" consultants do, making short visits to peripheral countries and, based on that visit, make recommendations. This work of formulating a document for UNIDO emerged unpredictably and with such a short time frame (six weeks) that I doubted if it could be up to ICISD's expectations. I did not offer myself to do the job. They asked me to work in a situation that wasn't all that short of an emergency.

Gabriel: In my doctoral thesis I explore this topic a bit, and speculate that if Papanek's vision of design in peripheral countries had somehow prevailed or deserved greater prominence in face of the peripheral governments at that time, this would have further deepened the dependency relationships. In this sense it would have been more than appropriate to use the term "pirouettes of neo-colonialism" (that gave title to your criticism of Papanek's book), given the risks involved in the eventual adoption of the practices recommended by the author. Do you agree with the view that there would be a deepening of dependency if that course were followed?

Bonsiepe: It is possible to ponder about possible consequences if that had been the case, nothing more. I cannot speculate on the consequences of a hypothetical possibility. Papanek has the merit of having drawn attention to design problems outside the mainstream of the profession. My criticism was directed against what seems to me an anti-industrial romanticism enchanted by the simple solutions with the aroma of do-it-yourself, very attractive particularly to young designers who maintain a certain reservation or even aversion against the range of products that constitute the current professional practice. Papanek identified a problem, but the path he proposed to solve the problem seemed wrong to me. When I was teaching for a semester at Carnegie Mellon University in Pittsburgh in 1964, Victor Papanek invited me for a visit to North Carolina where he worked at that time. I respect the work he did, but later on, from my own experiences, I perceived the limitations of his approach to the development of design in peripheral countries. My respect for him did not stop me from writing a critical piece about his best-known book, in 1973.

Gabriel: In the interview you gave to James Fathers, you said that the Design for Need movement belongs to the past and should not be considered outside its historical context. However, recently (in 2010) Bruce Nussbaum started a controversy that sparked several articles (and their replications) questioning whether humanitarian design would not be a new form of imperialism. There is therefore a resurgence of this theme, and even Papanek's book (*Design for* *the Real World*) has piqued interest and rereads by young students. Alpay Er has pointed out to me that in the preface to the second edition of his book Papanek assumes and embodies the criticism that many design students (including myself) have made in their lectures in Latin America and other peripheral countries, that he would be at least patronizing, if not imperialistic, in his positions. However, Papanek decided not to change the contents of his book, to maintain its historical context—which caused the book to often be read without noting the self-criticism, except by a very attentive reader of the second edition preface, in which the author rejects much of his ideas and the content of the book. What do you think of these two issues: (1) the resurgence of the Papanek book as a reference for young designers; and (2) humanitarian design as neo-imperialism.

Bonsiepe: I find it encouraging that young designers cultivate the practice of reading (besides reading Facebook and Twitter messages). Even more so if what they read is a book by a colleague that I respect on the one hand, and who, on the other hand, I have a professional divergence with.

   With regard to the critique of humanitarian design, it seems to me unjust and even cynical to disqualify tout court attempts to contribute to the solution of emergencies and poverty as an imperialist operation, all of this in no less than a print journalism body such as *Business Week*, that represents the interests of the business establishment in the United States. This resembles a deceptive maneuver to hide from the real imperialism that works with other methods, much more efficient and dangerous: destabilization of governments that dare provoke the hatred of hegemonic powers, pressures, and even financial and economic bottlenecks, globally tuned media campaigns, low-profile wars, commercial punitive actions, and, if deemed necessary, more aggressive actions.

   I observe with some reserve the humanitarian actions of NGOs in peripheral countries, which are sometimes camouflaged to promote hegemonic interests. But I would not put all those humanitarian efforts in the same bag. You need to differentiate.

PART 3: DESIGN, DEVELOPMENT, GLOBALIZATION

Gabriel: In your essay Design and Democracy (Chapter 1.9), you quote Kenneth Galbraith and Vance Packard while writing about the economic function of design as a tool of power submitted to the market and to economic interests, whether from companies or nations, who use design to "boost exports and generate economies of value-added products rather than mere commodities." But you also discuss industrialization—and I take the liberty here to replace "industrialization" with "design role"—as "a means of democratizing consumption" and as a way of improving various aspects of everyday life. Are these two views of design usually mutually exclusive, or can they be complementary? In other words: Is it possible for the market to develop while answering both the interests of the industry and the broader interests of society, or will the market always be "manipulative," seeking only to increase productivity, profits, and consumption?

Bonsiepe: I return to my answer to an earlier question on industrialization. We cannot speak of industrialization in abstract, neutral terms. We must ask about the content and orientation of industrialization: What industrialization? Let me explain. At this moment in several Latin American countries, there are highly contaminating and water-consuming open-air explorations of mineral resources that harm local population and their agriculture. This kind of industrialization not only seems harmful, but also counterproductive to the interests of the local population. It only benefits a few local politicians and investment capital, and no one else.

When I mentioned investing in industrialization, I referenced the manufacturing industry and the assumption that only through manufacturing industrialization would it be possible to democratize consumption and reduce inequalities. Besides, the concept of consumption has changed. We must ask (today): Consumption of what objects? This leads to further questioning as to whether the market would be the most appropriate social institution to respond to those changing needs. Today the market is enthroned as the dominant institution to regulate social and environmental relations. There are doubts as to whether this institution is—and may be—the most appropriate for solving the serious problems humanity faces today. Right, there are antagonistic forces with an asymmetric distribution of power. However, the contradictions between industrial interests and broader societal interests are not necessarily insurmountable. One cannot demonize industry as such. Here too we must ask: What industry? Industrialization in the form of "make-up" factories seems to me a pseudo-industrialization, since it contains no ingredient of innovation; they are merely reproductive.

You mentioned the Design for Need conference and movement—a formulation that implicitly evokes the opposite: the no-need design. Obviously, the motto Design for Need was a reaction against the dominant landscape of professional design that could not meet needs located in an area beyond the mechanism of the market. That is why the movement necessarily contained a political, critical, and even utopian component—a concept that has been questioned and even declared obsolete, for understandable reasons. In times of restoration, a utopian posture bothers by daring people to think about the possibility of a different reality.

José Mauro: With globalization, the gap between the Northern and the Southern Hemispheres became smaller, in terms of both the circulation of ideas and people. The Design for Need movement, which ran from the 1970s to the 1980s, was developed at the height of the Cold War, a time when ideological polarization was strongest. How do you assess the impact of globalization on the possible change of agenda for this movement?

Bonsiepe: Beneath the surface of the end of the Cold War are continuing ideologies that demonize one another. Moreover, the asymmetry of power relations has not diminished. "Globalization" is a term that signals the expansion of capitalist relations with "Western" cultural values and lifestyles around the world. It tends to encompass everything, penetrates to the last corner of the planet, and so it is global. Already in the 1970s it was difficult to characterize the difference between design for needs and design for the opposite, the affluence. These difficulties still exist. What can be said is that environmental issues should not be separated from the most pressing social issues today. The term "globalization" becomes devoid of meaning if used to explain—and excuse—everything, but it can have meaning if appropriately employed to characterize the current socio-environmental problem (among others, the climatic change aggravated by massive human interventions) that is, indeed, global.

José Mauro: The beginning of the third millennium marks the rise of the Southern Hemisphere and the ascension of emerging powers (the BRICS) as new players in the global economy, all of which makes the world more complex and multifaceted. Take China and India, for example, where a simple polarity between developed North and underdeveloped South becomes problematic to preserve. I also think of large Korean companies (the so-called Chaebols, such as Samsung and LG), and Chinese companies (such as HTC and Huawei) that invest heavily in design as a differentiating element. In addition, companies like Apple transfer their production to China, which necessarily leads to the transfer of design practices from the "centre" to the "periphery." I would like to hear your thoughts on the rise of BRICS in the context of design practices, and how their economic rise impacts design practices.

Bonsiepe: Because of its geographical connotations, the North-South binomial was a term with limited explanatory value. That is why I preferred to use the term "peripheral countries." As India, China, Brazil, and Korea enter the scene, the landscape becomes more diverse, and thus will probably change. I do not allow myself speculation about the influence these countries can exert in the field of design. I cannot comment on the design know-how in China and India—because I've never visited China. Production know-how can possibly be something which one can quickly learn, but I'm not sure if the same is true about design know-how. As you say: the mentioned countries, and especially China and Korea, are making considerable efforts to create design potential on a broad basis distancing themselves from the traditional value of mimicry. If a teaching tradition places more emphasis on respect for tradition and conformity than to the willingness to break rules—an indispensable condition for innovation—that may be a barrier to a country becoming a player of importance in the field of design, as was Italy long before the release of the work of the Memphis group. In Italy, something we called "industrial humanism" was practiced. The archetype of this businessman was Adriano Olivetti, who was one of the leaders in this process, possibly the most important one. I confess I would like to see more examples of this attitude.

Selected Projects in Latin America

In chapters 34-40 a link to materiality of design is provided by a selection of seven projects developed in the 1970s and 1980s in different political contexts and institutional settings in Chile, Argentina, and Brazil. They illustrate a variety of problems addressed by design. The designs range from simple, socially relevant products like a spoon for health recommended dosage of milk powder to complex agricultural machinery to increase food production and reduce imports through to the design of the Opsroom for the emblematic Project Cybersyn. They also reveal the technological limitations for design activities in peripheral countries. Thus, in each case, the search for simplicity is mandatory. (GB, 2021)

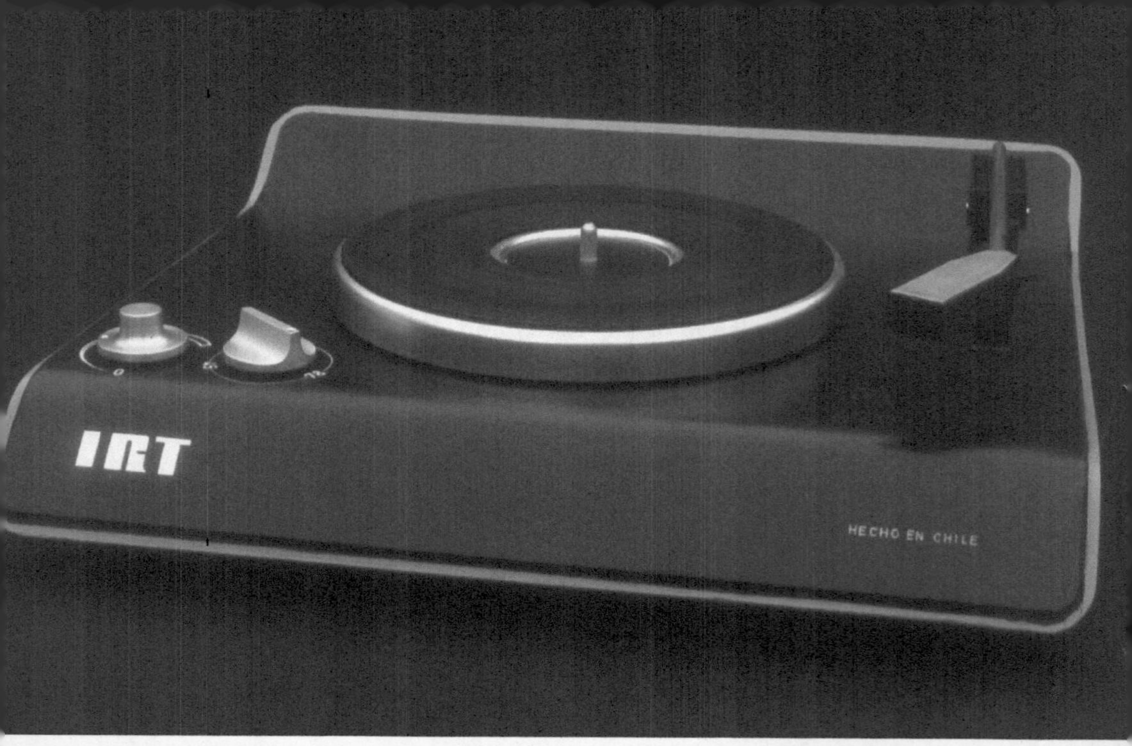
IRT
HECHO EN CHILE

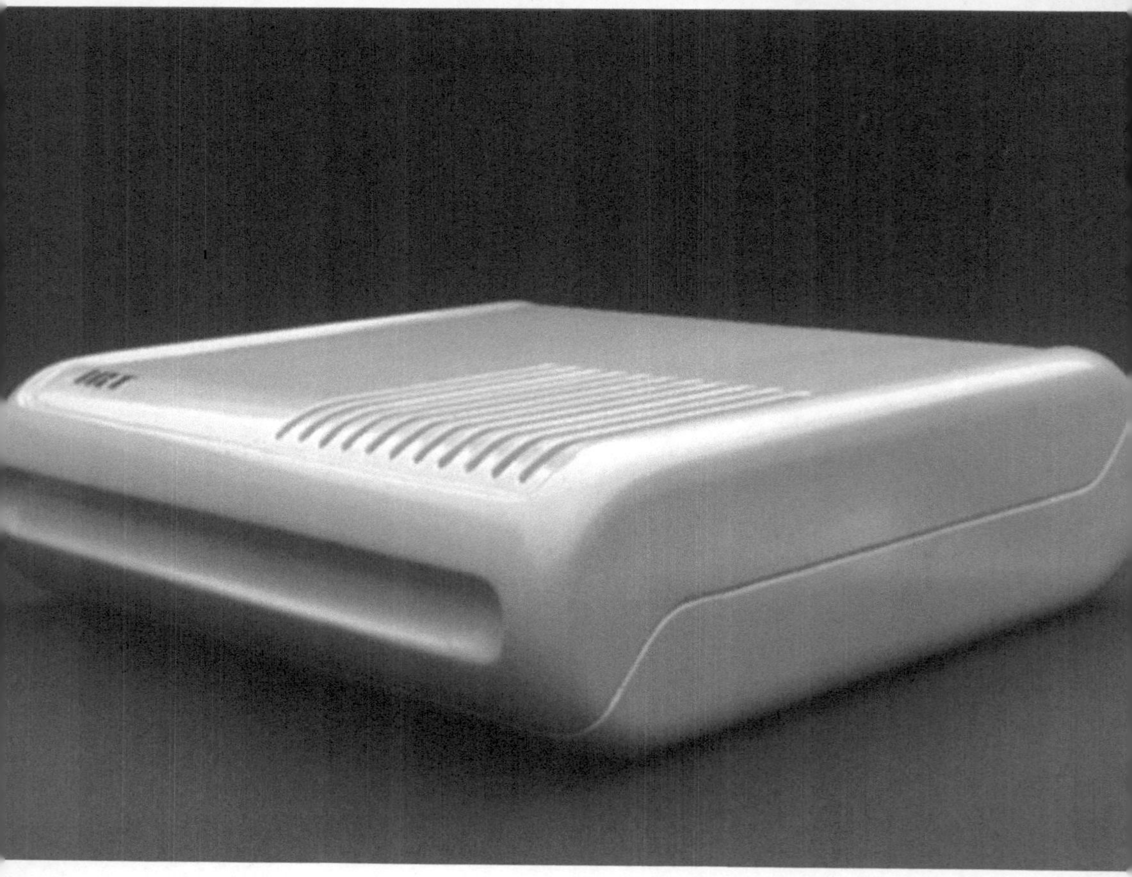
IRT

34 INEXPENSIVE RECORD PLAYER

(CHILE, 1972)

The development of this consumer good seemed unjustified under the supply difficulties that arose in Chile in 1972. However, it was intended to be used in the context of economic policies to siphon off circulating financial resources that exerted inflationary pressure, to be locally manufacturable and to replace far more expensive imported models.

Design: Grupo de Diseño Industrial, Instituto de Investigaciones Tecnológicas, INTEC
Santiago de Chile

The development of this product encountered two changing different objectives: first, a purely economic-financial objective—offering a product for the local market (of the middle class) that could be used to siphon/suck off local financial resources/savings in order to reduce the inflationary pressure; and second, offer a product for social use in community meeting rooms, kindergartens, and so on. I favored the second objective, but in a context where economists dominated the first objective was preferred.

The player was designed to be inexpensive and capable of local manufacture. The casing consists of two shells that are almost identical on the outside. The lower shell holds the mounting plate to which all electromechanical parts are attached. The upper plate carries the loudspeaker. As the player was conceived a portable device, a recessed grip was molded into the underside, The semicircular cross-section at the front edge and back of the turntable is derived from a variant in which an attempt was made to make the housing from two symmetrical halves (tool with interchangeable parts). The illustrations (Fig. 4, 5) show the final design of the player and how the two casings interlock.

FIGURE 4 Inexpensive record player designed for local manufacture.

FIGURE 5 Record player case.

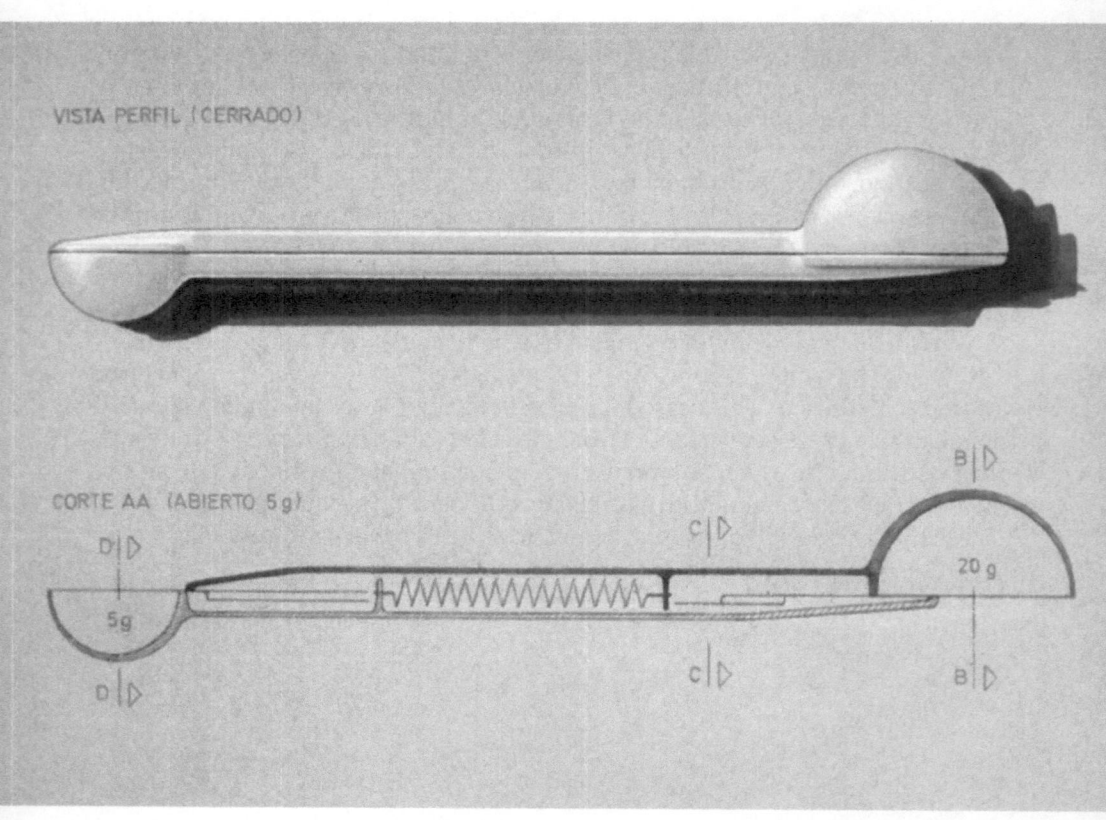

FIGURE 6 Drawings for the measuring spoon for milk powder distributed as part of a government-funded nutrition program for children.

35 NUTRITION PROJECT

Spoon for Milk Powder

(Chile, 1973)

One of the priorities of the Allende government was the attention given to guarantee the healthy nutrition particularly for the children of the poor population. One of the forty measures of the Allende government was the distribution of milk powder to children under fifteen years and breastfeeding mothers. In 1973, there were 3.6 million beneficiaries of the program.[1]

Design: *Grupo de Diseño Industrial, Instituto de Investigaciones Tecnológicas, INTEC, Santiago de Chile*

This project of a measure for powdered milk was part of a nutrition program for children in 1972–3 in Chile. The measuring device should ensure that children receive physiologically adequate milk that was neither too diluted nor too concentrated. The design should have the following characteristics:

— Be easy to handle
— Be hygienic
— Have low cost
— Be made of plastic material
— Measure portions of 5 and 20 grams

Based on the observation of customs of use, the design group developed a series of alternative proposals, among others, the delivery of small packages of 5 and 20 grams. This proposal was discarded due to the cost of purchasing new machinery. Alternatives with two parts and internal spring were also discarded because they were too complicated.

In another proposal, powder adherence to the walls of a measuring cylinder was used—a proposal that worked with a small amount but not with 20 grams. For this reason, the simplest—and most obvious—solution was chosen—a semicylindrical spoon with a reinforced handle with a T-shaped section (Fig. 6). The surface of the spoon is flat, allowing the excess powder to be easily removed. In short, this modest, far from spectacular project reveals the crucial weight of economic and technological factors that often conditions the work of the designer—which the team had to accept in favor of a simple and viable solution.

[1] "Medio Litro de Leche" para todos los chilenos: la historia de una política pública que se extiende por más de cuatro décadas. https://www.uchile.cl/noticias/152243/la-historia-detras-del-medio-litro-de-leche-para-todos-los-chilenos

FIGURE 7 Agricultural chopper during field test

36 AGRICULTURE PROJECT: CHOPPER

(Chile, 1973)

The government of Salvador Allende considered it as one of its priorities to guarantee food production and reduce imports of both agricultural products and agricultural machinery in order to alleviate the consequences of the shortage of foreign exchange. This project was developed with the second aim in mind—and as a test of how design might contribute to the redesign of imported products for local production.

Design: Grupo de Diseño Industrial, Instituto de Investigaciones Tecnológicas, INTEC, Santiago de Chile

The development of this agricultural machine (Fig. 7), starting from a redesign of an imported machine, served for increasing the productivity in agriculture. The requirements included a high and a low ejector for the cut plants, a maximum weight of 600 kilograms for the small width (120 centimeters) and 800 kilograms for broad width (150 centimeters).

Using components with modular coordination, the weight of the chopper has been reduced by 35 percent. The ejectors consist of bent metal sheet point soldered on a bent rim zone. The area of mayor mechanical stress (the frame consisting of strong U-profiles) has been visually emphasized by red color, whereas the "skin" has been painted with a clear paint.

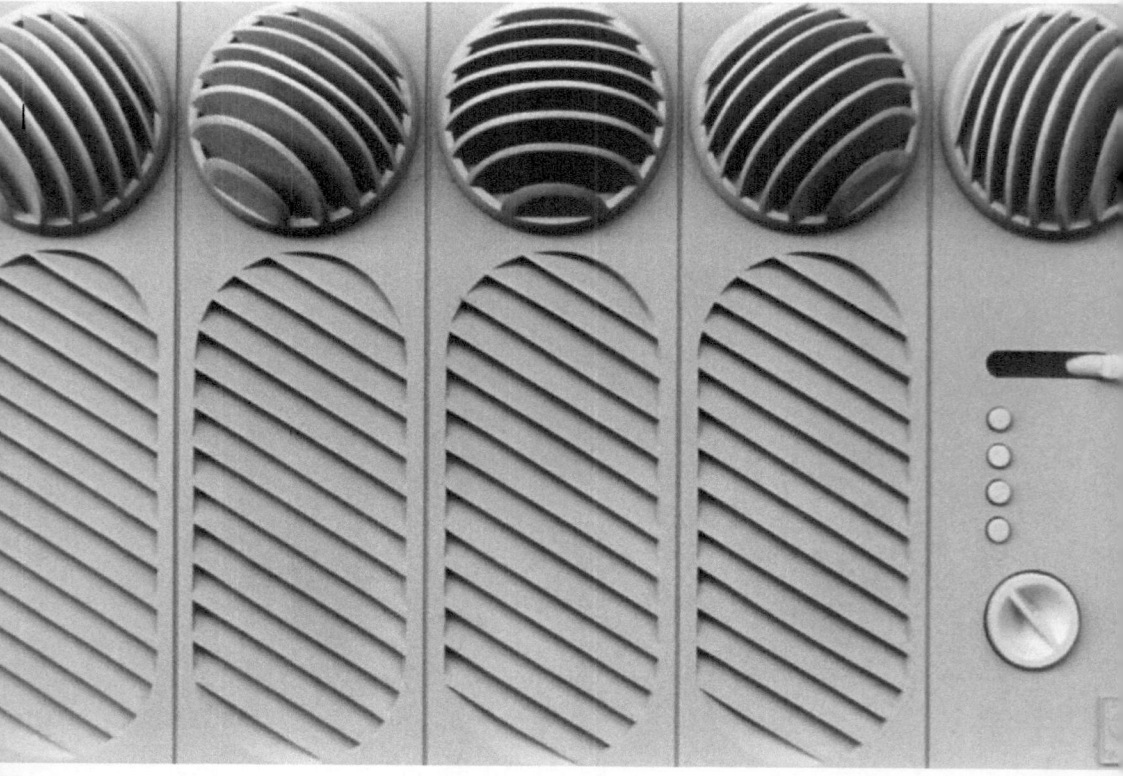

FIGURE 8 Front view of the model with the basic concept of the sphere-like openings for controlling the angle of the airflow. Subdivision of the front panel in a set of modules.

37 CONSUMER PRODUCT: AIR-CONDITIONING

(Argentina, 1980)

This project fell into the phase of neoliberal economic policy that began with the military coup in Argentina in March 1976. Fiscal manipulation of the exchange rate (overvaluation of the local currency and undervaluation of the US dollar) made it impossible for local producers to compete with the prices of external products. Thus, producers that once had produced their own products (and designs) mutated into importers and sellers of imported products. So, this design fell by the wayside.

—Gui Bonsiepe, 2021

Design: MM/B Diseño (Méndez Mosquera/Bonsiepe/Kumcher)
Buenos Aires
Team members: Designers, Hugo Legaria and Miguel Muro; Technical drawing, Héctor Taboada

The product analysis revealed several unsatisfactorily resolved details (lack of structural stability of the plastic frame, precarious fixing of the chipboard to the plastic frame by hinges, exceedingly complex solution of the air outlet consisting of narrow vertical and horizontal fins that are difficult to assemble).

Concerning the semantic dimension of this product, one can ask why disguise an electromechanical product that is not a piece of furniture as furniture. Starting from this questioning, the design of the front panel has been developed.

The front panel of this product is divided into three functional areas:

1) Cold air outlet: the airflow must be directed upward approximately 45 degrees and laterally (approximately 45 degrees at each side).
2) Air intake area covered with a chipboard that permits access to the plastic foam filters.
3) Control area (switch on/off, regulate the temperature).

Instead of the complex solution with fins in form of a grid, the chosen scheme was for a control element in form of a spherical calotte with inclined subdivisions that rotates in a groove. The air intake has been designed with the sufficient depth to avoiding exposing the interior where the plastic foam is placed. Instead of hiding the controls of the airflow, they have been placed on the front area facilitating its operation. The front view of the model shows the detail of the basic concept of the sphere-like openings for controlling the angle of the air-flow and the subdivision of the front panel as a set of modules (Fig. 8).

Chapter 38 presents two projects that were conducted under the aegis of the LBDI—Laboratório Brasileiro de Desenho Industrial/the Brazilian Laboratory for Industrial Design based in the city of Florianopolis, Santa Catarina State, in the Brazilian South. The lab started in 1983 as a pilot project supported by Brazilian national research agency (CNPq) and as a partnership between several institutions including Sebrae (Brazilian Service of Support to Micro and Small Companies). The institute was administratively linked to the Federal University of Santa Cartarina, but it operated with full autonomy and was thus freed from obstructive, time-consuming academic and bureaucratic procedures.

Its location on a region marked by small and medium enterprises distant from the main industrial states (São Paulo, Belo Horizonte, Rio de Janeiro) in the Southeast of Brazil was intentional, as a strategy to offer intensive short-term practical design courses and to provide design consulting services to the local small and medium-sized companies.[1].

The Lab offered nonacademic short courses, which were not connected to a design degree and were aimed at working professionals in the areas of industrial design, engineering, and architecture. Teaching activities were complemented with product development activities for small and medium-sized companies.

Projects shown here illustrate the working method that aimed to promote a strong link between design and the technological-industrial base. Therefore, there is a decided denial of personalist design. In many projects, particularly the more complex projects, the "morphological box," or "Zwicky matrix," method[2] was used to list alternatives for solving subproblems. It is worth mentioning that the projects documented here date from the predigital period, without the possibility of digital rendering. The mock-ups were made of polyurethane, plaster, and corrugated cardboard.

---

[1] At that time the Brazilian government gave support to some 80000 scholarships with sometimes 5 years duration. That period was considered excessively long. Industry needed immediate assistance.

[2] The Swiss physicist Fritz Zwicky (1898-1974) invented this method to visualize combinations of new possibilities and the interaction between determining factors for the solution of a problem.

38 TWO PROJECTS FOR LOCAL INDUSTRY

(Brazil, 1984, 1986)

AIR COMPRESSOR

Design: LBDI
Santa Catarina, Brazil, 1984

This, for industrial designers, was a rather unusual project, and resulted from the experimental development of an air compressor by a local factory in Southern Brazil that produced the compressor in large scale—many hundreds units per day. The aim consisted in reducing noise and vibrations and simplifying the production process.

What could designers contribute to an artifact that is generally considered to be an exclusive engineering task? The approach and contribution from designers produced unexpected results in the concept of this experimental development. The hermetically sealed housing of mechanical components consists of two shells of metal sheet soldered through a horizontal seam—similar to a pot with a lid. By orienting the soldered seam about 45 degrees, the new design requires one step less in the stamping process (four instead of five steps). Formally, complex details such as the corner detail, where convex and concave radii meet under different angles and orientation, have been checked and resolved with the help of plaster models. (See Figure 9.)

This text is adapted and translated from the book Gui Bonsiepe, *Design como pratica de projeto* (Blücher, 2012), 126–9. By kind permission of Blücher publishers.

PIPING MACHINE

Design: LBDI. Short-term course project
Santa Catarina, Brazil, 1986

This project started by approaching the local association of producers of shoe machinery in Southern Brazil offering design services. As generally with the rest of managers of capital goods (machinery producers), the term "design"

FIGURE 9 Corner detail of the design of the compressor.

was not known, and it was unclear what design could do to improve the quality of the product.

From a visit to local shoe production companies in Southern Brazil resulted the suggestion to use different machines as a starting point for learning concretely the design process, improving the product structure and quality of use of this tool. The machine is used to lower (chamfer) the leather using a cylindrical rotating knife. It has three subfunctions: sharpen the razor with an emery, collect the chips, and exhaust air (dust separation).

The redesign consists of a self-supporting structure from bent and soldered metal sheet. It covers all moving parts, reducing the probability that the operator gets hurt. The two lateral "wings" serve for supporting a basket on each side, one for the unmachined pieces and the other for the machined pieces. In this way the workflow was improved. Furthermore, ergonomical standards were applied. (See Figure 10.)

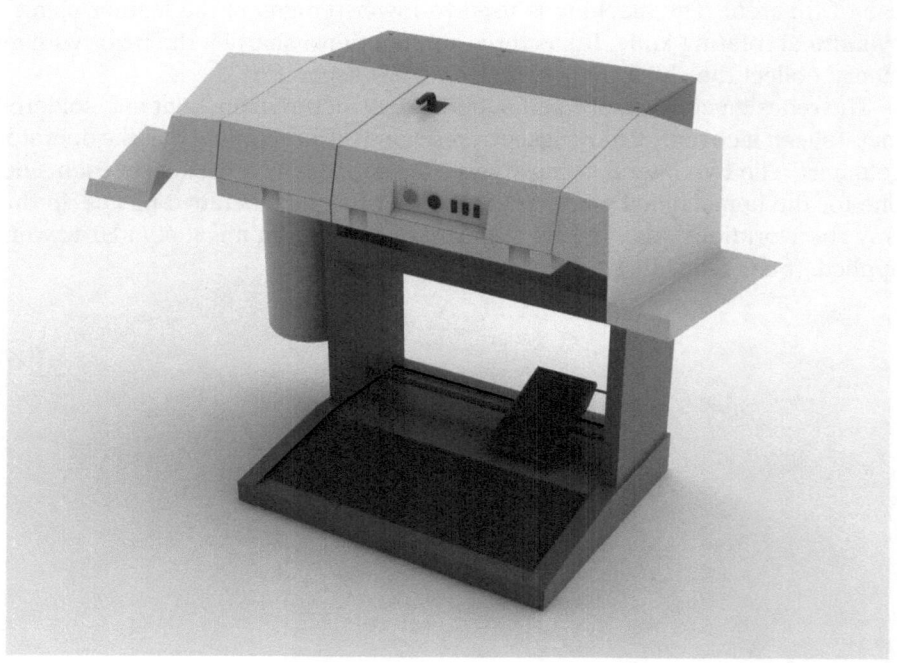

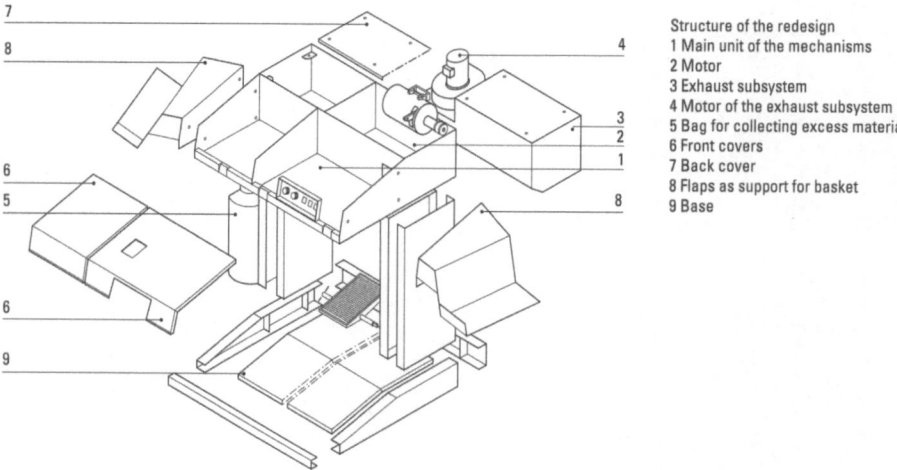

Structure of the redesign
1 Main unit of the mechanisms
2 Motor
3 Exhaust subsystem
4 Motor of the exhaust subsystem
5 Bag for collecting excess material
6 Front covers
7 Back cover
8 Flaps as support for basket
9 Base

FIGURE 10 Piping Machine: Redesign and exploded view of the structure of the redesign.

39 HEALTH CARE PROJECT

Needle for Blood Sampling

(Brazil, 1986)

Design: LBDI
Santa Catarina, Brazil

This needle serves for diabetics to take blood samples for glucose analysis. The small device works with a mechanism similar to ballpoint pens. The rectangular section was chosen because the user has to put the device between four fingers and press with the thumb to unblock the needle under pressure. The illustration below (Fig. 11) shows the top view with the push button to shoot the needle under pressure against the patient's skin

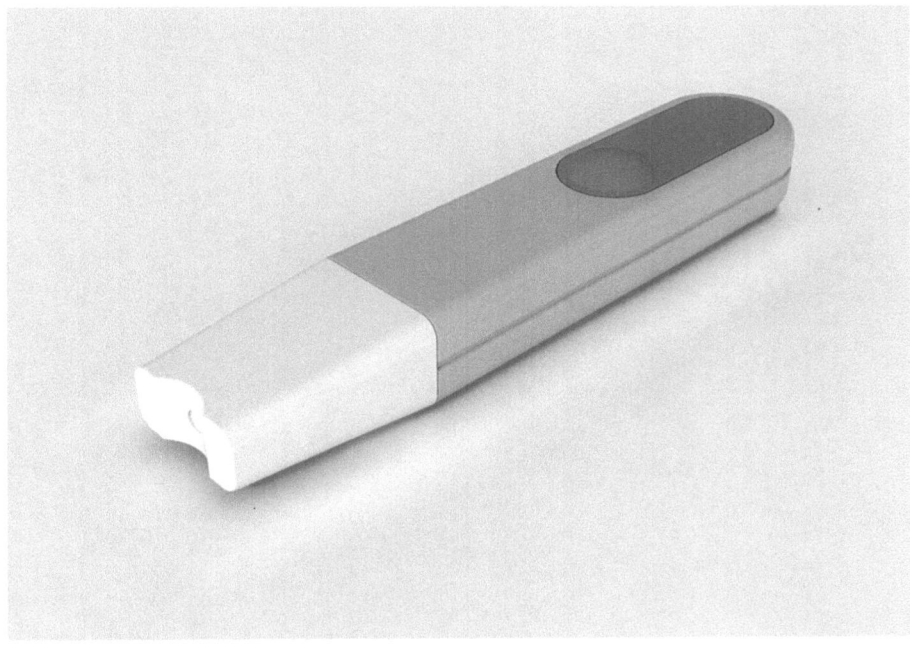

FIGURE 11 Needle for blood sampling. Top view with the push button to shoot the needle under pressure against the patient's skin.

39. HEALTH CARE PROJECT

Needle for Blood Sampling

(Brazil 1993)

CASE STUDY OF PROJECT CYBERSYN, CHILE

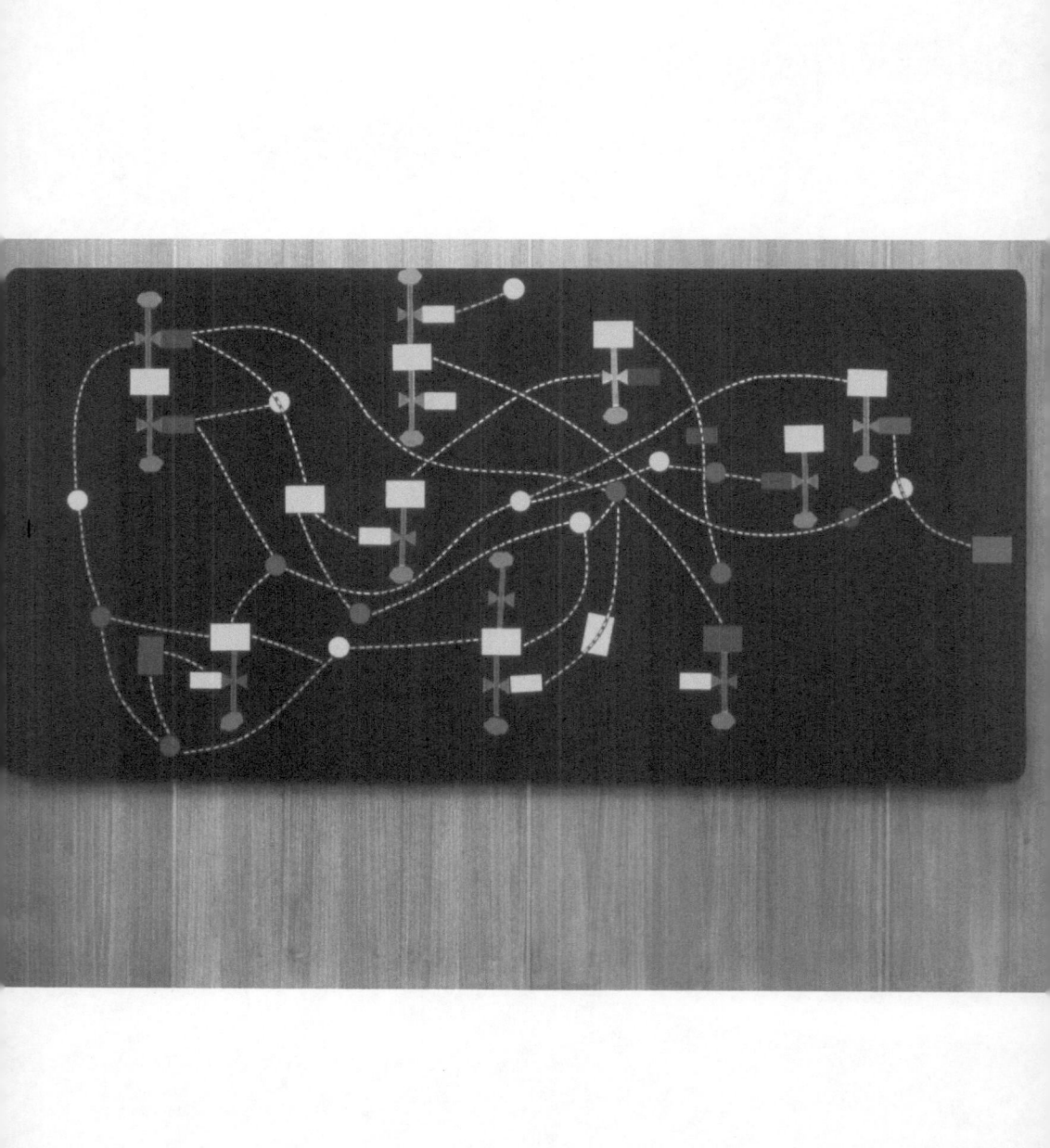

FIGURE 12 (previous spread) Cybersyn Opsroom: view of the room showing chairs and displays for presenting economic data and simulations. Chile, 1973.

FIGURE 13 Ops room: Information graphics. Panel of the Future. Metal panel with magnetic components and polarized light to simulate the future behavior of economic entities and the consequences of decisions.

FIGURE 14 Ops room: Information graphics. Diagrams of Input / Output relations between industrial branches in the private sector (APP) and social property sector (ASP) and in particular industries.

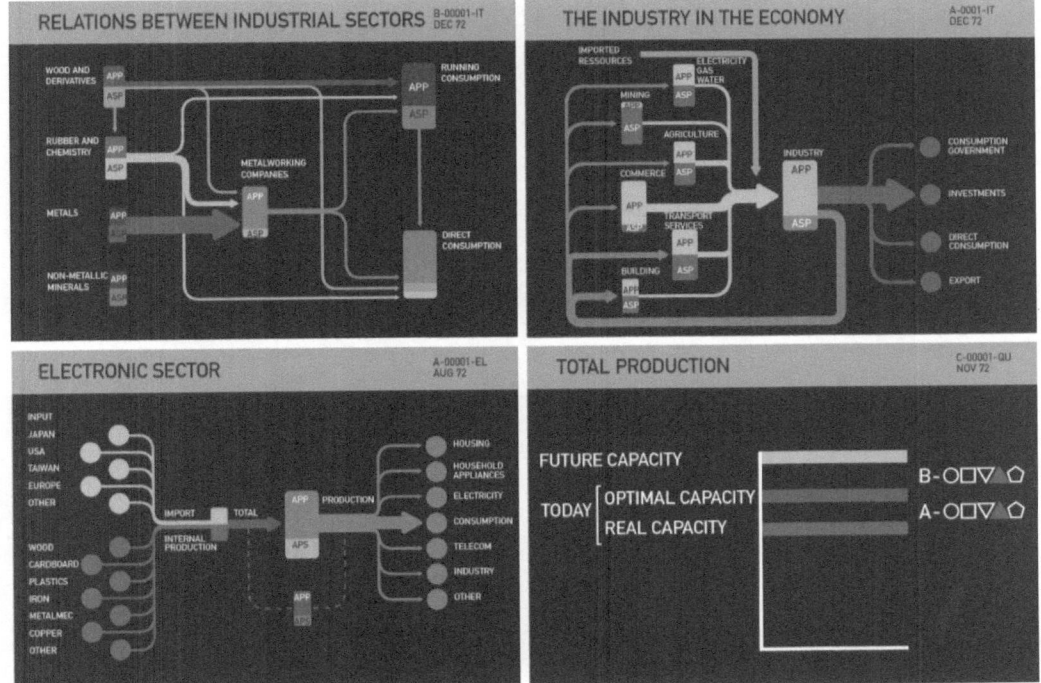

Editorial note: The Cybersyn project as a whole and the design of the opsroom have been already referred to a number of times in this volume in the context of Bonsiepe's work in Chile between 1968 and 1973. Given both the complexity of the project and the renewed interest in it in the last few years, it seemed to us worthwhile to develop a fuller case study of the project as a suitable conclusion to this section of the volume.

40 OPSROOM: INTERFACE OF A CYBERNETIC MANAGEMENT ROOM

Design: Grupo de Diseño Industrial, Instituto de Investigaciones Tecnológicas, INTEC, Santiago de Chile

(1972–73)

INTERSECTION BETWEEN SOCIAL PROGRAM, TECHNOLOGY, AND DESIGN

There are not many examples of the convergence between social responsibility, advanced technology, and design. An approximate balance is rarely reached between these three factors. For this reason, CYBERSYN (an acronym of Cybernetics and Synergy)—a cybernetic management center for the Chilean economy—can be considered an exception. The ambitious project in the political[1], social, and technological sense ended abruptly on September 11, 1973, due to the military coup against the democratically elected government of Salvador Allende. The project was conceived by the cybernetic management theorist and consultant Stafford

Beer, founder of cybernetic management and president of the World Organization of Systems and Cybernetics, and the then director of the Corporation de Fomento[2] Fernando Flores. In its theoretical and practical focus, the project can be seen as a bold anticipation of big-data management that considerably surpassed the technical resources available at that time, especially in a peripheral country like Chile. Despite the technical limitations, it was partially accomplished. Today it constitutes a rich research theme,

Originally published in Portuguese in Gui Bonsiepe, *Design como prática de projeto* (São Paulo: Blucher, 2012). Translation by Lara Penin. Courtesy Blucher.

[1] This is a political project, in the following sense of the concept: "Since the 1920s, the attribute 'political' has extended its meaning, indeterminately denoting any radical break, any suspension of consensus. . . . 'Politician' calls the desire for a beginning, the desire that finally a fragment of reality can be presented without fear and without law—solely as a result of human, artistic, erotic or scientific experience. The art-politics connection is incomprehensible if the political term is not understood in this broad and subjective sense." Alain Badiou, *Das Jahrhundert* (Zürich; Berlin: diaphanes, 2006), 183–4. In the meantime, publications have been launched, among others one from a political perspective: Susan Buck-Morss, "Historical surprises from the recent past," Chapter 2 in *Revolution Today* (Chicago: Haymarket Books, 2019), 13–16. And another from an art theory perspective. Karen Benezra, "Cybersyn: Style, Management, and the Object of Design," Chapter 4 in *Dematerialization – Art and Design in Latin America of Studies in Latin American Art* (Oakland: University of California, 2020), 133–67.

[2] This institution, created in 1939 by the government, had the function of supporting industrialization in Chile with public resources. It can be compared to that of an industry ministry. Fernando Flores proposed to hire Stafford Beer as a consultant in Chile.

especially for economists,[3] systems analysts, business administrators,[4] cyberneticians, technologists,[5] technology historians,[6] communication and media theorists,[7] and design historians[8].

CYBER MANAGEMENT

Stafford Beer and his collaborators (computer scientists, system engineers, economists, mathematicians, programmers) set out to develop a tool for flexible, decentralized, dynamic, and, above all, interconnected economic management, which differed from the hitherto known socialist techniques of centralized planning. It should also overcome the contradiction between autonomy and control (control in the English sense). To that end, Stafford Beer used a thesis by Norbert Wiener and Claude E. Shannon, according to which complex phenomena can be understood as systems, whose behavior can be extrapolated through the analysis of timelines, and can be applied in forecasts, regardless of whether they are weather forecasts, stock markets, or production statistics. Simply put, it was necessary to answer two questions: First, what information does a planning and management team need to make appropriate decisions for economic policy; second, how to communicate this information? In the literature on control systems prior to cybernetics, you can read:

Stimulated by Norbert Wiener after the Second World War, not only engineers but also economists, anthropologists and social scientists took up theories on feedback, stability and interconnected systems by applying this knowledge to any other area, from global ecology to urban traffic. What they shared was not

---

3   Sebastian Vehlken, "Environment for Decision – Die Medialität einer kybernetischen Staatsregierung. Eine medienwissenschaftliche Untersuchung des Projekts Cybersyn in Chile 1971-73," Master's thesis (Bochum: Ruhr Universität, 2004).

4   The management center St. Gallen directed by Fredmund Malik takes care—obviously under different socioeconomic and sociopolitical conditions—of the continuity of Opsroom and presents a virtual cockpit version on the network: "Operations Room – Total Control at your Fingertips." http: // www. malik-mzsg.ch/mcb/htm/1083/de/mcb.htm (accessed July 26, 2008).

5   Eden Medina, "Democratic Socialism, Cybernetic Socialism – Making the Chilean Economy Public," in *Making Things Public: Atmospheres of Democracy*, eds. Bruno Latour and Peter Weibel (Karlsruhe: ZKM Center for Art and Media; Cambridge, MA: MIT Press, 2005), 708–19.

6   Eden Medina, "Designing Freedom, Regulating a Nation: Socialist Cybernetics in Allende's Chile," *Journal of Latin American Studies*, no. 38 (2006): 571–606; Andrew Pickering, "Kybernetik und die Mangel: Ashby, Beer und Pask," in *Kybernetik und neue Ontologien* (Berlin: Merve, 2007), 87–125.

7   Claus Pias, "Unruhe und Steuerung. Zum utopischen Potential der Kybernetik," in *Die Unruhe der Kultur: Potentiale des Utopischen*, eds. Jörn Rüsen and Michael Fehr (Weilerswist: Velbrück Wissenschaft: 2004), 301–26; Claus Pias, "Der Auftrag. Kybernetik und Revolution in Chile," in *Politiken der Medien*, eds. Daniel Gethmann and Markus Stauff (Zürich; Berlin: diaphanes, 2004), 131–45. Accessible at: http: // www. uni-essen.de/~bj0063/texte.html (accessed July 25, 2008).

8   Hugo Palmarola Sagredo, "Productos y socialismo: diseño industrial estatal en Chi-le," in *1973 – La vida cotidiana de un año crucial*, ed. Claudio rolle (Santiago: Planeta, 2003), 225–95; Jane PaVitt and David Crowley, "The High-Tech Cold War," in *Cold War Modern – Design 1945 –1970*, eds. David, Crowley and Jane PaVitt (London: Victoria and Albert Museum, 2008), 162–92.

so much a particular methodology or theory, but the idea that different aspects of the world can be understood as systems and can be modeled as processes of flow, feedback and human-machine interactions. The terminology of feedback, control, communication and information proved to be very flexible and adaptable.[9]

As one of the assumptions of this project, it was considered important to involve the workers and employees of the companies in the management, through system feedback loops. At that time, this goal could not be achieved, due to insufficient technological means. It is important to highlight this aspect, because this attempt is not visible in the project that was carried out. Today, with the advancement of information technology, democratic management is technically possible.

However, as is well known, sociopolitical conditions are often lacking. Here a bifurcation between technological and social possibilities opens up. The project was not limited to the creation of a management and planning center, but to the application of this concept at different levels of the industrial organization, passing through the industrial sectors to the level of each company. Despite the technical limitations of the time, the idea of a nonhierarchical management of companies would constitute the core of a concept of concrete economic democracy. Depending on the political stance, this project will be classified as utopian or outdated. Utopian in the sense of not being achievable. Outdated in the sense of belonging to a long-gone era making it no longer possible to establish sociopolitical and socioeconomic alternatives, as if history were frozen in its current phase. Such attitudes can be criticized for revealing a limited worldview.[10] The project was not dominated by the intention of a rigid state-driven nationalization, since for managing the economy according to the social program of the Allende government; it was enough to keep 80 to 100 large companies under state control.

THEORY OF REQUISITE VARIETY

First, the project criticized the usual management information systems (MIS) that, in general, generate a counterproductive avalanche of data. In contrast, Stafford Beer was committed to applying his idea of cyber management—presented in numerous publications[11]—on a large scale for democratic purposes, using, on the one hand, the theory of requisite variety (formulated by

---

9   David A. Mindell, *Between Human and Machine – Feedback, Control, and Computing before Cybernetics* (Baltimore; London: The John Hopkins University Press, 2002), 316.
10 The transfer of utopian impetus to the area of science fiction and the dialectic between the imaginary and reality were commented by Fredric Jameson: "the more surely a given Utopia asserts its radical difference from what currently is, to that very degree it becomes, not merely unrealizable but, what is worse, unimaginable," Fredric Jameson, *Archeologies of the Future – The Desire Called Utopia and Other Science Fictions* (London; New York: Verso, [2005] 2007), xv.
11 Stafford Beer, *Brain of the Firm – The Managerial Cybernetics of Organization* (London: The Penguin Press, 1972); Stafford Beer, *Decision and Control – The Meaning of Operational*

W. Ross Ashby) and, on the other hand, his viable systems theory (VSM, viable system model).[12] The theory of requisite variety states that a system exposed to the destabilizing influences of the environment—which happens in any company, from large, medium, and even micro-companies—needs to be able to produce a variety higher than that generated by the environment, in order to guarantee its existence. Only the variety can force down the variety, according to Ashby's correct formulation.

For the management of a highly complex system such as the economy, the requirement of the necessary variety presupposes the availability of up-to-date information on the activities of companies, because appropriate decisions cannot be taken without them. The task of constructing a routine to quickly transmit relevant economic data, through a data transmission network, was assigned to the engineers. Previously, this data reached the governing bodies with a lag of several months, thus losing its practical value. On the other hand, the designers' team had the task of visually processing these data to facilitate the work of managers, distributed over five levels, according to the Stafford Beer[13] model for cyber management.

TECHNICAL DETAILS OF CYBERSYN

The system consists of four subsystems—Cybernet, Cyberstride, Checo, and Opsroom. They were described in relevant publications of the specialized literature.[14] Cybernet meant the networking of socialized companies through approximately 500 telex devices. Cyberstride consisted of a series of software programs to elaborate the production indexes. Checo was a simulator for virtually experiencing the consequences of economic decisions. The Opsroom finally formed the front end of the entire system that in this way became usable. The fifth component Cyberfolk that aimed at democratizing the decision process by including workers remained in the planning stage.

THE DESIGN OF THE INTERFACE

The interface—at that time this concept did not cover industrial design and graphic design—of the cybernetic management center was conceived, developed, and implemented in the years 1972–3.[15]

Research and Management Cybernetics (London; New York; Sydney: John Wiley & Sons, 1966).

[12]   Allenna Leonard, "The Viable System Model and its Application to Complex Organizations – The First Stafford Beer Memorial Lecture July 8, 2007," World Multiconference on Systemics, Cybernetics and Informatics, Orlando, 2007.

[13]   Stafford Beer, "Corporate Structure and its Quantification," in Brain of the Firm – The Managerial Cybernetics of Organization (London: The Penguin Press, 1972), 198–212.

[14]   Medina, Cybernetic Revolutionaries.

[15]   Cybersyn demonstrated its potential and importance as a tool to counter the effects of the "1972 October strike" by some 20,000 truck drivers in Chile, which—funded by the US government— aimed at overthrowing the popular government.

In the specialized literature, the interface became known as Opsroom, or Operations Room, using the terminology of strategic planning.

DESIGN WITHOUT BRIEFING

There was no detailed briefing in the form of a list of requirements and conditions to develop the Opsroom. In this respect, the project violated the traditional rules of systems sciences and design in general. The design team was entrusted with the vague task of equipping a space with chairs and tables for about ten people, in which economic data were presented on a projection screen, so they could make political and economic decisions. The work covered, on the one hand, the design of the space with the relevant equipment and, on the other, the formulation of rules (design algorithms) for the visualization of the information presented in the management center.

LOUNGE ATMOSPHERE

Based on the first sketches, the ideas expressed verbally by Stafford Beer were being clarified. He dismissed the first attempt—it did not coincide with his idea. Thus, a new attempt was made, which came closer to Beer's ideas. The drawings of the second phase show a space with armchairs arranged in a circle in which there is no preferential situation (see Figure 12). In keeping with his deeply unconventional anti-bureaucratic style, Beer wanted a relaxed atmosphere. On the walls of the space with a circular layout, economic data would be projected using slide projectors located outside the room in a service area (Figures 13, 14, and 15). The data would be accessed through a central control panel mounted on a pulpit.

In the pilot project, the central control panel was replaced by a keyboard integrated into the armrest of the chairs (Figures 18 and 19). There were two reasons for this: on the one hand, it is more comfortable for people to have the keyboard at their fingertips in their chair instead of standing up and walking to the center of the room to operate the control panel. On the other hand, the conversation among the participants become more fluent due to the decentralized keyboards.

DATA COMPRESSION THROUGH DIAGRAMMATIC VISUALIZATION

Extracts from the project report were saved and thus preserved from destruction. Technical information about information design is likely to be of historical interest only. Today, they would obviously be replaced by efficient and much faster digital technologies. At that time, as is known, these technologies were not available and for this reason the visual material was elaborated using artisanal methods. The diagrams were drawn by hand using colored cards, templates, and typewriters for the texts. These diagrams were then photographed with a camera (35 millimeter slides), and the slides were available the next day for the following session. That was the maximum achievable in "real time" at that time (Plate 8).

Although information design has evolved a lot with the introduction of information technology, nothing has changed in the problem of how to summarize and visualize complex economic data. What was said about maps or cartographic representations in general, which are the best procedures for visualizing data (Krzysztof Lenk), may also be applied to diagrammatic representations. They also work well as data visualization tools.

COMPONENTS OF THE PHYSICAL AND SEMIOTIC INTERFACE

The room contains the following displays:

1. The datafeed module with a large panel (80 x 120 centimeters) and three small panels (40 x 60 centimeters).
    (Figures 16, 19, 20).
2. The "staffy" or viable system model (VSM) module—an animation display for the five-level model of cyber management.
3. The module of two overhead projectors (100 x 100 centimeters) for the projection of additional information from the staffy.
4. The algedonic module (see the following explanation) divided into two panels with different windows (Figure 20).
5. The future panel (a metal sheet of 100 x 200 centimeters, with magnetic symbols to simulate the effects of decisions) (Plate 8).

The room was equipped with seven seats with a built-in keyboard to access the datafeed. A hexagonal floorplan was used for the room due to architectural constraints (Figure 16).

Displays for visual presentation of information could be installed on five of the six walls of the room, while the sixth side was reserved for the entrance door (Plate 6, Figure 17).

In the center of the room were seven 270-degree swivel chairs. On the right armrest, keyboards with ten keys were placed to call the datafeed slides. The keys were arranged in three rows:

— The three square keys in the top row were used to select one of the three displays.
— The five keys in the central row with different contours corresponded to a code to identify the slides.
— In the bottom row were the Reset and Enter keys. When pressed, this key would light up, as well as in the other seats, to indicate "in use." On the left armrest, there was an ashtray and a bas-relief for glasses (Plate 7, Figure 18).

ALGEDONIC MONITORS

The term "algedonic" is not well known in the specialized cybernetic literature. It is explained by Stafford Beer as follows:

> Algedonic . . . refers to regulation in a non-analytical way. For example, we can teach a person how to solve a task, analytically explaining the 'how' and the 'why'. But we can also proceed algedonically using a system of punishments and prizes that does not offer such explanations.
>
> Algedonic loop: a circuit for algedonic regulation that can be used to block an analytical control circuit. For example, a stinging pain can prevent us from completing a task that we fully understand and want to complete. Error-proof devices can be used to block a section of a factory when a certain critical value is breached, without knowing the cause of this incident.[16]

To better differentiate the different types of information, they were presented on backgrounds of different colors:

— Sky blue for exceptional situations;
— Green, yellow, red for contingent problems, according to the degree of recursion.

To indicate the degree of danger, that is, the time that passed without the company being able to solve the problem, different intermittent side windows (slow, medium fast, and fast) were used to indicate the urgency of solving the problem. Typographic information was printed on white acrylic strips (company name, indicators, arrows to indicate the behavior of an indicator: decreasing, growing, constant). The small square display window on the left showed a red light that switched on when the staff communicated an algedonic situation. (See Figure 20.)

"STAFFY"

It was an animation display (215 centimeters high, 133 centimeters wide, and 50 centimeters deep) imported from England. This display was used to visualize the flows (including the degree of intensity: low, normal, high) between the five different levels of the cyber management model. This display was considered by Stafford Beer as the main piece for the visualization techniques, although not all experts with knowledge of the system were able to recognize its practical value at first sight.

---

[16] Beer, *Brain of the Firm*, 305.

FUTURE PANEL

It was a metallic sheet to simulate the consequences of economic decisions. It was covered with brown fabric to avoid uncomfortable reflections. Thin magnetic plates with different colors and contours (icons) as used in flow charts could be placed on the 100 x 200 centimeter surface. The flow lines consisted of segments of a flexible and magnetized plastic profile covered by a material that allowed, under polarized light, to dynamically simulate the flows (See Plate 8.).

MATERIALS, FINISH, COLORS

The display cases and chairs were made of white fiberglass-reinforced polyester. The walls of the room were covered with eucalyptus wood. The ceiling was painted white, and the floor was covered with brown carpet. The room also had two sets of adjustable light fixtures: a channel for indirect light and six spots close to the corners of the hexagon.

MAINTENANCE

The area around the room was used for maintenance. The sixteen slide projectors were placed on support structures. The overhead projectors for algedonic displays were also operated in this maintenance area, through an intercom that communicated the room with the service area.

INFORMATION DESIGN: DESIGN ALGORITHMS

For visualization of economic data, as many visual analog codes (colors and icons) as possible were used instead of numeric codes. For this, there was an argument that Otto Neurath proposed back in 1920: Complex information or macro-information presented using a nondiscursive qualitative code is more easily readable and understandable than numerical tables with statistical data whose degree of precision is often too high to be used efficiently by system users.[17] The mathematical foundations for the manufacture of flow diagrams were determined in a technical document from the computer department of CORFO.[18]

---

[17] "Visual statistics is part of an attempt, which has existed for at least 200 years, to extract new knowledge from numerical statistics using visualization techniques and to make that knowledge accessible to a majority of people, through an appropriate design." Sybilla Nikolow, "Kurven, Diagramme, Zahlen- und Mengenbilder. Die Wiener Methode der Bildstatistik als statistische Bildform," *Bildwelten des Wissens - Kunsthistorisches Jahrbuch für Bildkritik* 3, no. 1 (2005): 20–33.

[18] Humberto P. Gabella, *Técnica de la flujo-gramación cuantificada para efectos del control en tiempo real* (primera versión) (Santiago: Corporación de Fomento de la Producción, 1972).

A small manual was created containing the rules in the form of design algorithms to produce the originals with respect to the following:

— Distribution of graphic elements on the surface
— Size, type, and position of typographic elements
— Color codes
— Line thickness for diagrams
— Shape and size of arrows
— Type of photographs.

The application of these rules guaranteed a coherence of the different visual materials starting from sketches generated by the different teams of the industrial information system. The visual grammar to encode the information reduced the variety of visual material and avoided the danger of a proliferation of ad hoc solutions. Due to the semantic complexity of the system, it was not possible to develop a systematic chromatic grammar. Chromatic rules referred only to a group of diagrams and varied between different types of information.
     Graphical representations were divided into six groups:

— Flow diagrams
— Taxonomic diagrams
— Panels with text
— List of slides for the datafeed
— Photographs
— Special cases

To guarantee an aesthetic-formal consistency of the diagrammatic representations, templates were established with exact measurements for each type of information.

LINE AND ARROW THICKNESS

To visualize the different flow intensities, five different thicknesses were used in two variants (depending on the available space).

COLOR CODES AND SYMBOLS

Certain meanings were visualized using standardized symbols, as used in diagrams. (Circle = Input, Output. Vertical rectangle = Amount of capital, company).

TEXTS AND TYPOGRAPHY

The originals were typed and reproduced photographically (black-and-white negatives). The negative served as a slide. Emphasized words could be highlighted using colors.

TAXONOMIC DIAGRAMS

With these diagrams, the indicators "Potentiality," "Ideal capacity," and "Real level" are displayed. Stafford Beer defines these concepts as follows:

> Real: it is what is being done now, with the available resources and under the existing conditions.
>
> Ideal: it is what could be done now (at this moment), with the available resources and under the existing conditions, provided there was the will to do so.
>
> Potential: it is what should be done by developing resources and removing restrictions, operating in a range considered feasible.[19]

HISTORY OF TECHNOLOGY

A few days before the coup, the operating room should have been installed in the government palace at the request of President Allende. As a doctor, Allende had very well understood the scope of the project and its theoretical foundations, among others, neurophysiological. After the coup, the memory of the pioneer Cybersyn project was eliminated. Today, Cybersyn constitutes a chapter of the history of technology in Latin America yet to be written or, in a broader context, an example in the experimental field covered by ruins and tombs of democratic and social relations, oriented toward autonomy, and that is an unfinished history.

[19] Beer, *Brain of the Firm*, 207.

FIGURE 15 Basic concept of the preliminary project. Armchairs distributed in a circle without preferential position. Central control panel to load the slides that will be projected on the displays on the walls.

FIGURE 16 Outline of the concept of the central control panel and walls for projecting information.

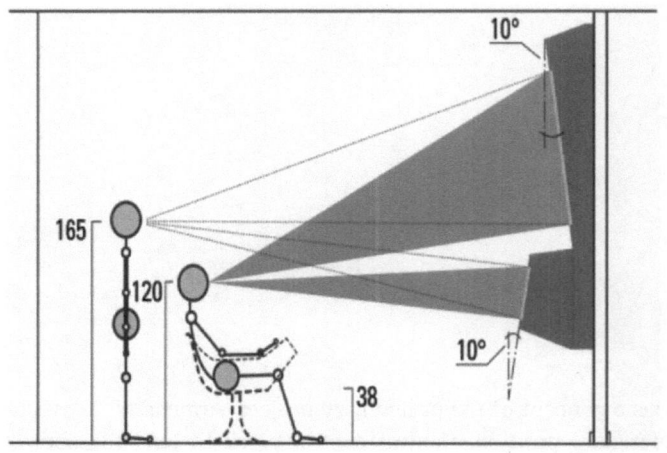

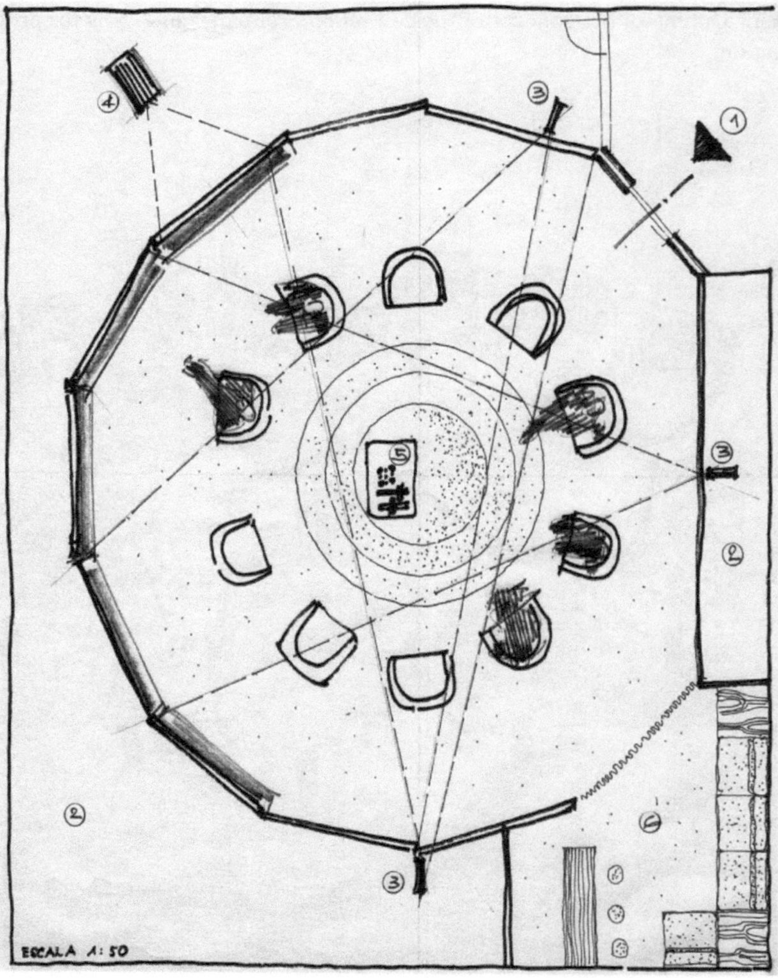

ESCALA 1:50

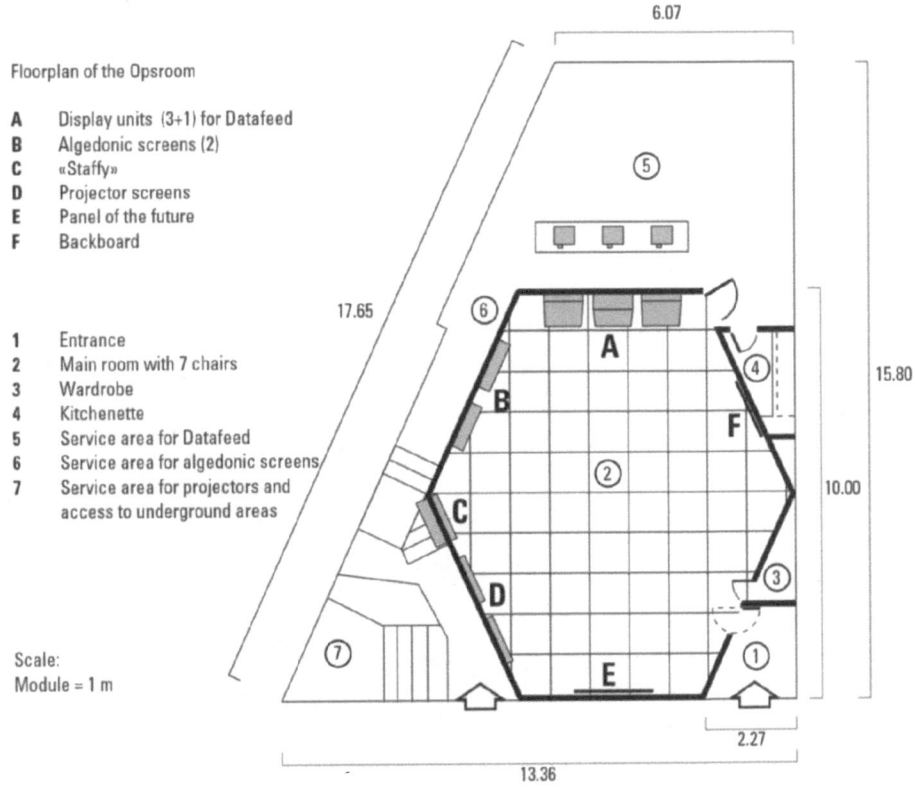

Floorplan of the Opsroom

A   Display units (3+1) for Datafeed
B   Algedonic screens (2)
C   «Staffy»
D   Projector screens
E   Panel of the future
F   Backboard

1   Entrance
2   Main room with 7 chairs
3   Wardrobe
4   Kitchenette
5   Service area for Datafeed
6   Service area for algedonic screens
7   Service area for projectors and
     access to underground areas

Scale:
Module = 1 m

FIGURE 17 Ergonomic design sketch of the two positions for viewing data in the operations room.

FIGURE 18 Sketch of the polygonal floor plan with projectors installed outside the room itself

FIGURE 19 Floor plan for the final version of the Cybersyn Opsroom.

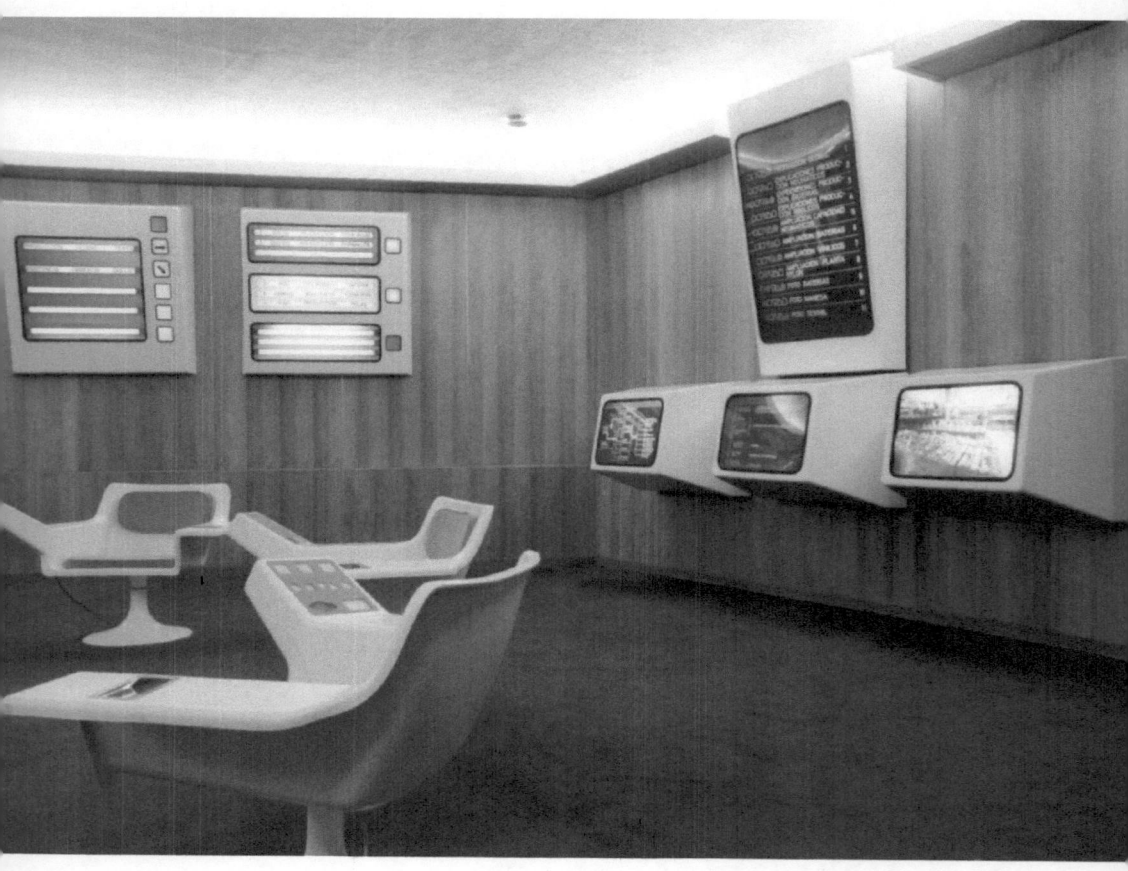

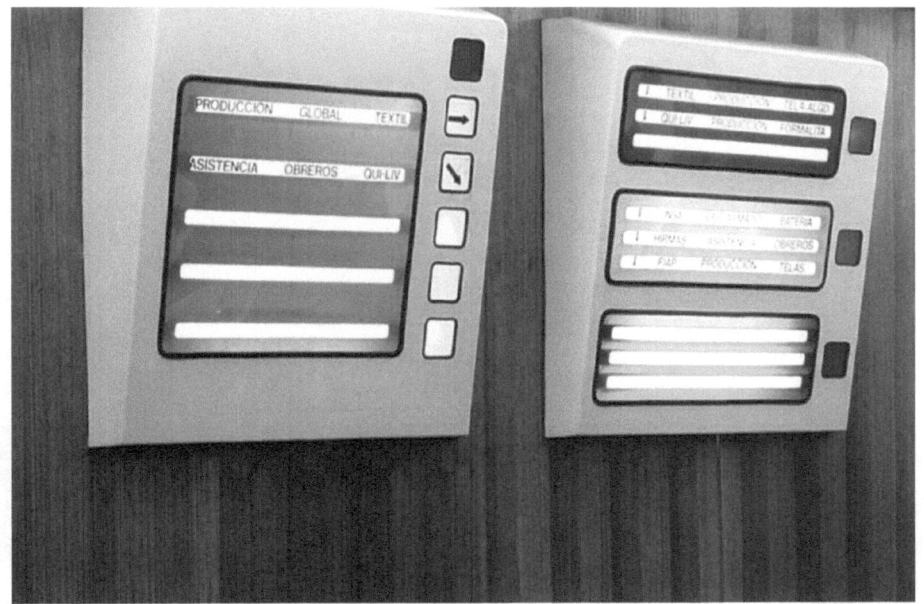

FIGURE 20 Opsroom: Close-up of chairs and datafeed screens.

FIGURE 21 The algedonic screens from the Cybersyn operations room.

FIGURE 22 Datafeed for the presentation of economic data.

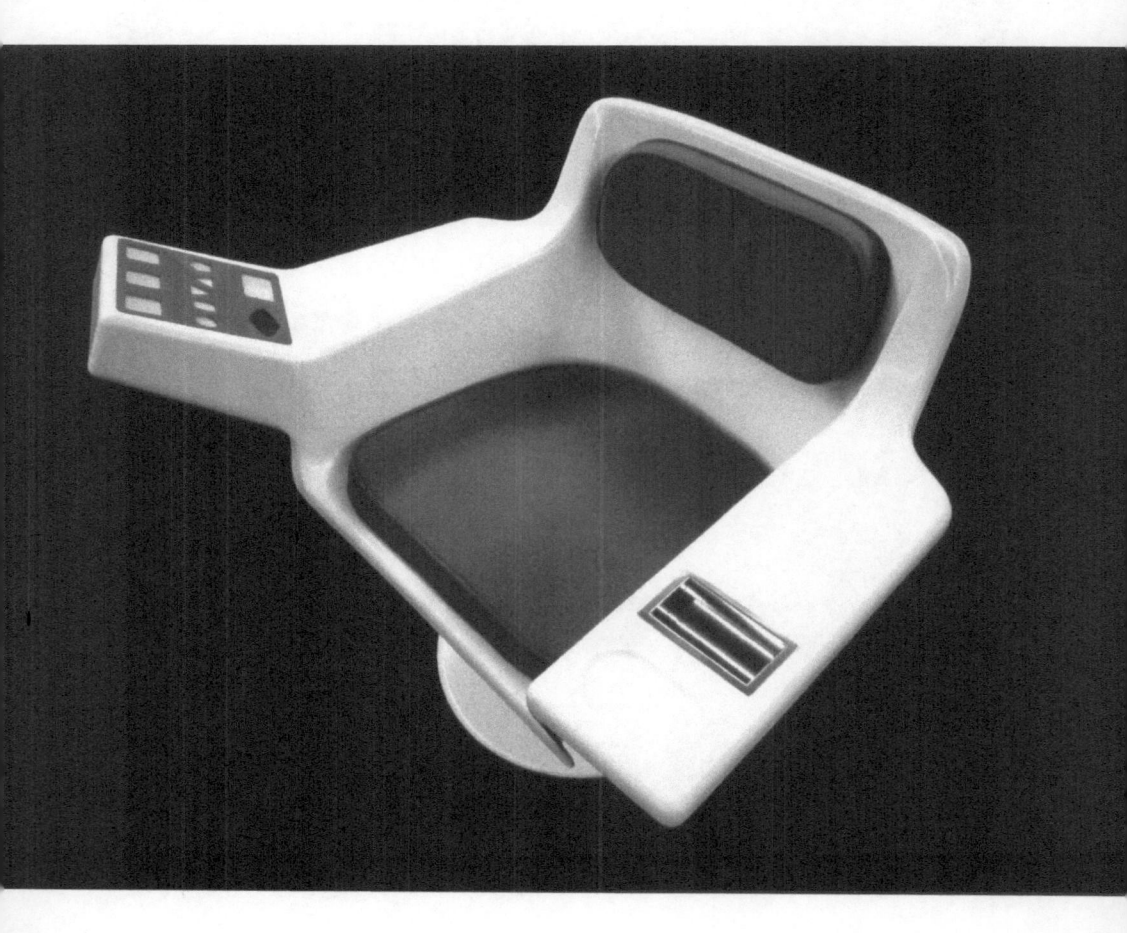

FIGURE 23 Swivel armchair with punch controls built into the armrest..

FIGURE 24 A design sketch for the armrest of the operations room chair showing the geometric "big hand" buttons.

FIGURE 25 The operations room chair showing the armrest.

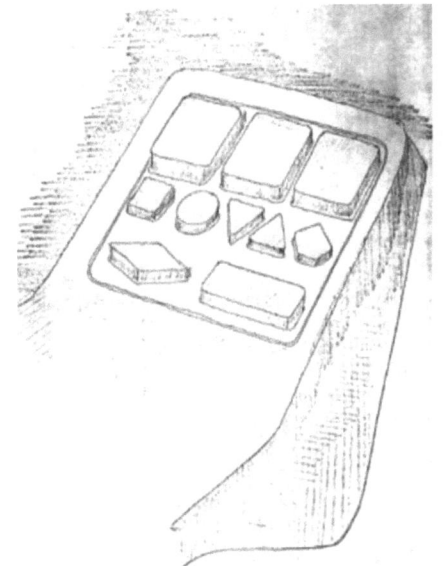
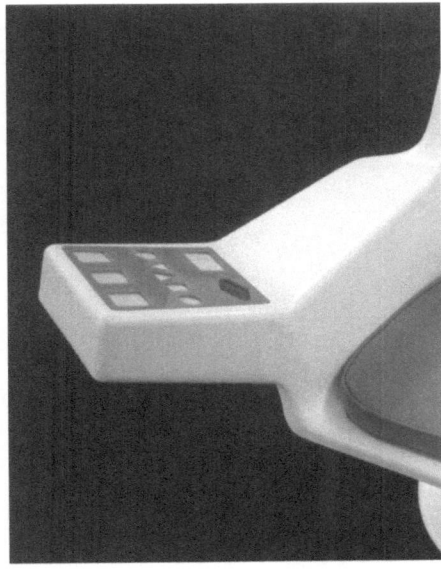

Editorial note: This chapter is an extract from Eden Medina's book, Cybernetic Revolutionaries: Technology and Politics in Allende's Chile *(Cambridge: MIT Press, 2011, pp. 108–128). It was added to this volume by the editors because it offers an extremely clear account of the wider social and political and economic context of the Cybersyn project. As Bonsiepe notes in Chapter 1.9, Medina was one of the first researchers to work on recovering the detailed story of the Cybersyn project. It is in no little part to her pioneering work that there is renewed interest in the Cybersyn project and in the economic policies of the Allende government before the 1973 coup.*

SOCIALISM BY DESIGN

(2011)

Eden Medina

SOCIALISM BY DESIGN

Of the four subprojects that composed Project Cybersyn, the operations room best captured the vision of an alternative socialist modernity that the project represented. The futuristic design of the room, and the attention it paid to its human user, would never have come about if the State Technology Institute had not had its own team of professional designers that it could assign to the project. Because this team did not exist before Allende's election, it is worth taking a moment to describe how it came to be and the role industrial design played in the creation of Chilean socialism.

The industrial production of goods for mass consumption constituted one of the central goals of CORFO under Allende. Beginning in 1971, the agency pursued a number of programs to "augment the production capacity of goods for popular consumption," including plans for the design and manufacture of low-cost automobiles, bicycles, motorcycles, sewing machines, household electronics, and furniture, among other items. For example, Citroën of Arica began constructing a new "automobile for the people" at the government's request, a Chilean version of the German Volkswagen.[1] Using funding and technology from its parent company, the Chilean Citroën plant drew up plans for a utility vehicle modeled after the Citroën Baby Brousse, a jeeplike conveyance that the French manufacturer had designed for public transportation in Vietnam. Citroën christened the new design Yagán, after a Chilean Indian tribe indigenous to Tierra del Fuego. Cristián Lyon, then director of Citroën Arica, remembered that the designers "wanted to see [a vehicle] that was native like the Yagáns."[2] Another example was the manufacture of low-cost televisions for popular consumption produced between 1971 and 1972 by the mixed-area

---

[1] In 1971, Minister of the Economy Pedro Vuskovic ordered the manufacture of a utility vehicle akin to the jeep that would cost less than $250 to produce. The Yagán was Citroën's response. Automobiles were among the most highly politicized technologies of the UP era. Although the Popular Unity program banned government workers from using automobiles for private use, government workers received priority in the distribution of new cars produced by Chilean factories, which sparked charges of favoritism. In 1971, CORFO created the Automotive Commission in an effort to coordinate the distribution of Chile's limited automobile supply in the face of rising demand and to bring Chilean automobile production in line with the goals of the UP. Among the core objectives was the production of utility vehicles and automobiles for mass consumption. *Primer mensaje del Presidente Allende ante el Congreso Pleno, 21 de Mayo de 1971* (Santiago: Departamento de Publicaciones de la Presidencia de la República, 1971), 119.

[2] Interview of Cristián Lyon, "Creando El Yagán," 2003, formerly available at the Web site of Corporación de Televisión de la Pontificia Universidad Católica de Chile, now defunct.

enterprise Industria de Radio y Televisión S.A., or IRT.[3] The IRT Antú was a black-and-white unit with an eleven-inch screen. Production of the Antú meant that television, previously obtainable only by well-to-do Chileans, became available to the masses for the first time.

Projects such as the Antú television and the Citroën Yagán paralleled UP policies for income redistribution and represented a "diversification and decentralization" of property, distribution patterns, and commercialization practices within Chilean industrial firms.[4] As a result of these efforts, poor Chileans and members of the working classes gained access to products and services previously reserved for the elite, a maneuver that raised levels of popular support for the UP, particularly during 1971 and early 1972.

The State Technology Institute also wanted to change Chilean material culture to reflect the goals of Chilean socialism. In an interview with *Science* magazine reporter

Nigel Hawkes, the deputy director of the State Technology Institute explained, "it is important for Chile to be selective about the technologies it adopts, because in the long run they may determine social values and the shape of society—as the automobile has in the United States, for example."[5] In addition to fostering the manufacture of low-cost, durable goods for popular consumption, Popular Unity's technological goals included decreasing Chilean expenditures on imported technologies and foreign patents, using science and technology to satisfy the specific biological and social needs of the Chilean people, producing a greater number of consumer and capital goods domestically, and improving both education and the dissemination of technical knowledge at Chilean universities, industries, and research institutes.

The State Technology Institute created the Industrial Design Group to assist with these efforts. In her study of the history of design education in Latin America, Silvia Fernández writes that Chile during the Popular Unity period was "the most advanced example in Latin America of design successfully integrated into a political-economic project in support of a social program."[6] The state support for design during the Allende years and its place in the Popular Unity program resulted from a series of coincidences and personal connections—although in hindsight design clearly forms part of a larger set of political, economic, social, and technological changes that were linked to Chile's revolutionary process.

Gui Bonsiepe, the head designer of the operations room, had studied at the Ulm School of Design (Hochschule für Gestaltung Ulm) in Germany beginning in the mid-1950s. One of the most influential design schools in Germany,

---

[3]   CORFO controlled 51 percent of the company, while RCA maintained the minority share of 49 percent. CORFO, "Comité de las Industrias Eléctricas y Electrónicas," CORFO en el gobierno de la Unidad Popular, November 4, 1971, Santiago, Chile.
[4]   CORFO Relaciones Públicas, "Rol de CORFO en los propósitos de cambios," CORFO en el go-bierno de la Unidad Popular, November 4, 1971.
[5]   Nigel Hawkes, "Chile: Trying to Cultivate a Small Base of Technical Excellence," *Science* 174, no. 4016 (1971): 174.
[6]   Silvia Fernández, "The Origins of Design Education in Latin America: From the Hfg in Ulm to Globalization," *Design Issues* 22, no. 1 (2005): 10.

perhaps second only to the Bauhaus, the Ulm School began in 1953 as a center for design education in industrial design, visual communication, industrialized architecture, and information design. From its inception, the school melded design education and practice with the social and political goals of European postwar reconstruction, including the promotion of democracy. The Ulm School also argued that design should be integrated into industrial production processes, where it would improve the production and use of material artifacts ranging from "the coffee cup to the housing estate."[7] The Ulm School moved design closer to science and technology and melded the visual aspects of design with scientific ideas, mathematical analyses, and user studies. Cybernetics, semiotics, systems theory, operations research, analytic philosophy of language, and Gestalt psychology all influenced the design methodology practiced at the school. The regular arrival of new guest instructors and visiting lecturers, such as Norbert Wiener and R. Buckminster Fuller, made student education in this range of areas possible.

Bonsiepe studied in the Design of Information Department. He first encountered cybernetics there. After he graduated from the program, he continued to work in a research and development group at the Ulm School and designed one of the first interfaces for an Olivetti mainframe computer.[8] Ulm professor and fellow designer Tomás Maldonado was Bonsiepe's intellectual mentor, and he made his first trip from Germany to Maldonado's home country, Argentina, in 1964 to work on design projects. Bonsiepe returned to Latin America for four months in 1966 as a consultant for the United Nations International Labor Organization (ILO). During this time, he gave a seminar on packaging design and developed a curriculum for an Argentine school of design.[9] "In Latin America I discovered the political dimension of design," Bonsiepe said, "not in the sense of political parties, but in the sense that professional work [in this area] can have a social dimension."[10]

In 1968, Bonsiepe accepted a more permanent position with the International Labor Organization to work with Chile's State Development Corporation to introduce industrial design in small- and medium-sized Chilean industries.[11] (His departure coincided with the closing of the Ulm School.[12]) Industrial

---

[7] Max Bill made this statement on July 5, 1954, to commemorate the first phase of construction of the Ulm School. "History: From the Coffee Cup to the Housing Estate," HfG Archiv Ulm, www. hfg-archiv.ulm.de/english/the_hfg_ulm/history_4.html.

[8] Bonsiepe said, "Under the direction of Tomás Maldonado, I designed a complete pictogram system in analogy with language, differentiating pictograms for 'verbs,' 'adjectives,' and 'nouns' that could be combined for the large control panel. In 1960 we didn't have the technical term interface, much less human-user interface." Gui Bonsiepe, interview by author, May 21, 2008, La Plata, Argentina.

[9] Although Bonsiepe completed the curriculum, it was never implemented.

[10] Bonsiepe interview.

[11] Bonsiepe signed a one-year contract initially, which he subsequently extended until 1970.

[12] There are multiple readings of why the Ulm School closed. Some historical analyses cite internal conflicts and financial difficulties as the main reasons. Bonsiepe also credits political conflicts between the school and "conservative government circles" as leading reasons to the closure of the institution: "For the conservative environment the school was too irritating because it did not maintain the silence of the cemetery that is so highly cherished by the forces of the status

design was a new field in Chile, and at the University of Chile it was being developed by a core group of undergraduate students who lacked a formal mentor. Fernando Shultz, Alfonso Gómez, Rodrigo Walker, and Guillermo Capdevilla were students at the College of Applied Art at the University of Chile, which had advertised a program in design. Only after they arrived on campus did they learn that the program existed in name only, there was no curriculum, and they were among its first students.[13] Since the college did not have a good understanding of design, the students bore the burden of forming their own program. The new design students faced a considerable challenge: none of the faculty at the College of Applied Arts specialized in design or had a design background (most worked in the fine arts or architecture). The university "didn't know what design was, and we didn't have a clear idea either," Shultz said. "But we [the design students] knew that there was something else; that there was another alternative. And that was what we were looking for, to be designers."[14] They pushed the university to create a design department with programs in textile and garment design, landscape design, interior design, graphic design, and industrial design.[15] The students found faculty from various parts of the university to teach classes in all these areas but one: industrial design, the area that most interested them. They realized they needed to look beyond their home institution for the education they wanted.

After meeting Maldonado at a 1968 UNESCO-sponsored conference in Buenos Aires, the students learned of Bonsiepe's impending arrival in Chile. When Bonsiepe's boat arrived at the port city of Valparaíso, the four students were there to meet him. They convinced him to take a role in their education, and he, in turn, became a demanding taskmaster who pushed them to read widely and cultivate competencies in a range of areas, including engineering, economics, the social sciences, and design.

In 1970, Bonsiepe accepted an offer to teach design at the School of Engineering of the Catholic University. Bonsiepe's move presented new opportunities for this particular group of design students. They began working as teaching assistants for engineering classes at the Catholic University, even though they were officially enrolled as students at the University of Chile, a rival institution. Teaching engineers led them to appreciate the benefits of combining design with engineering. "The engineers had the know-how," Shultz noted, but in his opinion they were like catalogs that contained a rigid set of

quo. The fact that members of the school organized a protest march against the war in Vietnam did not fare well with the population in Ulm." Gui Bonsiepe, e- mail to author, August 19, 2010.

[13] Chilean universities began offering courses in design as early as 1966, the first at the University of Chile in Valparaíso. Fernando Shultz said that he did not know about the course offerings in Valparaíso when he was deciding where to study. Fernando Shultz, interview by author, September 9–10, 2008, Mexico City, Mexico.

[14] Ibid.

[15] The Department of Design was an interdisciplinary endeavor from the outset. Most professors were from the Faculty of Fine Arts and the Faculty of Architecture, two campuses located in different parts of the city. Moreover, the department depended administratively on the Faculty of Engineering, located in yet another part of the city. Since the university couldn't find a way to bring the department under the control of a single faculty, "they put us there like an island," Shultz said. "This gave us a certain internal autonomy." Shultz interview.

solutions. In contrast, designers looked for different solutions but lacked the technical expertise the engineers possessed.

While at the Catholic University, Bonsiepe extended his role as teacher and mentor to a group of four graphic design students from the School of Communications. Unlike the industrial design students and the majority of students at Catholic University's engineering school, the four graphic design students—Eddy Carmona, Jessie Cintolesi, Pepa Foncea, and Lucía Wormald—were all female (Figure 22). "In the school where we studied [the School of Communications], there were almost no men," said Foncea. "So, rightly, we were girls, just like they [the industrial design students] were the ones who worked with hard things, materials." In Foncea's opinion, this gender divide was "part of a [social] reality that, in a certain form, still exists today in Chile," where science and engineering are male-dominated fields.[16] These two groups of students, the industrial design students and the graphic design students, would contribute to the design and construction of the Cybersyn operations room.

In 1970, Flores was still the director of the engineering school at the Catholic University, and he met Bonsiepe through a mutual friend. Years later Flores confessed to Bonsiepe that he did not have a high opinion of the design profession until he visited Bonsiepe's home and saw one of Stafford Beer's books on Bonsiepe's bookshelf. As Bonsiepe tells it, Flores remarked, "There were probably only two people in Chile who knew this book at that moment [Bonsiepe and Flores], and I thought that if a designer reads Stafford Beer, the design profession must have something serious in it."[17] Bonsiepe credits Flores for promoting industrial design education in Chile when it was still in its infancy. "This also happened in Brazil and Argentina," Bonsiepe noted. "Engineers with decision-making power created the conditions for the field of industrial design. This is not a well-known historical fact."[18]

When Allende came to power, Flores used his positions as both general technical director of CORFO and president of the board of the State Technology Institute to create the first state-sponsored industrial design group, which was to be housed at the State Technology Institute and led by Bonsiepe. The four industrial design students also moved to the institute. For Shultz, the move meant not finishing his undergraduate degree at the University of Chile, a sacrifice he was willing to make. Higher education was far less attractive to him than the possibility of contributing to the Chilean road to socialism. Shultz noted that at the time finishing a degree was seen as bourgeois, or akin to having a "title of nobility," which was not appealing to the young design student. In addition to

[16] Carmen (Pepa) Foncea, interview by author, July 25, 2006, Santiago, Chile. However, this is not the whole story. Although Foncea remembered that the graphic design students at the Catholic University were predominately female, the students in graphic design at the University of Chile were predominately male. Understanding the gender dynamics present in Chilean design education in the late 1960s and early 1970s is beyond the scope of this book and warrants greater analysis.
[17] Bonsiepe interview.
[18] Ibid.

FIGURE 26 Graphic design students who worked on the opsroom project (left to right):
Pepa Foncea, Lucía Wormald, Eddy Carmona, and Jessie Cintolesi. Personal archive of
Pepa Foncea. Image used with permission from Pepa Foncea.

FIGURE 27 The State Technology Institute (INTEC) Industrial Design Group. Front row
(seated, from left): Rodrigo Walker, Gustavo Cintolesi, and Fernando Shultz Morales.
Second row: Alfonso Gómez. Back row (seated, from left): Gui Bonsiepe, Pedro
Domancic, Werner Zemp, and Guillermo Capdevila. Not pictured: Michael Weiss.

Capdevilla, Walker, Shultz, and Gómez, Bonsiepe assigned additional designers and mechanical engineers to the Industrial Design Group, including three Ulm School graduates (Figure 22). Outside the institute Bonsiepe continued to work with the four graphic design students from the Catholic University; the four women contributed to several institute projects from 1970 to 1973, including the design of the institute's logo. Although the State Technology Institute benefited from the contributions of the graphic designers, the four women were not formally invited to join the institute. Foncea believes this was because graphic design had a less obvious connection to improving Chilean production capabilities than the field of industrial design.[19]

From 1971 to 1973 the State Technology Institute developed nearly twenty products, including inexpensive cases for electronic calculators; agricultural machinery for sowing and reaping that furthered the agrarian reform by raising the productivity of the land (see Chapter 4.6, Plate 3, Figures 5, 6, and 7); spoons for measuring rations of powdered milk given to children through the National Milk Plan (see Chapter 4.5, Plate 2, Figure 4); a collection of inexpensive, durable furniture for use in public housing projects and playgrounds; and a record player inexpensive enough for popular use (see Chapter 4.4, Plate 1, Figure 3).

These goods were simple in design, easy to construct, inexpensive, and of good quality, all important considerations for the majority of Chilean consumers. These products also illustrated the political dimensions of design. A piece of agricultural machinery that cut grass to feed livestock (see Chapter 4.6) was Bonsiepe's favorite product "because it was directly related to the production of food—in this case, milk," and would raise levels of Chilean nutrition.[20] Taken together, these projects illustrate a shift in the definition of industrial success and the considerations driving technological innovation. Instead of giving priority to the production of capital-intensive goods and the maximization of profit, as private companies had in the past, the government emphasized accessibility, use value, and the geographic origin of component parts. These new considerations reflected the economic policies of Popular Unity and the social goals of the Chilean revolution. Far from being neutral, the technologies described here intentionally reflected the philosophy of the Allende administration and became tools for revolution.

BUILDING THE OPSROOM

The Cybersyn operations room fit with the political mandate of the Industrial Design Group, but it was unlike anything else it created. While its other projects were closely tied to the day-to-day life of the Chilean people, the room was more of a futuristic dream (Plate 6, Figure 17). However, it did incorporate elements characteristic of the Ulm School of design and reflected the merging

[19] Foncea interview.
[20] Bonsiepe interview.

of engineering and design that had taken place at the Catholic University. The designers paid great attention to ergonomics and concerned themselves with such questions as the best angles for a user to read a display screen (Figure 14). They studied aspects of information visualization and wondered how they could use color, size, and movement to increase comprehension or how much text could be displayed on a screen while maintaining legibility. The operations room offered a new image of Chilean modernity under socialism, a futuristic environment for control that meshed with other, simultaneous efforts to create a material culture that Chileans could call their own.

Beer gave the design team general instructions about the type of control environment he wanted to create. He asked Bonsiepe to create a relaxing environment, akin to a British gentlemen's club. The designers drew up plans for a "relax room" that used indirect lighting to simulate a "saloon" atmosphere.[21] The plans included space for a bar where room occupants could make pisco sours, a popular Chilean cocktail. The design also represented the future of Chilean socialism. "I believe that the original idea was Stafford's own," said designer Fernando Shultz, "that we are looking toward the future," and creating an aesthetic that would break with the way things were done in the past.[22] Rather than replicating old designs of control rooms or gentlemen's clubs, the designers gave the room a futuristic flair. For example, they proposed that the chairs and screen cases be made out of fiberglass, a relatively new construction material that lent itself to organic curved shapes that were difficult to achieve with more traditional building materials.

In April, Bonsiepe sent Beer sketches of a circular room with ten chairs placed around a single control mechanism (Figures 12 and 13). The circular arrangement meant the seating arrangement could not be hierarchical, and the central control mechanism determined which data sets appeared on the wall displays. One wall contained a representation of Beer's five-tier Viable System Model. A series of slide projectors placed behind a wall projected slides of economic data onto acrylic screens, which Beer called "datafeed." These back projections created the effect of a high-tech flat-panel display.

By mid-June, the team had located a small space (approximately 24 feet by 12 feet) where the room could be housed. The small dimensions required the industrial design group to rethink its original layout. Among the changes they made, the designers put the screen for the five-tier Viable System Model on a rail so that it could easily be moved out of the way; they reduced the number of chairs from ten to no more than four; and they nixed the bar. These changes concerned Beer, who described the new space as claustrophobic and unable to accommodate enough people in decision-making. Moreover, he felt that the smaller space did not do an adequate job of selling the project. "We already have a selling problem in principle," Beer wrote. "This [small room] aggravates

---

[21]   Gui Bonsiepe, memo to Stafford Beer, "Sketches for the Op- Room," April 21, 1972, box 62, Beer Collection; INTEC, "Sala de operaciones anteproyecto," April 1972, box 62, Beer Collection.
[22]   Shultz interview.

it."[23] As Beer saw it, Project Cybersyn aspired to fundamentally change management practices in the enterprises and government offices. People would need to be convinced of the superiority of this cybernetic approach, and he hoped the modern-looking control room would offer an effective form of visual persuasion.

In August, the team finally located a more suitable space for the operations room, an interior patio of a downtown building that previously had been used to display automobiles. The space offered a number of advantages, including 400 square meters (4,303 square feet) of open space with no columns, the opportunity to construct a ceiling at any height necessary, and a central Santiago location in a building several stories tall "so nobody will be able to actually see us working" in the patio area.[24] As an added perk, the National Telecommunications Enterprise (ENTEL) owned space in the same building and had wired it with telecommunications capabilities. CORFO arranged for the room's construction, and Bonsiepe began working with the architect on the room's design, which could now accommodate a greater number of people and display screens than would have been possible in the tiny space the team had found back in June (Figure 16).

The new design allowed for seven chairs arranged in a circle in the middle of the room. Putting an uneven number of individuals in the room meant there would be no tied votes.

In deciding on this number, the team also drew from the influential 1956 paper "The Magical Number Seven, Plus or Minus Two: Some Limits on Our Capacity for Processing Information," by Princeton psychologist George A. Miller. Miller suggested that human beings could best process five to nine information channels, seven on average.[25] The team felt that limiting the number of occupants to seven would allow a diversity of opinion but still permit each voice to be heard. Paper was explicitly banned from the room, and the designers did not provide a table or other area for writing. Beer believed the use of paper detracted from, or even prevented, the process of communication; writing was strictly prohibited in the operations room.

The designers originally wanted to make the room circular as well, but when this proved difficult, they opted for a hexagon, a configuration that permitted five distinct wall spaces for display screens plus an entrance.[26] Upon entering the room, a visitor would find that the first wall to the right opened up into a small kitchen. Continuing to the right, the second wall contained a series of four "datafeed" screens—one large and three small—all housed in individual fiberglass cabinets (Figures 17, 20, and 21). The large screen was positioned above the three smaller screens and displayed the combination of buttons a user needed to push on the armrest of his chair to change the data and images displayed on the three screens below. The armrest also included

[23] Stafford Beer, telex to Gui Bonsiepe, June 27, 1972, box 66, Beer Collection.
[24] Roberto Cañete, letter to Stafford Beer, August 14, 1972, box 66, Beer Collection.
[25] George A. Miller, "The Magical Number Seven, Plus or Minus Two: Some Limits on Our Capacity for Processing Information," *Psychological Review* 63 (1956): 81–97.
[26] Bonsiepe interview.

a hold button that, when pushed, gave that user control over the displays until the button was released (Figures 18 and 19). Although the dimensions of the room had changed, the new space still placed a series of slide projectors behind the wall and used them to back-project slide images onto the datafeed screens, thus simulating flat-panel displays. The armrest buttons sent signals to the different projectors and controlled the position of the slide carrousel. Slides displayed economic data or photographs of production in the state-run factories[27] (Plate 8).

Rodrigo Walker, a member of the Industrial Design Group who worked on the design and construction of the operations room, said the user's ability to create his own path through the data was "like a hypertext" but one that preceded the invention of the World Wide Web by more than twenty years. While the parallel with the Web is not exact, the room did offer a nonlinear way of seeing the Chilean economy that broke from the presentation of data in traditional paper reports. The three screens contained a mix of flow diagrams, graphs of actual and potential production capacities, and factory photographs, an intentional mix of quantitative and qualitative data designed to give the occupant a "physical relationship" to the enterprise being discussed.[28]

The third wall held two screens for recording Beer's algedonic signals, which would warn of trouble in the system. The screens displayed the overall production trends within different industrial sectors and listed urgent problems in need of government attention. A series of red lights appeared on the right-hand side of each screen and blinked with a frequency that reflected the level of urgency that a given problem posed.

The fourth wall held a board with a large reproduction of Beer's Viable System Model and two large screens that could show additional information of use to the occupants. Beer insisted that the Viable System Model appear in the room to help participants remember the cybernetic principles that supposedly guided their decision-making processes. However, interviews revealed that few team members—let alone factory managers and CORFO employees not directly involved in the project—truly understood the Viable System Model. Some found it strange that such a theoretical representation appeared in a room dedicated to concrete representations of data and decision-making. The board was so closely associated with Stafford Beer that the project team referred to it as "Staffy."

Occupying the final wall was a large metal board covered in fabric (Plate 8). Here users could change the configuration of magnets cut in various iconic forms, each of which represented a component or function of the Chilean economy. This physical model served the same basic purpose as the model being developed by the CHECO team; both offered policy makers an opportunity to play with their policies and visualize different outcomes, but unlike CHECO the metal board was the epitome of low tech.

[27]   For a detailed description of the operations room design, see Grupo de Proyecto de Diseño Industrial, "Diseño de una sala de operaciones," *INTEC* 4 (1973): 19–28.
[28]   Rodrigo Walker, interview by author, July 24, 2006, Santiago, Chile.

The British company Technomation completed four screens for the datafeed display. However, import licenses were difficult to acquire from the Central Bank: "I have had the [word] IMPOSSIBLE written in red tape and with flashing lights on every step of the bureaucratic way," Cañete complained, alluding to the flashing red lights in the operations room that signaled trouble.[29] He thus conceived of an elaborate plan to smuggle the screens into Chile marked as donations from "Artorga," a reference to the British cybernetic investment club ARTORGA (the Artificial Organism Research Group) to which Beer belonged. But at the eleventh hour the Central Bank came through with the import licenses, and the screens reached Chile in September.[30]

The Chilean government dedicated some of its best resources to the room's completion. Its futuristic design, which borders on science fiction, was unlike anything being built in Chile at the time (Plate 6, Figure 17). It is often compared with the style of design found in Stanley Kubrick's classic movie *2001: A Space Odyssey* (1968), although the designers vehemently dispute that they were influenced by sci-fi films. "There was no reference point for this project," asserted Rodrigo Walker. "If I told you, 'Let's go build a movie theater,' you would have a reference point, you could begin to imagine what it would look like. But there was no operations room [in Chile], there was nothing that we could look at."[31] So they looked at design styles elsewhere and found inspiration in the work of Italian designers who used unorthodox materials, such as plastic and fiberglass, to create furniture with a sleek organic form. Only a few people in Chile knew how to work with fiberglass, and it had previously been used to construct swimming pools, not furniture, but the designers felt the material gave them the practical and stylistic elements they desired. "I think the room looked the way that it did because of the materials that we used . . . polyester with fiberglass, an organic material that allows you to do anything that you want," Walker noted.[32] Using these new materials allowed the designers to project a new image of socialist modernity that rivaled science fiction.[33]

The operations room also gave the designers opportunities to form new working relationships, which they viewed through the lens of socialist change. For example, the designers wanted to attach the fiberglass form of the seat to a metallic base that swiveled. However, the swivel mechanism they envisioned was not manufactured in Chile, and the designers could not import the mechanism because of the government's shortages of foreign credit and the invisible blockade. The designers consulted with workers in their metal shop, who devised an alternative design that used grease alone

[29] Cañete to Beer, August 14, 1972.
[30] Roberto Cañete, telex to Stafford Beer, August 16, 1972, box 66, Beer Collection.
[31] Walker interview.
[32] Ibid.
[33] Stories differ on where the chairs were constructed. Roberto Cañete remembers driving to the countryside to pick up the fiberglass chairs from a workshop and, upon arriving, finding a chicken perched on top of one of the futuristic seats. Cañete interview.

and allowed the upper part of the chair to move without friction. Thus, Chilean socialism not only inspired the use of new materials but forced Chileans to develop innovative ways of working with old materials. Ideas that originated on the shop floor mixed with those of the professional designers, and, in the context of the Chilean road to socialism, this mixing had new significance. One designer, Fernando Shultz, said that Chilean socialism opened up a new awareness of worker participation that was "very subtle" but still part of the government's program. For Shultz, asking for workers' suggestions to improve the design team's work was not a simple act but rather the result of "a mental process, a process of conscience and commitment" that was set in motion by the Popular Unity government.[34] In the area of industrial design, Cybersyn thus resulted in more inclusive and participatory design practices.

Work on the slides showing production data and factory photographs was supposed to start in August, and the team secured one of the top photographers in Chile to assist with their production. But the team was not sure how to create a clear, homogeneous representation of factory data that managers and government administrators could easily understand. This uncertainty delayed production of the slides, and the team worried that the photographer would be otherwise engaged by the time they were ready for him.

The slides provided a way for the design team to update the data displayed in the operations room. But the team did not use a computer to generate these visual displays of data, as they would today. Instead, Bonsiepe enlisted the four female graphic design students from the Catholic University to create, by hand, camera-ready versions of the flow charts and graphs that the photographer could convert into slides. The graphic designers completed the first flow charts showing production activities in September; these gave an overview of production in several nationalized textile enterprises.[35]

Although the operations room presented a sleek, futuristic vision of socialist modernity in which an occupant could control the economy with the touch of a button, maintaining this illusion required a tremendous amount of human labor. In this case, it required some of Chile's best graphic designers to draw by hand every graph and chart the room displayed. These images needed to change regularly to permit the form of dynamic control Beer imagined, yet there were no plans to automate this process in the future. Although Allende believed that Chile would have a revolution with "red wine and empanadas," this assertion failed to account for the actual complexity that the process entailed. In the same way, the clean, futuristic appearance of the control room obscured the vast network of individuals, materials, expertise, and information required to make economic management appear simple.

[34] Shultz interview.
[35] Gui Bonsiepe, memo to Stafford Beer, August 28, 1972, box 56, Beer Collection.

DESIGN FOR VALUES

Beer and the Industrial Design Group were well aware that design could reflect social values. For example, Beer found the early design sketches for the operations room, which placed a single control mechanism in the middle, to be lacking because the design inhibited democratic participation. As a result, Bonsiepe sent Beer a new set of sketches that put the mechanism for controlling the content of the datafeed display screens in the armrest of each chair. Occupants could thus change the data displayed by pushing different combinations of geometric buttons. This new design gave all occupants equal access to the data and allowed them to control what was displayed inside the room. The geometric buttons also made the room more inviting by replacing a more traditional mechanism, the keyboard. Beer imagined that the individuals sitting in the operations room would be either members of the government elite or factory workers, individuals who did not know how to type—a skill typically possessed by trained female secretaries. With little instruction, occupants could use the large "big-hand" buttons on each armrest. Participants could also "thump" these buttons if they wished to emphasize a point. Beer claimed that an interface of large, geometrical buttons made the room more accessible for workers and prevented it from being a "*sanctum sanctorum* for a government elite." Through this design decision, the system allowed for worker participation.[36]

While politics favoring class equality influenced the design decision to use the buttons, this design decision was also gendered. Beer stated that the decision to eliminate the need for a keyboard literally eliminated the "girl between themselves and the machinery" and thus brought the users closer to the machine.[37] He was referring to a literal woman, a typist who would navigate the keyboard interface on behalf of the bureaucrats or factory workers occupying the operations room chairs. Other gendered assumptions also entered into the design of the control environment. In addition to eliminating female clerical work, the room was explicitly modeled after a gentlemen's club. It also encouraged a form of communication that bears a closer resemblance to masculine aggression ("thumping") than to a form of gender-neutral or feminine expression. Bonsiepe later acknowledged that "in hindsight I can see a gender bias" in the room's design.[38]

The characteristics ascribed to the room's future occupants reveal assumptions about who would hold power within the Chilean revolution and who constituted a "worker." Generally speaking, factory workers and bureaucrats would have the ability to make decisions affecting the direction of the country; clerical workers, women, and those operating outside the

---

[36] Beer, *Brain of the Firm*, 270.
[37] Beer said the occupants of the operations room would view a keyboard as "calling for a typing skill, and [would] want to insinuate a girl between themselves and the machinery. . . . It is vital that the occupants interact directly with the machine, and with each other." Stafford Beer, *Platform for Change: A Message from Stafford Beer* (New York: J. Wiley, 1975), 449.
[38] Bonsiepe interview.

formal economy would not.[39] The operations room also offers a valuable counterexample in the history of technology, a field filled with examples that link female labor to the routinization of work and unskilled labor. Here we see an opposite but no less interesting phenomenon: Beer and the designers viewed female clerical work as *too* skilled; it therefore needed to be eliminated to make the room accessible.

The design of the operations room illustrates that even futuristic visions of modernity carry assumptions about gender and class. Moreover, the design of this control space shows how cultural and political givens limit technological innovation. By treating the design of the operations room as a historical text, we can see how the Allende government framed its revolutionary subjects and ultimately limited the redistribution of power within Chile's socialist revolution.

[39] The design of the room supports a common criticism of the Popular Unity government, namely that it held an ambivalent attitude toward women and that the Chilean left focused on the largely male groups of industrial workers and, to a lesser extent, rural peasants. For example, during the Popular Unity period, Chilean social scientist Lucía Ribeiro criticized existing bodies of social theory for failing to address the experiences of Chilean women and privileging production outside the home over reproduction and related domestic activities. Years later, historian Sandra McGee Deutsch concluded, "Existing studies have stressed the left's inability—of which there are numerous examples—to conceive of female participation in the struggle for socialism." In a similar vein, the historian of Chile Thomas Miller Klubock writes, "The popular fronts, with the support of labor and the Left, built their political hegemony on the foundation of a gendered political ideology that defined the rights and benefits of national citizenship in terms of the male worker and head of household and the female housewife." See Lucía Ribeiro, "La mujer obrera chilena: Una aproximación a su estudio," *Cuadernos de la realidad nacional*, no. 16 (1973); Sandra McGee Deutsch, "Gender and Sociopolitical Change in Twentieth- Century Latin America," *His-panic American Historical Review* 71, no. 2 (1991): 297–8; and Thomas Miller Klubock, "Writing the History of Women and Gender in Twentieth-Century Chile," *Hispanic American Historical Review* 81, nos. 3–4 (2001): 507.

PARACHUTING INTO THE FUTURE

Marcos Martins and Zoy Anastassakis

RATIONALISM, AUTONOMY, PROGRESS

The currents that drove Gui Bonsiepe's theoretical and practical project in the second half of the twentieth century do not blow the same way in contemporary times. Even considering the obvious need to contextualize his texts in their historical circumstances, it is difficult to read them today without persisting comparisons getting in the way. This challenge can be dealt with through an effort to take up the essential aspects of Bonsiepe's thinking, detaching them from positive expectations regarding globalization, which today are problematized by decolonialism, or from a project for a future that history will apparently have buried.

However, such an exercise excludes the possibility that the contrast may allow us to better see certain movements of our own time that we tend not to see. More than that, there is the risk of hastily decreeing the end of an era and its replacement by another, when they actually coexist. It is possible to let Bonsiepe's thought reach us not by means of a linear causal historical explanation, but through flashes of the past, according to the much cited Walter Benjamin's concept of history. Unable to recover the past "the way it really was," we can at least let it precipitate into the present as a "memory as it flashes up in a moment of danger."[1]

Looking back at Bonsiepe's wide range of fields and theoretical elaborations as presented in these texts, three positive values stand out as recurring motifs and themes: "rationalism," "autonomy," and "progress." These three terms are organized in an argumentative logic in defense of a certain conception of design, in the form of a teleological sequence. For Bonsiepe, it is only through rationalism that design practices can achieve their autonomy, in terms of both its technical vocation and its social reach, freeing themselves from the dangerous association with a "second rank applied art" (Chapter 17, p. 169) or with "sophisticated and expensive luxury goods for the five percent economically privileged in the population" (Chapter 13, p. 141).

From this perspective, design activity can play a prominent role in the processes of industrialization and modernization through rationalism, more specifically the one developed at HfG Ulm, a school "distinguished by a

---

[1] Walter Benjamin, "On the Concept of History," in *Selected Writings. Volume 4*, eds. Howard Eiland and Michael W. Jennings (Cambridge, MA: Belknap Press of Harvard University Press, 2006), 389–400, 390.

view of design as an activity on a technical foundation - an activity that was to be practiced with a basically rational approach" (Chapter 13, p. 141). It is this autonomy that would guarantee design a prominent place in industrial processes, as these processes would constitute the only way for progress and, therefore, for the future. The latter connects with the question of autonomy in the field of Design in two ways.

First, this notion of the future guarantees the survival of the activity itself. "Today more than ever," says Bonsiepe, in 2003, "it should be clear that professions that do not produce their own specialized knowledge have no future" (Chapter3, p. 27). Secondly, designing, once its autonomy is guaranteed, becomes the main vector of progress, the "driving force of modernity" itself. Crediting the strength of this program to Tomás Maldonado, who allowed, "to re-interpret the world from the viewpoint of design rationale," Bonsiepe argues that "to be radically modern is to invent the future, to design and arrange it, and that includes the future of that same modernity" (Chapter 1, p. 13).

This logic that connects rationality to autonomy and the future does not just reveal a certain vision and program for a design framework. It replicates itself for projects about the future and progress that Bonsiepe establishes for "peripheral" countries. The rationalism of the Ulm school would find a "latent need" (Chapter 11, p. 119) in these countries, Bonsiepe senses, precisely because there would already be an aspiration for independence and autonomy in them. Here, this autonomy would also come through the implementation of a design discourse, without which these countries would not be able to take possession of their own future: "What characterizes the peripheral world is the lack of a design discourse. That is why these countries have not, so far, had a future—for the future is where design unfolds. Only through design is it possible to appropriate the future" (Chapter 1, p. 13).

In this sense, Bonsiepe positions the field of Design—as formulated at HfG Ulm—as a direct heir of the Enlightenment project: "in terms of the history of ideas, the HfG followed in the tradition of the Enlightenment, a movement which has lost nothing of its relevance. A look at the current global political situation is enough to persuade one that our present epoch suffers not from too much, but from too little Enlightenment" (Chapter 1.4, p. 35).

DESIGNING FACING GAIA

Today, Enlightenment's certainty in human development based on rationality and progress is, at least apparently, discredited. The narratives of an "end of the world"[2] already overshadow the belief in scientific progress. In

---

[2]   Déborah Danowski and Eduardo Viveiros de Castro, *The Ends of the World* (Cambridge: Polity Press, 2017).

particular, science, conceived in modernity as the maximum value of human rationality, has been gradually eclipsed by the predatory policies and politics of neoliberalism. Confronted with the current climate crisis, we are faced with a sad dilemma: must we regret the barbaric decline of the belief in reason and insist on a recovery of the possibilities of progress through reason? Or should we question the very idea of the enlightenment's rationality so that we can look for other ways of living beyond developmentalism?

Today, beyond the field of Design, the formulation of the problem moves, in its statement, from a "yes or no" to human reason, to a reduction of both its centrality and its role in planning a future. In the field of philosophy of science, Isabelle Stengers, in the book *Catastrophic Times: Resisting the Coming Barbarism*,[3] asks questions on what it means to live, in today's world, in the midst of this state of intense exhaustion and hesitation. According to Stengers, from now on, a new era arises, bringing with it an inconvenient truth: global climate change, until recently imagined as a remote possibility, is now emphatically in place. Stengers calls this event "the intrusion of Gaia." But Gaia is not the concrete planet; it is the intrusive event itself: a new situation in which nature not only asks to be protected; it is able to intrude into our lives. Now, the "natural" world, which we thought to be passive and inert, is violently interfering with the plans of industrial developmental progress.

Gui Bonsiepe, in 1994 (Chapter 3, p. 27), already discusses the impacts of the environmental crisis on industrial production and industrialism, pondering then on the practical, sociopolitical and ethical dimensions of design. However, his proposal to review industrialism in relation to the field does not imply a disengagement from the domain of industrial production, understood as a fundamental cultural manifestation for a modernity still desired and not fully realized. Bonsiepe insists that it is modernity that creates "an obligation to make this planet more worth living on" (Chapter 1, p. 13), arguing that a "renunciation of the project of enlightenment seems [to be] the expression of a quietist, if not conservative attitude—an attitude of surrender that no designer should be tempted to cherish" (Chapter 9, p. 87).

Stengers, on the other hand, ponders that now, in the face of the intrusion of Gaia, it is necessary to reconsider not only the modes of industrial production, as Bonsiepe proposes, but also industrialism itself. It is not a question, then, of mobilizing design as a means to reformulate or rationalize the modes of industrial production in the face of the challenges posed by the climate crisis, safeguarding its conceptual pillars. More radically, Stengers proposes, the current moment demands a reconsideration of *all* our Enlightenment heritage. In her opinion, only by questioning the whole of the modern framework and our commitments to it would we be able, then, to respond to the intrusion of Gaia which, in imposing itself on us, bothers us, and demands the cultivation of unusual responsive skills.[4]

3   Isabelle Stengers, *In Catastrophic Times: Resisting the Coming Barbarism* (Open Humanities Press, Meson Press, 2015).
4   Donna Haraway, *Staying with the Trouble: Making Kin in the Chthulucene* (Durham and London: Duke University Press, 2016).

In our book for the *Designing in Dark Times* series, we give a tangible example of this pressure for reconfiguration of ways of living that Stengers points out, in which elements of the environmental crisis intertwine with political, economic, and financial dimensions. We recount some reports of a specific moment in the history of the Superior School of Industrial Design (Escola Superior de Desenho Industrial, ESDI), in Rio de Janeiro, Brazil, which found itself challenged amid a situation of sudden material and institutional precariousness. It is opportune to make a comparison between the Brazilian school and the HfG-Ulm, both because of their historical ties and also because of the contrast between today's EDSI—urged to develop the "new skills" Stengers mentioned—and the modernist project of future that gave rise to the school.

A CASE FOR OPENNESS

Founded in 1962, ESDI was the first Brazilian design school to offer a bachelor's degree in Industrial Design. The first plans for its implementation date back to the country's industrialization between the 1950s and 1960s, a time of progressive ideas shared both by enthusiastic intellectuals and government's "powerful men." At that moment, the idea of planning was adopted as a motto by the federal government, which invested intensively in industrialization as a vector for national development. In this context, it would be necessary, then, to form a body of professionals capable of designing for the industry. Hence, the first moves to create industrial design courses in the country.

ESDI came into being with the support of a group of designers identified with the Bauhaus and the HfG-Ulm, such as the Swiss Max Bill and the Argentinean Tomás Maldonado. In 1960, Bill suggested that two former students from Ulm joined the ESDI faculty, the German product designer Karl Heinz Bergmiller and the Brazilian graphic designer Alexandre Wollner. Intensifying the flow of design ideas between Germany and Brazil, the presence of Wollner and Bergmiller at ESDI consolidated strong kinship ties between the design conceptions formulated in Ulm and the genealogy of the Brazilian school's structure.[5] This alignment of proposals was recognized by the "matrix." In 1965, in a letter to Bergmiller and Wollner, Maldonado wrote that "the Ulm School and the one in Rio de Janeiro are, at the moment, the only ones in their specialty that present a progressive structure."[6] It is important to note, however, that this alignment has never meant that ESDI has followed the German school model in a homogeneous and unquestioned way. Similarly, in the case of the HfG-Ulm, what became known as the "model" never represented, as is well known, a uniform ideological block that, since its closure, has become stereotypically associated with its name.[7]

[5]   Pedro Luiz Pereira de Souza, *Esdi, biografia de uma ideia* (Rio de Janeiro: EDUERJ, 1996), 30.
[6]   Ibid., 16.
[7]   Ibid., 51.

With these caveats, we can, however, state that the rationalism from HfG-Ulm, as well as the proposal that design could become a factor of education and enlightenment of society, are remarkable features of ESDI since its genesis. This influence gains a new inflection with the arrival of Gui Bonsiepe's ideas. His writings in the *Ulm* magazine were already known and discussed at ESDI, but his influence became more tangible when the school's students came into contact with transcripts of lectures Bonsiepe had given in Valparaiso in 1970.[8] They enthusiastically identified with his proposals for a design that was connected with public policies; instrumental for the autonomy and independence of national industry; marked by interdisciplinarity; and averse to regionalisms.

A 1972 text written for the participation of ESDI in the International Design Biennial, in Rio de Janeiro, clearly reflected Bonsiepean ideas: "The linking of the designer to economic, technological and social reality is only possible, integrating it into an interdisciplinary team directly linked to production. [...] Universities and specialized schools should contribute with the creation of project and research groups in industrial design." The article ends by stating that these ideas could only escape a "utopian failure, to the extent that they were formulated together with the deliberative organizations of the country."[9]

Written by young students from a peripheral design school, these words seem to corroborate Bonsiepe's idea that there would be a "latent need," in peripheral countries, for a design education project linked to industrial growth and social improvement. These themes resonated, at ESDI, with the concerns of one of the founders of the school, Aloisio Magalhães, who, with his more spontaneous approach, discussed how to mobilize design as a tool for autonomous and culturally oriented development. The last sentence of the pamphlet for the 1972 Design Biennial, by conditioning the project to the adherence to it by the spheres of power, perhaps indicates a possible reason for its disarticulation.

The government's neglect of public education institutions in Brazil today imposes a great distance between a design school and "deliberative instances." In fact, part of Brazil's public policy regarding education has been not one of collaboration, but of attack, considering public universities as a burden to be eliminated. At ESDI, this scenario has imposed challenges entirely different from those faced by who took care of its birth.

Almost fifty years after its foundation, in 2016, when we were both elected as Deans, the school was suddenly threatened with closure. Funded by the State of Rio de Janeiro, because of extravagant spending on the 2016 Olympic Games and a series of scandals of misappropriation of public funds, corruption, and money laundering that together decimated the state budget and the funding for the university to which ESDI belongs, the school was practically unable to function. The crisis that resulted from a series of delays and interruptions in the transfer of funds for payment of salaries, scholarships, and basic infrastructure

[8]   Ibid., 233.
[9]   Ibid.

lasted for more than a year. In an unusual way, the intensification of external pressure on the school ended up leading the community to rehearse ways of living in such a precarious environment, different from the usual institutional routine.

The movement to collectively recover the sense of community and care for the school has led us to unusual experiments. Among them, we highlight one, for radically promoting the meeting between the school's rationalist tradition and other cosmologies. During this intense and troubled times, a group of Indigenous artists and researchers visited us, such as Alberto Álvares, Daiara Tukano, Denilson Baniwa, Francy Fontes, Ibã Sales Huni Kuin, Inê Kuikuro, Jaider Esbell, Sandra Benites, and Wally Kamayurá. Their encounter with teachers and students confronted the school's segregationist elitism, which insisted in ignoring the contribution of affirmative action policies, which for almost twenty years now have helped diversifying the student body which today is made up of about 40 percent of Black and poor students.[10]

The power of these "Afro-Indigenous relations,"[11] when recognized in the classroom of a design school with a Eurocentric matrix, repositions some important issues. Amid these encounters, it was possible to explore the idea that the rationalist, progressive, and universalist apparatus can always be redirected according to ways of perceiving and experiencing the world that are not at the service of homogenization, but, rather, of recognizing spaces for coexistence while respecting their differences. At that time, we realized that design practice can also be reconsidered as a tool for pluriversal[12] correspondences,[13] which are situated between the hegemony of Western universalist ontology and the pluriverse of socionatural configurations. Which is to say, as Caroline Gatt and Tim Ingold ponder: "design, in this sense, does not transform the world. It is rather part of the world's transforming itself."[14]

This brief account presents a moment in which a project of progress based on the meeting between an industrial rationality and progressive public policies has proved unsustainable, opening up new possibilities. The contrast between today's ESDI and the ideas that underpinned its creation in 1962, points to the broader issue of understanding design as a future-oriented project activity. But what is the difference between the future at this beginning of the twenty-first century and that of the mid-twentieth?

[10]   Zoy Anastassakis, "Redesigning Design in the Pluriverse: Speculative Fabulations from a School in the Borderlands," in *Design Struggles: Intersecting Histories, Pedagogies and Perspectives*, eds. Claudia Mareis and Nina Paim (Amsterdam: Valiz, 2020).

[11]   Marcio Goldman, "Beyond Identity. Anti-Syncretism and Counter-Miscegenations in Brazil." Lecture to the Workshop "Brazil at the Crossroads: Looking Beyond the Current Crisis," sponsored by TORCH. The Oxford Research Centre in the Humanities, University of Oxford, Oxford, UK (May 28, 2018). Available at: https://www.academia.edu/43368178/Beyond_Iden tity_Anti_Syncretism_and_Counter_Miscegenations_in_Brazil

[12]   Arturo Escobar, *Designs for the Pluriverse: Radical Interdependence, Autonomy, and the Making of Worlds* (Durham and London: Duke University Press, 2018).

[13]   Tim Ingold, *Correspondences* (Cambridge: Polity Press, 2021).

[14]   Caroline Gatt and Tim Ingold, "From Description to Correspondence," in *Design Anthropology: Theory and Practice*, eds. Wendy Gunn, Ton Otto and Rachel Charlotte Smith (London; New York: Bloomsbury, 2013), 139–58.

UNSOLVING PROBLEMS

In an interview given in 2002, Bonsiepe comments on an occasion in which he is referred to as the "parachutist from Ulm" (Chapter 11, p. 119). The term, says the interviewer, would correspond to the idea of Western intervention in a developing country. Rejecting the "missionary" (Chapter 11, p. 119) label, Bonsiepe argues that, back in 1976, in an interview with Sonia Newby, he called himself a parachutist in an ironic way, confessing his ignorance about the country in which he had just "landed," and where, precisely, he would dedicate himself to the cause of local autonomy through the creation of design policies. Bonsiepe's "landing" aimed at the promotion of the autonomy of countries threatened to remain forever under the yoke of domination by others who already had set in motion a project of economic development. His parachutes, therefore, can be thought of as a metaphor for the arrival at a satisfactory level of independence and social progress, achieved through a *programmed* future.

In his book *Ideas to Postpone the End of the World*,[15] Ailton Krenak develops a quite different and provocative vision of "postponement," in which the future divests itself from any possibility of conquest, victory, appropriation, possession, and landing. The allegory is that of a colored parachute: "Why do we hate the sensation of falling?" asks Krenak, recognizing that "that's all we've been doing of late. Falling, falling, falling.... Let's put our creative and critical capacity to use making some colourful parachutes to slow the fall, turn it into something exciting and edifying—after all, we like nothing more than fun and games, enjoying our time on this earth."[16]

Although this is an already-much-cited passage by now, hurriedly taken as a panacea, we believe that, by putting it side by side with Bonsiepe's story, we can see more deeply complex historical indexes pointed by Krenak's apparent simplicity. His parachutes are not primarily a device to help landing, but rather a device to *perpetuate* the fall as long as possible, like an endless story. In this fall, everything falls and, as David Kopenawa and Bruce Albert tell us in the book *The Falling Sky: Words of a Yanomami Shaman*,[17] even the sky can fall. With no sky above us nor place to land, we fall. Falling, we get in touch with all the worlds we had left aside while holding hope for future landings. Thus, we gain time, at last, to look around and pay attention to what surrounds us. This form of expanded perception, liberated from any function or purpose, is expressed in another attribute of this allegory: Krenak's parachute is colored. Color, here, is neither romantic, decorative, nor functional: it relates to the Indigenous technologies of knowledge production based on the expansion of perception, as is the case with dreams.

The same artifact, the same invention, *one singular design product* offers us two understandings, two apparently irreconcilable approaches

---

[15]   Ailton Krenak, *Ideas to Postpone the End of the World* (House of Anasi Press, 2020).
[16]   Ibid., 33.
[17]   Davi Kopenawa and Bruce Albert, *The Falling Sky: Words of a Yanomami Shaman* (Harvard University Press, 2013).

and cosmologies concerning the future. One aims to cushion the impact of contact with the ground, and has, at the moment of contact, the expectation of departure toward new conquests. The other, by adding color and subtracting the floor from the "parachute system," is prepared to not find a place of arrival, suspending us in a state of fall. So, while Krenak's parachute addresses an "end of the world," Bonsiepe's one points to the beginning of another.

However, beyond these differences, there seems to be a contact zone[18] between Bonsiepe's ideas and contemporary emerging issues as formulated by Ailton Krenak and Isabelle Stengers and exemplified in the account of the most recent events at ESDI. Bonsiepe insists on the need for a reconsideration of the inevitably political character of design, claiming that design must play a role compatible with its time: "questions regarding the political and social implications of design have now shifted into ecology" (Chapter 3, p. 27). This political call to action, in commitment with the ecological concern, Bonsiepe adds, "includes a component of hope—a dream, however vague, with the outlines of the society we want to live in."[19] The articulation between the political character of design and the vagueness of dreams seems to indicate a possible synching between the two approaches to the parachute and allows us to formulate the question: Accepting our condition of perpetual fall, would it be possible to reconsider designing as dream catching[20] the constant search to bring back to the materiality of the present time an imagination that we never manage to grasp?

---

[18] Mary Louise Pratt, "Arts of the Contact Zone," in *Profession* (1991), 33–40.
[19] Ibid.
[20] Tim Ingold, *Making: Anthropology, Archaeology, Art and Architecture* (London; New York: Bloomsbury, 2013).

Introduction to the Appendices

The appendix contains 4 documents of the end phase of the Ulm school, published in the last issue, number 21 (April 1968) of the journal of this institution. Two are my authorship (Appendices 1 and 2), and the other two are the resolution of the plenary assembly of the staff and students of HfG in February 1968, which effectively dissolved the school (Appendix 3), and a poster issued by the students of HfG commenting on the situation and drawing the relevant historical parallel (Figure 24).

It should come as no surprise that the institutional critique of the politically and culturally conservative forces determining the fate of the school, which is the focus of Appendices 1 and 2, contradicts a later, and particularly in Germany, favored semiofficial narrative according to which the reason for the termination of this cultural experiment is ultimately to be found in the intransigence of the members of the HfG, whereas the politicians are relieved of any responsibility for this sad story.

Admittedly, the HfG was an irritating institution. And it has remained so. A single-track reading of this project—its successes and failures—leads nowhere. Once an institution has suffered the misfortune of being apostrophized as a myth, the field of eager myth destructors opens up. Without wanting to take up old, albeit still-open controversial issues, the question today is how to tap the cultural heritage of this aborted, unfinished experiment of cultural innovation.

It is to be hoped that a new generation of historians will know how to exploit the potential of immaterial heritage.

APPENDICES

Three Notes on the Closure of HfG Ulm (1968)

1. THE SITUATION OF THE HFG: A RESIGNATION LETTER[1]

On February 23, 1968, the members of the HfG decided to terminate their activities at this institution with effect from September 30, 1968, if the government and parliament of the Land of Baden-Württemberg persisted in their previously published plans and conditions for continuing the HfG. As we go to press, it is not certain whether and in what form the HfG will continue its existence at Ulm or elsewhere. It is mainly due to the efforts of certain journalists and friends of the HfG that there are in fact any alternatives at all and that we are not presented with a fait accompli. They succeeded in bringing home to the public that the government's measures do not do justice to the HfG. These plans, however, were conceived entirely in the spirit of those measures which as far back as 1963–4 betokened an encroachment on the autonomy of the HfG and curtailed still further its already seriously restricted freedom.

Freedom is first and foremost economic freedom. And in this respect the position was never very favorable right from the start. It was a mistake to assume that an educational institution could and should be financed by earnings from industrial commissions. Education cannot be run out of its own resources. The HfG had therefore to rely on public funds and thus became dependent on the goodwill and understanding of elected representatives. Neither could be taken for granted. On the responsible committees, grants were often authorized only by narrow majorities in the teeth of stiff opposition. As the international reputation of the HfG continued to grow, the means whereby the demands arising from such a reputation might be met dwindled because the utterly inadequate funds made a mockery of its aims and commitments. After the HfG had eked out a day-to-day existence on an economy budget year after year and then on an emergency budget in a country whose representatives blithely pose as members of a developed industrial society, the HfG unanimously turned at bay in a resolution which was couched in no uncertain terms [see the following text]. "Disrespectful" was the word used by liberal-minded middle-of-the-road men to describe the hard and indeed harsh language of the manifestos, which did not hesitate to call a spade a spade. For it is pertinent to ask who stands to gain from pandering to the power of ignorance and the ignorance of power. Certainly not the HfG.

Arguments about the continuation of the HfG began even before a brick was laid. Apart from the politically motivated hostility engendered by the avowed

---

[1] Published in *ulm 21*, the final issue of *ulm: Zeitschrift der Hochschule für Gestaltung/Journal of the ulm school of design*, 1968, pp. 5–14.

anti-nazism of the HfG, the institution also had to contend with adverse opinions rooted in sheer provincial ignorance and cultural conservatism. The reason was that it did not fit into the traditional cultural scheme in which no provision is made for environmental design. It transcends a conception of culture where the focus is on the cultivation of the economically independent individual and social aspects are ignored. Culture in its bourgeois form does no harm; it jeopardizes nothing and nobody, least of all a society which can afford such a culture. Beyond the political line, it leaves everything precisely as it was. Admittedly Hegel answered the criticism that philosophy can never tempt a dog from behind the stove by saying that it was not its purpose to do so. The same might be said of culture. Yet it tends to become a passively accepted proceeding unless it does in fact attempt the next to impossible. And that such things are possible, that is, that philosophy can do far more than tempt a dog from behind the stove, is amply demonstrated by some events in world history.

A concept of culture which takes the environment to be its province cannot overlook the society living there; indeed, it is made constantly and urgently mindful of it. To be sure, there is no straight path from a well-designed advertisement or a well-turned doorknob to a better society. And although there was an instinctive consciousness at the HfG of the relationship of design to society, it was not actually embodied in its curriculum in a pondered form. The sociopolitical elements of the HfG were relegated in dilution to vague speeches about the cultural responsibility of the designer. Certainly, it would be naïve to expect an improvement of social conditions to result from a qualitative improvement of the world of signs and objects, although efforts to make the world a slightly more pleasant place to live in must undoubtedly be regarded as legitimate. By taking a leap into the pragmatic, one might rid oneself of doubts and find what is right by a subjective approach. For, since the environment is created and will continue to be created as a super-prothesis, it will not be the least of the factors deciding whether and how a society of whatever system will live and survive. Yet this noble demand that man, the crustacean, should be surrounded by a shell worthy of his humanity is inseparable from certain contradictions resulting from changes in historical circumstances.

In the mid-1950s, when the HfG was opened, there was still virtually no realization in industry of the necessity of design. The first missionary endeavors were therefore directed at the owners of the means of production with a view to convincing them that design and business are not irreconcilable opposites and may therefore contract a harmonious marriage. It was found then, and has frequently been confirmed since, that a product satisfying the designer's criteria of quality can also be a moneymaker. Today, design is part and parcel of industry, and only a hypocrite could complain about the improvement in the functional and aesthetic standard of the products. An industry with designers is an advance over a designless industry or an anti-design industry. During this process of assimilation, however, antinomies have sprung up between the satisfaction of needs and consumerism, although this cannot be turned as an argument against design itself. Whereas in theory the designer was to be the representative and interpreter of the interests of society as a whole, design was changed into an element of strategy in modern entrepreneurial policy in which the price war between competitors has been replaced by a design-based differentiation of products. This change in the situation has very special repercussions on the schools of design where kind professionalists can no longer be produced as a matter of course as in times gone by.

The technical rationalism advocated wholeheartedly by the HfG constituted a progressive element particularly during the earlier years of its existence. Previously opposed, it has now gained acceptance everywhere. The sociopolitical factor associated with this rationalism is, however, less welcome; for it cannot be fitted snugly into the productive and reproductive process of society. Industrial societies need intelligence to remain alive. One brand of intelligence in particular is favored. Instrumental intelligence is taken into service, but critical intelligence is desired to a lesser degree or not at all. Evaluated willy-nilly in terms of output, the HfG—particularly for the last five years or so—has had to prove its right to existence by becoming a production center of qualified designers. It became a designer factory which endeavored with shamefully slender funds to fulfill one part of its program, namely the training of designers.

After the HfG had cooperated in turning the unsolved problem of training designers for today into a solvable one, two other parts of its program would have had to be implemented: development projects and, more particularly, research in the field of environmental design. In any case, the mimetic process of training (pedagogics copying practice) would have had to be abandoned. Today, industrial practice is more advanced than pedagogics, whereas ten years ago it was the other way round. Now if training is not to become an insignificant appendage of industry, it must create its own models and patterns so as to give future practice its bearings; otherwise, training will be merely duplication. And thus, it would be unable to give a stimulus to practice in industry.

Internal discussion on a revision of what was taught and how it was taught, including a reorganization of the HfG, had already begun during the academic year 1966–7. In view of the apparently more-weighty considerations bound up with the threat to the very existence of the HfG, these discussions became somewhat unreal. Yet although the necessity of overhauling the HfG could no doubt be postponed, it could not be dropped altogether. If the HfG had been relieved of these external constraints, the realization that a monolithic concept of design is no longer tenable today would have produced its fruit. For the view that the problems of design can be solved primarily, if not exclusively, by designing has been shaken. The relationship between the designer and the sciences must be thought out afresh. So far designers have clung to the role of consumers of science hoping that someone somewhere will produce a piece of knowledge which they will apply and utilize if they come across it more or less by accident. Today, there is no future in this receptive attitude; it must be converted into a productive one. This can be achieved if the design schools do not train their students merely to make design objects but also to create design knowledge and design organization. In the last analysis, design is more than the creation of three-dimensional forms. The activities of the designer will become differentiated. There will be designers who work on the drawing board; there will be designers who research; and there will be designers who organize and plan. These are the lines along which we shall have to proceed in the future, and at the same time the eclectic attitude toward the sciences will have to be abandoned. Design, which might claim to organize and leave its imprint on a highly artificial and in future extremely complicated environment, needs the creation of a science of design as a branch of a future science of environment.

Viewed from the future, the HfG might appear to be a transitional institution that attempted to conjoin science and design but only succeeded in the initial stages of the synthesis. As a newcomer among the classical seats of learning, the HfG could not live independently of them. Its functional dependence on the production centers of knowledge became increasingly apparent. But since the latter are themselves bogged down in a serious crisis of a political nature and must ponder their own relationship to society, they can do little to help design out of a crisis which is rooted in the nature of the subject.

In view of the urgency and the rapidly increasing proportions of the problem confronting the occupants of a world environment, it would be hopeless to wait for the universities to reform their organization and their activities. Similarly, the organizational form of industry—and this applies in particular to capitalist industry—will not allow it to tackle and solve problems affecting society at large, that is, problems including such fields as community design. This touches on the immense sector of public use as against private consumption. Today a town, a hospital, and a school make up a hotchpotch of individual and part products which do not form a system or at best only a jerry system. To deal with the problems looming up there, it would be necessary to create new, versatile institutions where environmental design could be studied on a broad and interdisciplinary basis. Here would be a field of experiment for that collaboration between sociologists, psychologists, economists, engineers, doctors, and designers, which has so often been aimed at and so seldom attained. And at the same time this would spell the end of the obsolete arrangement whereby designers and architects are "advised" by scientists.

Trials could be made with new didactic ideas according to which each student is no longer the competitor of the others. Certificates of attendance as the expression of a repressive principle of performance, and indeed any didactic system which operates with the threat of minimal frustrations would be replaced by an emancipating form of instruction. Lectures, which are a highly uneconomic way of imparting knowledge unless it is new, would drop out and be replaced by teaching programs in which existing knowledge is concentrated. Heuristically oriented instruction would be replaced by instruction in which the solution of a problem is the focus of attention. The members of working groups might team up on the basis of their motivations and interests rather than be assembled according to the fortuitous criterion of their date of registration. The learning process would become productive instead of reproductive.

Perhaps the HfG could have stripped these speculations of their tentative character, although it must be remembered that experience shows that regeneration does not come about of its own accord or arise spontaneously from the matter itself but that taking a revolutionary grip on things must create it. But for this the HfG would have needed a freer climate and not had to solicit in fear and trembling the favor of the elected representatives whose hands control the moneybags and who have never found the HfG's desire for innovations and experiments congenial.

The HfG is therefore almost at an end. It is to be hoped that it will not suffer the same fate as the Bauhaus, that is, to be rendered harmless and put on show as an exhibit in the museum of cultural objects. Nor should the resolution of the members of the HfG be decked out as a heroic gesture. It was not the end of the HfG that was heroic but the hope presiding at its inception. The HfG is not to be gauged by what it achieved but by what it was prevented from achieving.

Gui Bonsiepe (31.3.68)

2. COMMUNICATION AND POWER: MARGINAL NOTES TO THIS ISSUE[2]

The number of voices critical of visual communication is growing and suggests that the training of the visual designer is running into difficulties. Too much importance is attached (they say) to formal and syntactical aspects and to secondary problems of communication. Techniques for prettifying the peripheral were taught but not methods for solving communication problems of truly greater importance. Interest (they claimed) was focused on the marginal areas of visual communication and not on the centers where the destinies of communication are decided, namely in the mass media of films and television, in the editorial offices of weeklies, the popular press, and broadcasting stations as centers of distributing the information merchandise.

Admittedly there is little evidence of realization in training institutions that the communications industry is a consciousness industry, whether it is concerned with the engendering of truth or untruth in consciousness, with enlightenment or ideology. The more visual designers concentrated on the aesthetic perfection of their designs, the more the communications industry was able to keep its power out of sight. The insistence on the aesthetic as one aspect of design is undoubtedly warranted and was capable of retaining its validity over the years. But the aesthetic cannot be maintained in unsullied and apolitical detachment from the social. Formerly the aesthetic figured as the anticipation of a state of affairs which implied liberation from the constraints of necessity. But the aesthetic met with a fate which could not have been foreseen. It was found that it could very readily be pressed into the service of repression. The forms of power have been sublimated. In the course of this sublimation the aesthetic—which was and still is a promise of the state of liberation of mankind—has been harnessed by the agencies of power and thus used to acquire and maintain power. No consequences have as yet been drawn from this change in the role of the aesthetic insofar as it affects either the theory or practice of training in visual communication.

Moreover, far too little attention has been paid to language in connection with visual communication. Restriction to the purely pictorial may be an advantage from the training point of view, but it does have the disadvantage that the visual designer as a specialist in the visual may be forced into a secondary role and that he is barred from access to the relevant communication problems. It is hardly surprising that when society throughout is in a process of rapid change—with new profession appearing and disappearing every day—the professional image of the visual designer is in need of adjustment just when one thought that it had been largely fixed by what had already been achieved.

[2]   Published in *ulm 21*, the final issue of *ulm: Zeitschrift der Hochschule für Gestaltung/Jounal of the ulm school of design*, 1968, p. 15.

FIGURE 28 "Exodus" poster, ulm HfG/Hochschule für Gestaltung, April 1968.

3. PLENARY ASSEMBLY OF THE STAFF AND STUDENTS OF HFG:
RESOLUTION OF THE HOCHSCHULE FUR GESTALTUNG, ULM[3]

The public was informed on February 22, 1968, of a decision made by the Cabinet of the Province Baden-Württemberg. According to this decision, the existence of the Hochschule für Gestaltung (Ulm School of Design) would be possible only under the condition that it be combined with Ulm State Technical School into a cooperative unit. The members of the Hochschule fur Gestaltung took the following position with regard to this decision:

1

The decision of the Council of Ministers is, for the undersigned members of the Hochschule für Gestaltung, not acceptable.

Arguments:

There are too strong contradictions contained in the Cabinet proposal for it to be considered convincing. On the one hand, the HfG is promised "relative" autonomy; on the other hand, the absurdity of the proposal is brought out by the fact that the rector of the HfG, if there would be such a position at all, is subordinate to another director. Our interest is not a puppet-autonomy but a real autonomy.

Within this proposed construction it is excluded that the HfG can complete its pedagogical mission and develop dynamically since administrative and pedagogical autonomy are inseparable.

It is argued that a transfer of the HfG as an autonomous institution under state administration would be more expensive than combining it with other institutions. Recent information obtained from authorities in the Ministry of Culture, charged with investigating this case, states that this is untrue.

The allegation that the HfG is incapable of autonomous self-administration, which is attempted to be proven by its 1.4 Million DM deficit, is a slander. It can be shown that the HfG has conducted its administration in an unobjectionable manner. Debts were contracted by the GSS foundation, the only admitted administrative structure in control of the HfG. This foundation not only misconducted financial politics but also failed to disclose the contents of its documents to the HfG and aid it to participate in its affairs.

The suggested hope of an all-encompassing Ulm University is deceptive in the foreseeable future since there is no objective basis for the realization of such a plan. If the project of an all-encompassing Ulm University was seriously intended, there would be no need of a detour via a cooperative unit HfG/Ulm State Technical School.

There is only one form to further the future existence of the HfG: an autonomous institution at university level under state administration. The

---

[3]   Published in *ulm 21*, the final issue of *ulm: Zeitschrift der Hochschule für Gestaltung/Journal of the ulm school of design*, 1968, p. 15.

final decision on this matter has to be given to the Parliament in a session at the beginning of March 1968. If the Parliament ratifies the proposal of the Cabinet, the members of the HfG declare the following:

The undersigned members of the HfG—as stated in former declarations—are not willing to accept the dictates of the government.

The members of the HfG have decided that the HfG as an international center of teaching, research, and development in environmental design will cease to exist in Ulm as an institutional organism on September 30, 1968.

The members of the HfG are decided to conclude their activity at the HfG on September 30, 1968, and in no manner whatsoever will continue the "HfG" under the dictated conditions.

The members of the HfG declare that the true reasons for the death of the HfG are not of a financial but a political nature. The members of the HfG protest against the cynicism which accuses of suicide, while really a cultural political murder is taking place. Responsible for the death of the HfG is its legal and financial supporting institution, the GSS foundation, inasmuch as year after year it conducted arbitrary and for the HfG disastrous financial politics.

Responsible for the death of the HfG are conservative opinion-forming circles within the city of Ulm, which have the interest of making the HfG into a pawn for local political ambitions.

Responsible for the death of the HfG is the government and many members of the Baden-Württemberg Parliament because of their fundamentally negative attitude toward the HfG and their ineradicable lack of comprehension.

Responsible for the death of the HfG is the federal government in Bonn, which followed with suspicious readiness the proposal of Tröger (constitutional expert who maintains the position that the federal government should not interfere with the educational matters of the provinces), which in reality does not apply to the HfG. For the HfG is a common responsibility of the federal and provincial governments, if only because of its 50 percent of foreign students.

These four groups, despite the recognition and support the HfG has found nationally and internationally, have finally achieved their aims and are now responsible for the liquidation of the Hochschule für Gestaltung.

Ulm, February 23, 1968

Accepted by the plenary assembly of the HfG (with four student abstentions) and signed by staff and students of the HfG

GUI BONSIEPE

A Brief Biography

German-born (1934), Argentina/Brazil-based information designer, educator, and author. Studied and taught at the *Hochschule für Gestaltung* (HfG) ulm until 1968. He participated under the direction of Tomás Maldonado, among others, in the development of a complex sign system for the interface of the Olivetti computer ELEA 9003.

Since 1969, worked mainly in Latin America providing consultancy and design services for public and private institutions. From 1973 to 1975, vice president of ICSID. From 1987 to 1989, specialized in interface design in a software house in Emeryville California. From 1993 to 2003, professor for interface design at the University of Applied Sciences, Cologne. Lives and works in Brazil and Argentina. Special interests: design in peripheral countries, critical design discourse, visual rhetorics.

Honorary degrees, among others, from the Universidade do Estado de Rio de Janeiro, Universidad Tecnológica Santiago de Chile, Universidad Autónoma Metropolitana Mexico, Universidade de Lisboa.

Author of several books published in Germany, Italy, Switzerland, Netherlands, Argentina, Brazil, México, Spain, and Korea:

Dall'oggetto all'interfaccia (1995).
Interface – An Approach to Design (1999).
Entwurfskultur und Gesellschaft (2009).
Diseño y Crisis (2012).
Trilogy of *Design, Cultura e sociedade* (2011), *Design como prática do projeto* (2012), *Do material ao digital* (2015).
http://www.guibonsiepe.com

CONTRIBUTORS

ZOY ANASTASSAKIS

Zoy Anastassakis is a Brazilian designer and anthropologist. Between 2016 and 2018, she was director at Superior School of Industrial Design, at the State University of Rio de Janeiro (ESDI/UERJ), where she works as associate professor. She coordinates the research group "Design and Anthropology Laboratory" (LaDA), where she experiments on means of composition between Anthropology and Design. In 2018, she was a guest researcher at the Department of Anthropology of the University of Aberdeen, Scotland, as part of the project "Knowing from the Inside: Anthropology, Art, Architecture and Design," coordinated by Tim Ingold. Since 2020, she is an associated researcher at the Center for Research in Anthropology (CRIA), in Lisbon, Portugal. Together with Marcos Martins, she is coauthor of a book on ESDI's experimentations during the period 2016–17, at Bloomsbury's series "Designing in Dark Times."

CONSTANTIN BOYM

Constantin Boym was born in Moscow, Russia, in 1955, where he graduated from Moscow Architectural Institute. In 1984–5, he earned the degree Master of Design from Domus Academy in Milan. In 1986, he founded Boym Partners Inc in New York City, which he runs together with his partner Laurene Leon Boym. Boym Partners bring a critical, experimental approach to a range of products and environments that infuse humor and wit into the everyday. The studio's designs are included in the permanent collection of the Museum of Modern Art in New York. Boym Partners was a winner of the National Design Award in 2009. The studio is a recipient of eight I.D. Magazine Annual Design Awards, including the Best of Category in 2000.

A book devoted to the work of Boym Partners, *Curious Boym: Design Works*, was published by Princeton Architectural Press in 2002. Boym's most recent book, *Keepsakes: A Design Memoir*, was published by Pointed Leaf Press in 2015. Since 2015, Boym has served as Professor and Chair of Industrial Design at Pratt Institute in New York.

FREDERICO DUARTE

Frederico Duarte is a Portuguese design critic and curator. He studied communication design at the University of Lisbon and worked as a designer in Portugal, Malaysia, and Italy. He has, since 2006, written articles and essays, contributed to and edited books and catalogues, given lectures and workshops, organized events and curated exhibitions on design, architecture, and creativity. In 2010, he graduated from the Design Criticism MFA at the School of Visual Arts in New York, and in 2014 started a collaborative PhD research project on contemporary Brazilian design at Birkbeck College (University of London) and the Victoria and Albert Museum. In 2017, he curated the exhibition "How to pronounce design in Portuguese: Brazil Today" for MUDE, Lisbon's design and fashion museum.

HUGH DUBBERLY

Hugh Dubberly studied environmental design at the University of Colorado, and graphic design at Rhode Island School of Design (RISD) and Yale. At Apple Computer (1986–94), Dubberly was a creative director, managing graphic design and corporate identity; he also produced the technology-forecast film Knowledge Navigator presaging the Internet and interaction via mobile devices. At Netscape (1995–2000), he became vice president of Design managing groups responsible for the design, engineering, and production of Netscape's web services. He cofounded Dubberly Design Office (2000), a software, service, and systems design consultancy. He has served on AIGA's national board and the SIGGRAPH Conference Committee, and chaired ACD's "Design for the Internet" Conference. He has taught at San Jose State, Art Center, CMU, Stanford, IIT/ID, Northeastern, and CCA. He edited a column "On Modeling" for ACM's journal, *Interactions*, and has published more than fifty articles on design methods. He was elected to the ACM CHI Academy, and is an AIGA fellow.

ETHEL LEON

Ethel Leon is Professor and a researcher of Design History. He holds a PhD in History of Architecture from the Faculty of Architecture and Urbanism at the University of São Paulo. As a curator, he has organized several exhibitions on Brazilian design. He is coeditor of *Agitprop*, online design magazine (2008–15). He is the author of the following books: *Brazilian Design who did who does* (Senac Rio, 2005); *João Baptista da Costa Aguiar Desenho Gráfico* (Senac / SP, 2006); *Memórias do Design Brasileiro (Brazilian Design Memoirs*, Senac SP, 2009); *IAC, primeira escola de design do Brasil (IAC, the first design school in Brazil*, Edgar Blucher, 2014); *Canasvieiras, um laboratório para o design brasileiro (Canasvieiras, a laboratory for Brazilian Design, UDESC/ FAPESC, 2014)*; author, with Marcello Montore, of the Brazilian chapter of *Historia del Diseño en America Latina y el Caribe*, organized by Silvia Fernández and Gui Bonsiepe. (Edgar Blucher, 2008), editor and coauthor of *Michel Arnoult Design e Utopia* (SESC, 2016), as well as coauthor of *Marcenaria Baraúna* (Olhares, 2017).

MARCOS MARTINS

Marcos Martins is a designer and associate professor at Universidade do Estado do Rio de Janeiro's (Rio de Janeiro State University) Escola Superior de Desenho Industrial (ESDI), where he was deputy director from 2016 to 2018. As a designer, he has worked in various fields, including, graphic and interactive design, computer animation, video, and exhibition design. His academic research is aimed at historicizing and analyzing the contemporary status of digital interfaces and environments while critically approaching the field of interaction design. In 2018, he conducted postdoctoral research at Princeton University investigating possible genealogical traits of today's social media interfaces. He advocates for the opening of the field of design to intersections with other domains such as art, philosophy, psychoanalysis, education, and decolonial studies, in order to interrogate the current status of design in face of challenging and unstable times. Together with Zoy Anastassakis, he is coauthor of a book integrating Bloomsbury's series, *Designing in Dark Times*, about educational experimentations during a period of economic and political crisis at ESDI, in 2016–17.

EDEN MEDINA

Eden Medina is Associate Professor of Science, Technology, and Society at MIT. Her research uses the history of science and technology as a way to understand processes of political change, especially in Latin America. She combines history, science, and technology studies, and Latin American studies in her writings. Her book, *Cybernetic Revolutionaries: Technology and Politics in Allende's Chile*, won the Edelstein Prize for outstanding book in the history of technology, the Computer History Museum Prize for outstanding book in the history of computing, and the Book Prize of the Recent History and Memory Section (honorable mention) of the Latin American Studies Association. Her coedited volume *Beyond Imported Magic: Essays on Science, Technology and Society in Latin America* received the Amsterdamska Award from the European Society for the Study of Science and Technology.

LARA PENIN

Lara de Sousa Penin is an associate professor of Transdisciplinary Design at Parsons School of Design, the New School, where she is former director of the *Transdisciplinary Design* graduate program (2015–2018) and currently coleads the *Graduate Minor in Civic Service Design*. She is a cofounder of the Parsons DESIS Lab, an action research laboratory that works at the intersection of service and strategic design, management, and social theory. Her work centers at the intersection of service and strategic design, participatory design, and social justice, seeking to explore the agency of design to effect positive social change. Lara is the author of *An Introduction to Service Design. Designing the Invisible* (Bloomsbury, 2018). Originally from São Paulo, Brazil, where she graduated in Architecture and Urbanism from the University of São Paulo, she has a PhD in Design from the Milan Polytechnic University.

INDEX

Page numbers followed with "n" refer to footnotes.